Springer Proceedings in Mathematics & Statistics

Volume 211

Springer Proceedings in Mathematics & Statistics

This book series features volumes composed of selected contributions from workshops and conferences in all areas of current research in mathematics and statistics, including operation research and optimization. In addition to an overall evaluation of the interest, scientific quality, and timeliness of each proposal at the hands of the publisher, individual contributions are all refereed to the high quality standards of leading journals in the field. Thus, this series provides the research community with well-edited, authoritative reports on developments in the most exciting areas of mathematical and statistical research today.

More information about this series at http://www.springer.com/series/10533

María A. Cañadas-Pinedo
José Luis Flores · Francisco J. Palomo
Editors

Lorentzian Geometry and Related Topics

GeLoMa 2016, Málaga, Spain, September 20–23

Editors
María A. Cañadas-Pinedo
Department of Algebra, Geometry,
Topology
University of Malaga
Málaga
Spain

Francisco J. Palomo
Department of Applied Mathematics
University of Malaga
Málaga
Spain

José Luis Flores
Department of Algebra, Geometry,
Topology
University of Malaga
Málaga
Spain

ISSN 2194-1009 ISSN 2194-1017 (electronic)
Springer Proceedings in Mathematics & Statistics
ISBN 978-3-030-09767-7 ISBN 978-3-319-66290-9 (eBook)
DOI 10.1007/978-3-319-66290-9

Mathematics Subject Classification (2010): 35J93, 35M10, 53A01, 53A05, 53A10, 53B25, 53B30, 53B50, 53C22, 53C25, 53C26, 53C40, 53C42, 53C50, 53C60, 53C80, 83C20, 83C57, 83C75

Printed on acid-free paper

This Springer imprint is published by Springer Nature
The registered company is Springer International Publishing AG
The registered company address is: Gewerbestrasse 11, 6330 Cham, Switzerland

Preface

Lorentzian Geometry was born as the geometric theory on which General Relativity could be mathematically stated. Nowadays, it constitutes a very active area of research with its proper specific weight in Differential Geometry and Mathematical Relativity. Many mathematical techniques are involved in this field (geometric analysis, functional analysis, partial differential equations, Lie groups and Lie algebras…), making any text focused on it of great interest to a broad audience.

Fifteen years ago, several Spanish researchers interested in Lorentzian Geometry and related mathematical topics, launched the biennial meetings on Lorentzian Geometry in Benalmádena 2001 (Málaga). This first meeting was developed in a very friendly atmosphere and had a vocation to be the first of a following suit on the same topic. As a consequence, a fruitful series of international meetings on Lorentzian Geometry started since that moment: Murcia 2003 (Spain), Castelldefels 2005 (Spain), Santiago de Compostela 2007 (Spain), Martina Franca 2009 (Italy), Granada 2011 (Spain), and Sao Paulo 2013 (Brazil). A special edition on "Lorentzian and conformal Geometry" was held in Greifswald (Germany) in 2014 in honour of Prof. Helga Baum. Along the years, the international character of these meetings has increased spectacularly. The excellent ambiance in which this series of meetings had been originated, recalling the first edition, led us to speak about the "Benalmádena's spirit".

The most recent edition was held at the University of Málaga, Spain, in September, 2016. This volume contains a picture of the trends in Lorentzian Geometry exposed in this VIII International Meeting on Lorentzian Geometry. Among others, it contains topics such as notable (maximal, trapped, null, spacelike, constant mean curvature, umbilical, isoparametric) submanifolds, causal completion of spacetimes, stationary regions and horizons in spacetimes, solitons in semi-Riemannian manifolds, relation between Lorentzian and Finslerian Geometries and the oscillator spacetime.

Let us provide a more specific summary about the contributions included here.

The Euclidean space and the Lorentz-Minkowski spacetime share the same underlying manifold. Therefore a natural question on spacelike hypersurfaces in the Lorentz-Minkowski spacetime arises. When a spacelike hypersurface has the same mean curvature as hypersurface of the Lorentz-Minkowski spacetime and as hypersurface in the Euclidean space? **Eva M. Alarcón, Alma L. Albujer** and **Magdalena Caballero** summarize results for the case when the two mean curvature functions are equal and constant. For the case when the mean curvature functions are equal but not necessarily constant, the authors generalize, to arbitrary dimension, some previous results on spacelike surfaces.

In their contribution, **Stephanie B. Alexander** and **William A. Karr** show that sectional curvature bounds of the form $\mathcal{R} \leq K$ are closely tied to space-time convex functions. Several consequences are obtained: a natural construction of such functions as well as an analogue of a theorem by Alías, Bessa and de Lira ruling out trapped submanifolds in new domains. In addition, they point several connections between totally independent researches by different authors.

Motivated by the very special role played by the Lorentzian oscillator group, **Giovanni Calvaruso** describes an explicit system of global coordinates for this Lie group, and uses it to compute its symmetries and solutions to the Ricci soliton equation.

Nastassja Cipriani and **José M.M. Senovilla** provide necessary and sufficient conditions for a spacelike submanifold of *arbitrary* co-dimension to be umbilical along normal directions. To this aim, they use the so-called *total shear tensor,* i.e. the trace-free part of the second fundamental form. They also show that the sum of the dimensions of the spaces generated by the total shear tensor and by the umbilical vector fields equals the co-dimension.

Ivan P. Costa e Silva presents a personal review of a number of results on the global geometric properties of stationary regions of spacetimes, both new and well-known (many of the latter with new proofs). He also discusses the general structure and regularity of the horizons associated with these regions. The analysis is largely carried out without assuming any field equations, asymptotic flatness/ hyperbolicity or dimensional restrictions, thereby emphasizing their independence of such extra, often physically motivated hypotheses.

J. Carlos Díaz-Ramos, Miguel Domínguez-Vázquez and **Víctor Sanmartín-López** are interested in isoparametric hypersurfaces in complex hyperbolic spaces, whose classification has recently been obtained by the authors. Concretely, they prove that given an isoparametric hypersurface in $\mathbb{C}H^n$, the principal curvatures of this hypersurface are pointwise the same, as the principal curvatures of a homogeneous hypersurface of $\mathbb{C}H^n$. Although this result can be obtained following the classification mentioned above, they prove it by a more direct approach, by lifting hypersurfaces in complex hyperbolic spaces to anti-de Sitter spacetimes.

The causal completion of spacetimes has succeeded as a tool to study different global properties of spacetimes. However, it is not always easy to extend the structure of the spacetime to the causal boundary. **Stacey (Steven) Harris** examines the extent to which various causal constructions and properties for spacetimes can be applied to the future completion of a strongly causal spacetime, considered as a topological space using the future chronological topology.

A very useful link between Lorentzian and Finslerian Geometries has been developed in the last years. In this volume, **Miguel Ángel Javaloyes** and **Miguel Sánchez** develop these relationships and several applications to spacetimes. In their contribution the purpose is twofold. On one hand, the authors provide a reference summary on the subject for Lorentzian geometers. On the other hand, they consider a big class of spacetimes admitting a time function t and characterize when the slices t = constant are Cauchy hypersurfaces.

Erdem Kocakuşaklı and **Miguel Ortega** study some general properties of translating solitons in the semi-Riemannian setting. In particular, they focus on the study of translating solitons which are invariant under the action of a Lie group of isometries of the ambient space, and they examine the behaviour near the singular orbit, whenever there exits, and at infinity. They also include several examples.

Wai Yeung Lam and **Masashi Yasumoto** investigate discretizations of surfaces with vanishing mean curvature (i.e., maximal surfaces) in the three-dimensional Lorentz-Minkowski space. In particular, the case of trivalent maximal surfaces is analysed. The authors derive a Weierstrass-type representation using discrete holomorphic quadratic differentials for these surfaces. Their contribution also includes a deep analysis of singularities of trivalent maximal surfaces.

The hypersurfaces with constant mean curvature in spacetimes are convenient initial data for the Cauchy problem of the Einstein equations. **Rafael López** provides a survey on this kind of hypersurfaces on the steady-state space. Three different models for this spacetime appear in his contribution. Each one is conveniently employed depending on the problems. The author focuses on Bernstein-type theorems by using as key tools the tangency principle and the Omori-Yau maximum principle.

Continuing with the celebrated Calabi–Bernstein theorem, **Rafael Rubio** presents a useful review about some of the classical and recent proofs of this theorem in the Lorentz-Minkowski spacetime for the two-dimensional case, as well as several extensions for Lorentzian-warped products and other relevant spacetimes. He also analyzes the problem of uniqueness of complete maximal hypersurfaces under the perspective of some new results.

Benjamín Olea focuses on the geometry of null hypersurfaces in Lorentzian manifolds. To this aim he uses the *rigging techniques,* one of the approaches to overcome the difficulties derived from the degeneracy of these objects. Concretely, he studies under which conditions the *rigged connection,* i.e. the Levi-Civita connection associated to the Riemannian metric induced by the rigging, coincides with the connection directly induced from the rigging, and gives some examples.

Finally, by means of the application of the Cauchy–Kovalevski theorem for partial differential equations, **Masaaki Umehara** and **Kotaro Yamada** construct all real analytic germs of zero mean curvature surfaces by using several concrete examples of such surfaces in $\mathbb{R}^3_1$ containing a certain light-like line. Besides they use a new approach to the subject, they obtain, as a consequence, new examples of zero mean curvature surfaces such that the causal type of one-side of the line is space-like and the other-side is time-like. Moreover, they provide several applications of these results.

Summing up, this volume constitutes a representative picture of the last progresses in the field of Lorentzian Geometry and related topics. Moreover, we think that the topics here included provide a nice approach to several very active research problems.

We would like to thank the careful work of the contributors, as well as the exhaustive revisions by the anonymous referees. We would also like to acknowledge Springer for its interest on this branch of knowledge and its staff for their friendly assistance.

Last but not least, we warmly thank all participants of the Lorentzian meeting for the excellent scientific level and pleasant atmosphere of this congress http://gigda.ugr.es/geloma/, as well as the support of the sponsors: the University of Malaga, the Vice-Rectorate for Research and Knowledge Transfer (UMA), the Department of Algebra, Geometry and Topology (UMA) and the Department of Applied Mathematics (UMA); the Spanish projects MTM2013-47828-C2-1-P, MTM2013-47828-C2-2-P and MTM2013-41768-P; the Academia Malagueña de Ciencias, the Sociedad Malagueña de Astronomía and the Real Sociedad Española de Física; and the Metro de Málaga.

Málaga, Spain
July 2017

María A. Cañadas-Pinedo
José Luis Flores
Francisco J. Palomo

Contents

Spacelike Hypersurfaces in the Lorentz-Minkowski Space with the Same Riemannian and Lorentzian Mean Curvature

Eva M. Alarcón, Alma L. Albujer and Magdalena Caballero

Abstract Spacelike hypersurfaces in the Lorentz-Minkowski space $\mathbb{L}^{n+1}$ can be endowed with two different Riemannian metrics, the metric inherited from $\mathbb{L}^{n+1}$ and the one induced by the Euclidean metric of $\mathbb{R}^{n+1}$. Consequently, we can consider two mean curvature functions naturally attached to any spacelike hypersurface, H_R and H_L. In this manuscript, we revise some known results for the case where $H_R = H_L$ is constant, and generalize to arbitrary dimension some recent results for spacelike surfaces with $H_R = H_L$ not necessarily constant obtained by Albujer and Caballero. Specifically, we prove that spacelike hypersurfaces with $H_R = H_L$ do not have any elliptic points. As an application of this result, jointly with a classical argument on the existence of elliptic points due to Osserman, we present several geometric consequences for the hypersurfaces we are considering. Finally, as any spacelike hypersurface in $\mathbb{L}^{n+1}$ is locally a graph over the hyperplane $x_{n+1} = 0$, our hypersurfaces are locally determined by the solutions to a certain partial differential equation called the $H_R = H_L$ hypersurface equation. The character of this equation is studied, and some uniqueness results for its related Dirichlet problem are given.

Keywords Mean curvature · Spacelike hypersurfaces · Elliptic points · Dirichlet problem

2010 Mathematics Subject Classification Primary 53C42 · Secondary 35J93 · 53C50

E.M. Alarcón
Departamento de Matemáticas, Campus de Espinardo, Universidad de Murcia, 30100 Murcia, Spain
e-mail: evamaria.alarcon@um.es

A.L. Albujer (✉) · M. Caballero
Departamento de Matemáticas, Campus Universitario de Rabanales, Universidad de Córdoba, 14071 Córdoba, Spain
e-mail: alma.albujer@uco.es

M. Caballero
e-mail: magdalena.caballero@uco.es

M.A. Cañadas-Pinedo et al. (eds.), *Lorentzian Geometry and Related Topics*, Springer Proceedings in Mathematics & Statistics 211,
DOI 10.1007/978-3-319-66290-9_1

1 Introduction and Background

A hypersurface in the Lorentz-Minkowski space $\mathbb{L}^{n+1}$ is said to be spacelike if its induced metric is a Riemannian one. We can endow a spacelike hypersurface in $\mathbb{L}^{n+1}$ with another Riemannian metric, the one inherited from the Euclidean space $\mathbb{R}^{n+1}$. Therefore, we can consider two different mean curvature functions on a spacelike hypersurface, the mean curvature function related to the metric induced by $\mathbb{R}^{n+1}$, that we will denote by H_R, and the one related to the metric inherited from $\mathbb{L}^{n+1}$, H_L.

A hypersurface in $\mathbb{R}^{n+1}$ is said to be minimal if its mean curvature function vanishes identically, that is $H_R \equiv 0$. Analogously, a spacelike hypersurface in $\mathbb{L}^{n+1}$ is said to be maximal if $H_L \equiv 0$. The study of minimal and maximal hypersurfaces is a topic of wide interest. One of the main results about the global geometry of minimal surfaces is the well-known Bernstein theorem, proved by Bernstein [5] in 1915, which states that the only entire minimal graphs in $\mathbb{R}^3$ are the planes. Some decades later, in 1970, Calabi [7] proved its analogous version for spacelike surfaces in the Lorentz-Minkowski space, the Calabi-Bernstein theorem, which states that the only entire maximal graphs in $\mathbb{L}^3$ are the spacelike planes. An important difference between both results is that the Bernstein theorem can be extended to minimal graphs in $\mathbb{R}^{n+1}$ up to dimension $n = 7$, as it was proved by Bombieri et al. [6], but it is no longer true for higher dimensions. However, the Calabi-Bernstein theorem holds true for any dimension as it was proved by Calabi [7] for dimension $n \leq 4$, and by Cheng and Yau [8] for arbitrary dimension.

It is interesting to note that any complete spacelike hypersurface in $\mathbb{L}^{n+1}$ is necessarily an entire graph over any spacelike hyperplane, see [4, Proposition 3.3]. Consequently, the Calabi-Bernstein theorem can also be expressed in a parametric way by asserting that the only complete maximal hypersurfaces in $\mathbb{L}^{n+1}$ are the spacelike hyperplanes. Its Riemannian analogue is not true since there exists a wide family of nonplanar complete minimal hypersurfaces in $\mathbb{R}^{n+1}$, even in the 2-dimensional case ($n = 2$).

As an immediate consequence of the above results, we conclude that the only complete hypersurfaces that are simultaneously minimal in $\mathbb{R}^{n+1}$ and maximal in $\mathbb{L}^{n+1}$ are the spacelike hyperplanes.

Going a step further, we can consider spacelike hypersurfaces with the same constant mean curvature functions H_R and H_L. In 1955, as a direct consequence of the classical divergence theorem, Heinz [12] proved that given a graph in $\mathbb{R}^3$ defined over a disk of radius R in $\mathbb{R}^2$ centered at the origin, $B_0(R)$, if $|H_R| \geq c > 0$ for a certain constant c, then $R \leq \frac{1}{c}$ necessarily. Some years later, Chern [9] and Flanders [10] simultaneously and independently extended this result to general dimension. Therefore, the only entire graphs with constant mean curvature H_R in $\mathbb{R}^{n+1}$ are the minimal ones. The Lorentzian version of this result is not true, there are examples of entire spacelike graphs with constant mean curvature H_L in $\mathbb{L}^{n+1}$ which are not maximal, for instance the hyperbolic spaces. However, taking into account the Calabi-Bernstein theorem, we conclude again that the only complete spacelike hypersurfaces in $\mathbb{L}^{n+1}$ with the same constant mean curvature functions H_R and H_L are the spacelike hyperplanes.

Without assuming any completeness hypothesis, Kobayashi [14] studied the problem for $H_R = H_L = 0$ in the 2-dimensional case. After presenting a classification of maximal ruled surfaces in $\mathbb{L}^3$, he showed that the only surfaces that are simultaneously minimal and maximal are necessarily ruled. And consequently, they are open pieces of a spacelike plane or of a helicoid in the region where the helicoid is spacelike. Recently, Albujer et al. [1, 2] have continued with the study of spacelike surfaces with the same mean curvature in $\mathbb{R}^3$ and in $\mathbb{L}^3$, not necessarily constant. Specifically, they have shown that those surfaces have non-positive Gaussian curvature with respect to the metric induced from $\mathbb{R}^3$ in all their points, and have obtained several interesting consequences about the geometry of such surfaces.

In general dimension, Lee and Lee [15] have recently presented nonplanar examples of simultaneously minimal and maximal spacelike graphs in the Lorentz-Minkowski space. Their examples can be seen as generalized ruled hypersurfaces, in fact they are a natural generalization of helicoids. However, there is no known classification of such hypersurfaces similar to Kobayashi's result.

Our main purpose in this manuscript is to generalize some of the results in [1, 2], providing some geometric properties of spacelike hypersurfaces in $\mathbb{L}^{n+1}$ with $H_R = H_L$. In Sect. 2 we present some basic preliminaries on spacelike hypersurfaces in $\mathbb{L}^{n+1}$ and their mean curvature functions with respect to the metrics inherited from $\mathbb{R}^{n+1}$ and $\mathbb{L}^{n+1}$. It is well known that any spacelike hypersurface can be locally seen as a graph over an open subset of a spacelike hyperplane, which without loss of generality can be supposed to be the hyperplane $x_{n+1} = 0$, see [16] for the proof in the two-dimensional case. Therefore, we also describe the normal vector fields and the mean curvature functions with respect to both metrics in terms of the differential operators of the function which locally describes the hypersurface. Finally, we recall a characterization result by Osserman [17] for hypersurfaces in $\mathbb{R}^{n+1}$ without elliptic points and we present its Lorentzian version for spacelike hypersurfaces in $\mathbb{L}^{n+1}$, see Theorem 2, which extends [1, Theorem 3].

In Sects. 3 and 4 we consider spacelike hypersurfaces in $\mathbb{L}^{n+1}$ such that $H_R = H_L$. Specifically, in Sect. 3 we prove that at any point of those hypersurfaces the principal curvatures cannot have all of them the same sign. From this theorem, as well as from Osserman's result, we get some geometric consequences to which the rest of the section is devoted.

In Sect. 4 we present the $H_R = H_L$ hypersurface equation. Any spacelike hypersurface is locally determined by a solution of this equation satisfying $|Du| < 1$, where D and $|\cdot|$ stand for the Euclidean gradient and Euclidean norm in $\mathbb{R}^n$, respectively. We prove the uniqueness of the Dirichlet problem associated to this partial differential equation under some appropriate boundary conditions. This is not trivial, since the equation is not always elliptic. We also consider rotationally invariant spacelike graphs with $H_R = H_L$, obtaining a uniqueness result for them.

Some of the proofs are analogous to the two-dimensional case, still we will include them for the sake of completeness.

2 Preliminaries

Let $\mathbb{L}^{n+1}$ be the $(n+1)$-dimensional Lorentz-Minkowski space, that is, $\mathbb{R}^{n+1}$ endowed with the metric

$$\langle , \rangle_L = dx_1^2 + \cdots + dx_n^2 - dx_{n+1}^2,$$

where $(x_1, \ldots, x_{n+1})$ are the canonical coordinates in $\mathbb{R}^{n+1}$, and let $|\cdot|_L$ denote its norm. It is easy to see that the Levi-Civita connections of the Euclidean space $\mathbb{R}^{n+1}$ and the Lorentz-Minkowski space $\mathbb{L}^{n+1}$ coincide, so we will just denote them by $\overline{\nabla}$.

A (connected) hypersurface Σ^n in $\mathbb{L}^{n+1}$ is said to be a spacelike hypersurface if $\mathbb{L}^{n+1}$ induces a Riemannian metric on Σ, which is also denoted by $\langle , \rangle_L$. Given a spacelike hypersurface Σ, we can choose a unique future-directed unit normal vector field N_L on Σ. Let ∇^L denote the Levi-Civita connection in Σ with respect to $\langle ; \rangle_L$. Then the Gauss and Weingarten formulae for the spacelike hypersurface Σ become

$$\overline{\nabla}_X Y = \nabla_X^L Y - \langle A_L X, Y \rangle_L N_L$$

and

$$A_L X = -\overline{\nabla}_X N_L,$$

respectively, for any tangent vector fields $X, Y \in \mathfrak{X}(\Sigma)$, where $A_L : \mathfrak{X}(\Sigma) \to \mathfrak{X}(\Sigma)$ stands for the shape operator of Σ with respect to N_L. The mean curvature function of Σ with respect to N_L is defined by

$$H_L = -\frac{1}{n} \operatorname{tr} A_L = -\frac{1}{n}(k_1^L + \cdots + k_n^L),$$

where k_i^L, $i = 1, \ldots, n$, stand for the principal curvatures of $(\Sigma, \langle , \rangle_L)$.

It is well known that there exists no closed (compact and without boundary) spacelike hypersurface in $\mathbb{L}^{n+1}$ [3, 4]. Therefore, every compact spacelike hypersurface Σ in the Lorentz-Minkowski space necessarily has nonempty boundary.

The same topological hypersurface can be also considered as a hypersurface of the Euclidean space, that is $\mathbb{R}^{n+1}$ with its usual Euclidean metric. For simplicity, we will just denote the Euclidean space by $\mathbb{R}^{n+1}$, the Euclidean metric and the induced metric on Σ by $\langle , \rangle_R$, and its norm by $|\cdot|_R$. In such a case, Σ admits a unique upward directed unit normal vector field, N_R. In an analogous way as in the Lorentzian case, let ∇^R denote the Levi-Civita connection in Σ with respect to $\langle , \rangle_R$. The Gauss and Weingarten formulae read now

$$\overline{\nabla}_X Y = \nabla_X^R Y + \langle A_R X, Y \rangle_R N_R$$

and

$$A_R X = -\overline{\nabla}_X N_R,$$

respectively, $A_R : \mathfrak{X}(\Sigma) \to \mathfrak{X}(\Sigma)$ being the shape operator of Σ with respect to N_R. The mean curvature function of Σ with respect to N_R is defined by

$$H_R = \frac{1}{n}\mathrm{tr}\, A_R = \frac{1}{n}(k_1^R + \cdots + k_n^R),$$

where k_i^R, $i = 1, \ldots, n$, stand for the principal curvatures of $(\Sigma, \langle, \rangle_R)$. Let us recall that a point $p \in \Sigma$ is said to be elliptic if all the principal curvatures of Σ at p have the same sign.

It is interesting to observe that the mean curvature functions have an expression in terms of the normal curvatures of any set of orthogonal directions. Specifically,

$$H_R = \frac{1}{n}(\kappa_{v_1}^R + \cdots + \kappa_{v_n}^R) \quad \text{and} \quad H_L = -\frac{1}{n}(\kappa_{w_1}^L + \cdots + \kappa_{w_n}^L), \tag{2.1}$$

where $\{v_1, \ldots, v_n\}$ and $\{w_1, \ldots, w_n\}$ are orthonormal basis of $T_p\Sigma$ with respect to $\langle, \rangle_R$ and $\langle, \rangle_L$, respectively.

A spacelike hypersurface is locally a graph over an open subset of the hyperplane $x_{n+1} = 0$, which can be identified with $\mathbb{R}^n$. Therefore, for each $p \in \Sigma$ there exists an open neighborhood of p, $\Omega \subseteq \mathbb{R}^n$, and a smooth function $u \in \mathcal{C}^\infty(\Omega)$ such that $\Sigma = \Sigma_u$ on this neighborhood, where

$$\Sigma_u = \{(x_1, \ldots, x_n, u(x_1, \ldots, x_n)) : (x_1, \ldots, x_n) \in \Omega\}.$$

It is easy to check that Σ_u is a spacelike hypersurface if and only if $|Du| < 1$, where D and $|\cdot|$ stand for the gradient operator and the norm in the Euclidean space $\mathbb{R}^n$, respectively. In this case, it is possible to get expressions for the normal vector fields N_L and N_R, as well as for the mean curvature functions H_L and H_R, in terms of u. Specifically, with a straightforward computation we get

$$N_L = \frac{(Du, 1)}{\sqrt{1 - |Du|^2}} \quad \text{and} \quad N_R = \frac{(-Du, 1)}{\sqrt{1 + |Du|^2}}. \tag{2.2}$$

And for the mean curvature functions we have

$$H_L = \frac{1}{n}\,\mathrm{div}\left(\frac{Du}{\sqrt{1 - |Du|^2}}\right) \quad \text{and} \quad H_R = \frac{1}{n}\,\mathrm{div}\left(\frac{Du}{\sqrt{1 + |Du|^2}}\right), \tag{2.3}$$

where div denotes the divergence operator in $\mathbb{R}^n$. Let us observe that

$$\cos\theta = \frac{1}{\sqrt{1 + |Du|^2}} \quad \text{and} \quad \cosh\psi = \frac{1}{\sqrt{1 - |Du|^2}},$$

where θ and ψ denote the angle between N_R and $e_{n+1} = (0, \ldots, 0, 1)$ and the hyperbolic angle between N_L and e_{n+1}, respectively. Moreover, from (2.2) it is immediate to get

$$\frac{\langle X, N_L \rangle_L}{\cosh \psi} = -\frac{\langle X, N_R \rangle_R}{\cos \theta}, \tag{2.4}$$

for any $X \in \mathfrak{X}(\Sigma)$, which is a global equality since it does not depend on u. Let us observe that, in the previous expressions, we are writing Du instead of $Du \circ \pi$, where π is the canonical projection of Σ_u onto Ω. On behalf of simplicity, we will continue using this identification along the manuscript.

According to Osserman [17], a hypersurface Σ in the Euclidean space $\mathbb{R}^{n+1}$ satisfies the convex hull property if every compact subset $D \subseteq \Sigma$ lies in the convex hull of its boundary. In [17, Theorem], he gave the following simple geometric condition characterizing those hypersurfaces.

Theorem 1 ([17, Theorem]) *A hypersurface Σ in $\mathbb{L}^{n+1}$ has the convex hull property if and only if there is no elliptic point in Σ.*

Theorem 1 also holds for spacelike hypersurfaces in $\mathbb{L}^{n+1}$. This yields from the following lemma contained in [1]. We expose the proof of the lemma for the sake of completeness.

Lemma 1 ([1, Lemma 2]) *Let Σ be a spacelike hypersurface in $\mathbb{L}^{n+1}$. Given $p \in \Sigma$ and $v \in T_p\Sigma$, let $\kappa_v^L(p)$ and $\kappa_v^R(p)$ denote the normal curvatures at p in the direction of v with respect to $\langle , \rangle_L$ and $\langle , \rangle_R$, respectively. Then,*

$$\frac{|v|_R^2}{\cos \theta(p)} \kappa_v^R(p) = -\frac{|v|_L^2}{\cosh \psi(p)} \kappa_v^L(p).$$

Proof Given $p \in \Sigma$ and $v \in T_p\Sigma$, let α be a smooth curve on Σ such that $\alpha(0) = p$ and $\alpha'(0) = v$. We will work at p, but for simplicity we will omit it. Then, by definition,

$$\kappa_v^R = \langle \overline{\nabla}_{t_R} t_R, N_R \rangle_R \quad \text{and} \quad \kappa_v^L = \langle \overline{\nabla}_{t_L} t_L, N_L \rangle_L, \tag{2.5}$$

where $t_R = \frac{\alpha'}{|\alpha'|_R}$ and $t_L = \frac{\alpha'}{|\alpha'|_L}$. We combine (2.4) and (2.5) to finish the proof. □

Consequently, κ_v^L and κ_v^R always have opposite signs. And so, all the principal curvatures of Σ with respect to $\langle , \rangle_R$ are positive if and only if all its principal curvatures with respect to $\langle , \rangle_L$ are negative, and vice versa. Equivalently, a point in Σ is elliptic with respect to the metric $\langle , \rangle_L$ if and only if it is elliptic with respect to $\langle , \rangle_R$. Hence, we have proved the Lorentzian version of Theorem 1.

Theorem 2 *A spacelike hypersurface Σ in $\mathbb{R}^{n+1}$ has the convex hull property if and only if there is no elliptic point in Σ.*

The above result is a generalization of [1, Theorem 3], which states that if Σ is a compact spacelike hypersurface in $\mathbb{L}^{n+1}$ not contained in the convex hull of its boundary, then it necessarily has an elliptic point.

3 Spacelike Hypersurfaces with $H_R = H_L$

We can now state and prove our first main result.

Theorem 3 *Let Σ be a spacelike hypersurface in $\mathbb{L}^{n+1}$ such that $H_R = H_L$. Then not all the principal curvatures have the same sign. That is, there is no elliptic point in Σ.*

Proof We are going to work locally, so we can assume that there exists an open subset $\Omega \subseteq \mathbb{R}^n$ and a smooth function $u \in \mathcal{C}^\infty(\Omega)$ such that $\Sigma = \Sigma_u$. We define Σ^* as the graph of u over the following open set

$$\Omega^* = \{(x_1, \ldots, x_n) \in \Omega : Du(x_1, \ldots, x_n) \neq 0\}.$$

Given $p \in \Sigma^*$, we consider its corresponding level hypersurface contained in $\mathbb{R}^n$ and we call its lifting to Σ, S_c. We are working in a neighborhood of p, hence we can assume that S_c lies on Σ^*. Since $Du \neq 0$ in Ω^*, its distribution is integrable, so we can consider the integral curve through $\pi(p)$. We denote by α its lifting to Σ^*. We observe that a vector field tangent to α is $\alpha' = (Du, |Du|^2) \circ \pi$. Therefore, we have two submanifolds defined on a neighborhood of p which are orthogonal at p for both $\langle , \rangle_R$ and $\langle , \rangle_L$. Now, let $\{e_1, \ldots, e_{n-1}\}$ be an orthonormal basis of $T_p S_c$ in $\mathbb{R}^{n+1}$. These vectors are also orthonormal in $\mathbb{L}^{n+1}$, and orthogonal to α' with respect to both metrics. Then, Lemma 1 gives us the following relationships, where we have omitted the point p on behalf of simplicity

$$\kappa_{e_i}^R = -\frac{|e_i|_L^2}{|e_i|_R^2} \frac{\cos\theta}{\cosh\psi} \kappa_{e_i}^L = -\sqrt{\frac{1-|Du|^2}{1+|Du|^2}}\, \kappa_{e_i}^L, \quad i = 1, \ldots, n-1 \quad \text{and}$$

$$\kappa_{\alpha'}^R = -\frac{|\alpha'|_L^2}{|\alpha'|_R^2} \frac{\cos\theta}{\cosh\psi} \kappa_{\alpha'}^L = -\left(\frac{1-|Du|^2}{1+|Du|^2}\right)^{\frac{3}{2}} \kappa_{\alpha'}^L.$$

By denoting $A = \sqrt{\frac{1-|Du|^2}{1+|Du|^2}}$, we rewrite the previous expressions as

$$\kappa_{e_i}^R = -A\, \kappa_{e_i}^L, \qquad i = 1, \ldots, n-1 \qquad \text{and} \qquad \kappa_{\alpha'}^R = -A^3\, \kappa_{\alpha'}^L. \tag{3.1}$$

As we are dealing with orthogonal directions at p for both $\langle , \rangle_R$ and $\langle , \rangle_L$, and we are assuming $H_R = H_L$, from (2.1) we get

$$-\kappa_{e_1}^L - \cdots - \kappa_{e_{n-1}}^L - \kappa_{\alpha'}^L = \kappa_{e_1}^R + \cdots + \kappa_{e_{n-1}}^R + \kappa_{\alpha'}^R,$$

which jointly with (3.1) implies

$$\kappa_{e_1}^L + \cdots + \kappa_{e_{n-1}}^L = -(A^2 + A + 1)\kappa_{\alpha'}^L,$$

and so

$$\left(\kappa_{e_1}^L + \cdots + \kappa_{e_{n-1}}^L\right) \kappa_{\alpha'}^L \leq 0.$$

On the other hand, we can express the normal curvature given by a unitary vector $v = \sum_{i=1}^n a_i e_i^* \in T_p\Sigma$, where e_i^*, $i = 1, \ldots, n$, stand for the principal directions of Σ at p, in the next way

$$\kappa_v^L = a_1^2 k_1^L + \cdots + a_n^2 k_n^L.$$

Then, if we suppose that all the principal curvatures have the same sign, we get a contradiction.

Consider now $p \in \Sigma \setminus \Sigma^*$. If $p \in \text{int}(\Sigma \setminus \Sigma^*)$, then Σ is locally a horizontal hyperplane around p, and so $\kappa_i^L = \kappa_i^R = 0$ for all $i = 1, \ldots, n$. Otherwise $p \in \partial\Sigma^*$, and the result follows from a continuity argument. □

We have just proved that a hypersurface Σ such that $H_R = H_L$ does not have any elliptic point, which jointly which Theorem 1 leads to some interesting geometric consequences, to which the rest of the manuscript is devoted. The first of them is immediate from both results.

Theorem 4 *Let Σ be a compact spacelike hypersurface with (necessarily) nonempty boundary such that $H_R = H_L$. Then Σ is contained in the convex hull of its boundary.*

Let us recall that any spacelike hypersurface is locally a graph Σ_u over an open subset $\Omega \subseteq \mathbb{R}^n$. From now on, we will focus on spacelike graphs.

First, we present a uniqueness result for graphs which are asymptotic to a spacelike hyperplane, where the term asymptotic is defined as follows. We say that two entire graphs Σ_u and Σ_v over $\mathbb{R}^n$ are *asymptotic* if for every $\varepsilon > 0$ there exists a compact set $K \subset \mathbb{R}^n$ such that $|u(x_1, \ldots, x_n) - v(x_1, \ldots, x_n)| < \varepsilon$ for every $(x_1, \ldots, x_n) \in \mathbb{R}^n \setminus K$. Observe that, without loss of generality, we can consider that those compact sets are Euclidean balls of a certain radius. If we define the *width* of a set in $\mathbb{R}^n$ as the supremum of the diameter of the closed balls contained in it, the concept of asymptotic graphs is not only well defined in the case of entire graphs, but also in the case of graphs over a domain of infinite width, that is, a domain which contains closed balls of any radius. Notice that this definition is a generalization of the classical concept of width for a convex body, see [18].

Theorem 5 *The only spacelike graphs Σ_u in $\mathbb{L}^{n+1}$ defined over an open subset $\Omega \subseteq \mathbb{R}^n$ of infinite width, with $H_R = H_L$, and asymptotic to a spacelike hyperplane, are (pieces of) spacelike hyperplanes.*

Proof Let us notice that Σ_u is a graph over any spacelike hyperplane, and in particular, over the hyperplane to which it is asymptotic. To prove it, it is enough to observe that if some timelike line intersects Σ_u twice, the plane with director vector $e_{n+1} = (0, \ldots, 0, 1)$ and containing that line cuts Σ_u in a curve that is timelike at some point, which is a contradiction.

Let us denote by Π the hyperplane to which Σ_u is asymptotic and let $v \in C^\infty(\Omega')$ be the function such that $\Sigma_u = \Sigma_v$, $\Omega' \subseteq \Pi$ being the domain of definition of v. Notice that the width of Ω' is also infinite.

For any $\varepsilon > 0$ there exists $(y_1, \ldots, y_n) \in \Omega'$ and $R > 0$ such that $|v(x_1, \ldots, x_n)| < \varepsilon$ for every $(x_1, \ldots, x_n) \in \Omega' \setminus \bar{B}_{(y_1,\ldots,y_n)}(R)$. By Theorem 4, we know that the graph of the restriction of v to $\bar{B}_{(y_1,\ldots,y_n)}(R)$ is contained in the convex hull of its boundary. Therefore, $|v(x_1, \ldots, x_n)| \leq \varepsilon$ for all $(x_1, \ldots, x_n) \in \bar{B}_{(y_1,\ldots,y_n)}(R)$, so this inequality holds globally on Ω'. Taking limits when ε approaches 0, we conclude that $\Sigma_u = \Omega'$. □

4 A Quasi-linear PDE Related to Spacelike Hypersurfaces with $H_R = H_L$

As we have already mentioned, any spacelike hypersurface is locally a graph Σ_u over an open subset $\Omega \subseteq \mathbb{R}^n$. Thanks to (2.3), if we consider the differential operator given by

$$Q(u) = \operatorname{div}\left(\left(\frac{1}{\sqrt{1-|Du|^2}} - \frac{1}{\sqrt{1+|Du|^2}}\right) Du\right),$$

those graphs are the solutions to the equation

$$Q(u) = 0, \tag{4.1}$$

satisfying $|Du| < 1$. We will refer to the above equation as *the $H_R = H_L$ hypersurface equation*.

Let us observe firstly that (4.1) is an elliptic quasi-linear partial differential equation, everywhere except at those points where $Du = 0$, at which it is parabolic. In fact, it is a tedious but straightforward computation to show that (4.1) can be expressed as

$$\langle B(Du), D^2u \rangle = 0,$$

D^2u being the Hessian matrix of u with respect to the Euclidean metric of $\mathbb{R}^n$ and $B(Du)$ a symmetric and positive definite matrix on Du, everywhere except at those points where $Du = 0$, where it vanishes.

It is well known that the solutions to a second order elliptic quasi-linear partial differential equation for an analytic operator Q are always analytic, see [13] for a proof of this fact. Therefore, if u is a solution of (4.1) with $0 < |Du| < 1$, it is necessarily analytic. However, in general the analyticity of the solutions of (4.1) cannot be guaranteed.

Along this section, let Ω be a domain of $\mathbb{R}^n$, that is an open and bounded subset of $\mathbb{R}^n$. Given a domain $\Omega \subset \mathbb{R}^n$ and $\psi \in C^0(\partial\Omega)$, the Dirichlet problem related to the $H_R = H_L$ hypersurface equation consists in finding a solution $u \in C^2(\Omega) \cap C^0(\overline{\Omega})$ to the boundary value problem

$$\left.\begin{array}{r} Q(u) = 0 \text{ in } \Omega \\ |Du| < 1 \text{ in } \Omega \\ u = \psi \quad \text{on } \partial\Omega \end{array}\right\}. \tag{4.2}$$

As a consequence of a uniqueness theorem for the Dirichlet problem associated to quasilinear elliptic operators [11, Theorem 9.3], we get our next result.

Theorem 6 *Let $\Omega \subset \mathbb{R}^n$ be a domain with smooth boundary and $\psi \in C^0(\partial\Omega)$ such that the Dirichlet problem* (4.2) *admits a solution u without critical points. Then, the solution is unique.*

Remark 1 It is interesting to observe that [11, Theorem 9.3] holds under four assumptions on the operator defining the equation, one of which does not hold in our case. Since we are assuming the spatially condition $|Du| < 1$, the coefficients of Q are not well defined on the whole $\Omega \times \mathbb{R} \times \mathbb{R}^n$, as it is required in the last hypothesis, but just on $\Omega \times \mathbb{R} \times B_0(1)$. However, studying in detail the proof of the cited theorem, we can realize that it is sufficient to consider the coefficients defined on $\Omega \times \mathbb{R} \times B_0(1)$.

It is still more interesting to put emphasis on the fact that the proof does not work if the ellipticity fails somewhere. Therefore, we can not omit the hypothesis on the gradient of u. However, as a consequence of Theorem 4, we get the following result on the uniqueness of the Dirichlet problem under appropriate boundary values.

Theorem 7 *The only solutions to the Dirichlet problem* (4.2) *with affine boundary value are the affine functions.*

Proof Let u be a solution of (4.2) and Σ_u its associated graph. From Theorem 4, Σ_u is contained in the convex hull of its boundary, which is contained in a hyperplane. Therefore, the spacelike graph Σ_u must be also contained in the same hyperplane, and consequently u is affine.

In the previous reasoning it is crucial to observe that Theorem 4 also works for C^2-hypersurfaces. □

4.1 On Rotationally Invariant Spacelike Graphs

From now on, let us consider rotationally invariant spacelike graphs with respect to a vertical axis. Therefore, we can assume without loss of generality that the graph Σ_u is determined by a function

$$u(x_1, \ldots, x_n) = f(r), \quad r = x_1^2 + \cdots + x_n^2, \tag{4.3}$$

where $f \in \mathcal{C}^\infty(I)$ for certain $I \subseteq [0, +\infty)$. As an immediate consequence of Theorem 7 we get the following uniqueness result for entire rotationally invariant spacelike graphs with $H_R = H_L$.

Theorem 8 *The only entire spacelike graphs Σ_u determined by a function u given by* (4.3) *such that $H_R = H_L$ are the horizontal hyperplanes.*

Proof Given a positive constant R, any entire solution to the $H_R = H_L$ hypersurface equation of the form (4.3) is a solution of the Dirichlet problem (4.2) over $B_0(R)$ with constant boundary value. By Theorem 7, the function u must also be constant in $\overline{B_0}(R)$. The result is proven taking limits when R approaches infinity. □

Let us observe that Theorem 8 works not only for entire graphs, but for graphs defined over a ball centered at the origin of $\mathbb{R}^n$.

Acknowledgements The first author has been supported by a PhD grant of the University of Murcia, Ayuda de Iniciación a la Investigación and she is partially supported by the Spanish Ministry of Economy and Competitiveness and European Regional Development Fund (ERDF), project MTM2013-47828-C2-1-P. The second author is partially supported by MINECO/FEDER project reference MTM2015-65430-P, Spain, and Fundación Séneca project reference 19901/ GERM/15, Spain. Her work is a result of the activity developed within the framework of the Program in Support of Excellence Groups of the Región de Murcia, Spain, by Fundación Séneca, Science and Technology Agency of the Región de Murcia. The third author is partially supported by the Spanish Ministry of Economy and Competitiveness and European Regional Development Fund (ERDF), project MTM2016-78807-C2-1-P.

References

1. A.L. Albujer, M. Caballero, Geometric properties of surfaces with the same mean curvature in $\mathbb{R}^3$ and $\mathbb{L}^3$. J. Math. Anal. Appl. **445**, 1013–1024 (2017)
2. A.L. Albujer, M. Caballero, E. Sánchez, Some results for entire solutions to the $H_R = H_L$ surface equation (preprint)
3. J.A. Aledo, L.J. Alías, On the curvatures of bounded complete spacelike hypersurfaces in the Lorentz-Minkowski space. Manuscripta Math. **101**, 401–413 (2000)
4. L.J. Alías, A. Romero, M. Sánchez, Uniqueness of complete spacelike hypersurfaces of constant mean curvature in generalized Robertson-Walker spacetimes. Gen. Relativ. Gravit. **27**, 71–84 (1995)
5. S.N. Bernstein, Sur un théorème de géomètrie et ses applications aux équations aux dérivées partielles du type elliptique, Comm. Soc. Math. Kharkov **15**, 38–45 (1915-17)
6. E. Bombieri, E. De Giorgi, E. Giusti, Minimal cones and the Bernstein problem. Invent. Math. **7**, 243–268 (1969)
7. E. Calabi, Examples of Bernstein problems for some nonlinear equations. Proc. Symp. Pure Math. **15**, 223–230 (1970)
8. S.Y. Cheng, S.T. Yau, Maximal space-like hypersurfaces in the Lorentz-Minkowski spaces. Ann. of Math. **104**, 407–419 (1976)
9. S.-S. Chern, On the curvatures of a piece of hypersurface in euclidean space. Abh. Math. Sem. Univ. Hamburg **29**, 77–91 (1965)
10. H. Flanders, Remark on mean curvature. J. London Math. Soc. **41**, 364–366 (1966)

11. D. Gilbarg, N.S. Trudinger, *Elliptic Partial Differential Equations of Second Order*. Classics Mathematics. (Springer, Berlin, 2001, reprint of the 1998 edition), xiv+517 pp
12. E. Heinz, Über Flächen mit eineindeutiger Projektion auf eine Ebene, deren Krümmungen durch Ungleichungen eingeschränkt sind. Math. Ann. **129**, 451–454 (1955)
13. E. Hopf, Über den funktionalen, insbesondere den analytischen Charakter der Lösungen elliptischer Differentialgleichungen zweiter Ordnung. Math. Zentralbl. **34**, 194–233 (1932)
14. O. Kobayashi, Maximal surfaces in the 3-dimensional Minkowski space $\mathbb{L}^3$. Tokyo J. Math. **6**, 297–309 (1983)
15. E. Lee, H. Lee, Generalizations of the Choe-Hoppe helicoid and Clifford cones in Euclidean space. J. Geom. Anal. **27**, 817–841 (2017)
16. R. López, *Constant Mean Curvature Surfaces with Boundary*. Springer Monographs in Mathematics. (Springer, Heidelberg, 2013), xiv+292 pp
17. R. Osserman, The convex hull property of immersed manifolds. J. Differ. Geom. **6**, 267–270 (1971)
18. R. Schneider, *Convex Bodies: The Brunn-Minkowski Theory* (Cambridge University Press, Cambridge, 1993), xiv+490 pp

Space-Time Convex Functions and Sectional Curvature

Stephanie B. Alexander and William A. Karr

Abstract We show that in Lorentzian manifolds, sectional curvature bounds of the form $\mathcal{R} \leq K$, as defined by Andersson and Howard, are closely tied to space-time convex and λ-convex ($\lambda > 0$) functions, as defined by Gibbons and Ishibashi. Among the consequences are a natural construction of such functions, and an analog, that applies to domains of a new type, of a theorem of Alías, Bessa and deLira ruling out trapped submanifolds.

Keywords Space-time · Convex function · Distance · Hessian · Trapped submanifold

1 Introduction

A study of the possible uses of convex functions in General Relativity was initiated by Gibbons and Ishibashi, according to whom: "Convexity and convex functions play an important role in theoretical physics ... [and] also have important applications to geometry, including Riemannian geometry ... It is surprising therefore that, to our knowledge, that techniques making use of convexity and convex functions have played no great role in General Relativity" [8].

Gibbons and Ishibashi introduce and mainly consider "space-time convex" functions on Lorentzian manifolds (M, g), or more generally, functions f satisfying

$$\overline{\nabla}^2 f \geq \lambda g, \quad \lambda > 0.$$

They find examples and nonexamples of such functions on regions in cosmological space-times and black-hole space-times. They show, for example, that such functions

S.B. Alexander (✉) · W.A. Karr
1409 W. Green St., Urbana, IL 61801, USA
e-mail: sba@illinois.edu

W.A. Karr
e-mail: wkarr2@illinois.edu

M.A. Cañadas-Pinedo et al. (eds.), *Lorentzian Geometry and Related Topics*,
Springer Proceedings in Mathematics & Statistics 211,
DOI 10.1007/978-3-319-66290-9_2

rule out closed marginally inner and outer trapped surfaces. Curvature bounds do not arise in their considerations.

The purpose of this note is to show that sectional curvature bounds of the form $\mathcal{R} \leq K$ are closely tied to space-time convex functions. Among the consequences:

- A natural construction of such functions.
- New domains that cannot support trapped submanifolds, namely a full neighborhood of a point q, rather than a neighborhood of q in the chronological future of q as has been considered previously, in particular by Alías et al. [2].

The bound $\mathcal{R} \leq K$, introduced by Andersson and Howard [4], extends Sec $\leq K$ from the Riemannian to the semi-Riemannian setting by requiring spacelike sectional curvatures to be $\leq K$ and timelike ones to be $\geq K$. Equivalently, the curvature tensor is required to satisfy

$$g(R(v,w)v,w) \;\leq\; K\big(g(v,v)\,g(w,w) - g(v,w)^2\big).$$

For $\mathcal{R} \geq K$, reverse the inequalities.

In addition, we indicate connections between investigations that have been pursued independently by various authors, including:

- Comparison theorems for Lorentzian distance on domains in the chronological future of a source point or hypersurface on which the source has no Lorentzian cut points, given timelike sectional curvature controls (see, for example, [2, 3, 6, 12]).
- Hessian comparisons on level hypersurfaces in exponentially embedded neighborhoods of a point or hypersurface, given a sectional curvature bound of the form $\mathcal{R} \leq K$ or $\mathcal{R} \geq K$ [1, 4].
- Space-time convex functions [8].

1.1 Outline of the Paper

Section 2 is an introduction to space-time convex and λ-convex functions, as defined in [8].

Section 3 summarizes certain theorems about Hessian and Laplacian comparisons on the Lorentzian distance function from a point or achronal spacelike hypersurface, under comparisons on timelike sectional curvature [2, 3, 6, 12].

Section 4 describes results from [1, 4] concerning the conditions $\mathcal{R} \geq K$ and $\mathcal{R} \leq K$ in semi-Riemannian manifolds. In particular, in [4] Andersson and Howard prove a comparison theorem for matrix Ricatti equations which applies to the second fundamental forms of parallel families of hypersurfaces under curvature comparisons. In [1], this theorem is adapted to tubes around points; as an application, the geometric meaning of the bounds $\mathcal{R} \geq K$ and $\mathcal{R} \leq K$ is found by introducing signed lengths of geodesics.

In Sect. 5, we use this framework to rule out trapped submanifolds in an exponentially embedded neighborhood of a point in a space-time satisfying $\mathcal{R} \leq K$.

2 Space-Time Convex Functions

Definition 2.1 Given smooth functions $f : M \to \mathbf{R}$ and $\lambda : M \to \mathbf{R}$ on a semi-Riemannian manifold (M, g), f will be called *λ-convex* if the Hessian $\overline{\nabla}^2 f$ satisfies

$$\overline{\nabla}^2 f \geq \lambda\, g, \tag{2.1}$$

or equivalently,

$$(f \circ \gamma)'' \geq (\lambda \circ \gamma)\, g(\gamma', \gamma') \tag{2.2}$$

for every geodesic γ.

Suppose M is Lorentzian. We say f is *space-time λ-convex* if f is λ-convex for some *positive* function λ, and $\overline{\nabla}^2 f$ has Lorentzian signature.

Note that this definition differs from the classical definition of convexity in that the right-hand sides of (2.1) and (2.2) need not be positive when $\lambda > 0$. Rather, controlled concavity is allowed along timelike geodesics, and is imposed in the definition of space-time convexity.

One of the simplest examples of a space-time λ-convex function is

$$f(\mathbf{x}, t) = \frac{1}{2}(\mathbf{x} \cdot \mathbf{x} - \lambda t^2), \quad (\mathbf{x}, t) \in \mathbf{E}_1^{n+1}, \tag{2.3}$$

on Minkowski space for some constant $0 < \lambda \leq 1$.

As pointed out in [8], the geometric meaning of space-time convexity is that at each point, the forward light cone defined by the Hessian $\overline{\nabla}^2 f$ lies inside the light cone defined by the space-time metric.

Definition 2.1 is consistent with current Riemannian/Alexandrov usage of "λ-convex" (see [14]); and also with the definition of "space-time convex" in [8] except that our λ is a positive function and Gibbons and Ishibashi take λ to be a positive constant. (However, Definition 2.1 differs from the usage in [1].)

In [8], Gibbons and Ishibashi begin an investigation of the geometric implications of space-time convex functions. For example, they show that a space-time with a closed marginally inner and outer trapped surface cannot support a space-time convex function.

Here a *marginally inner and outer trapped surface* Σ is a spacelike submanifold of codimension 2 whose mean curvature vanishes.

Seeking examples of space-time convex functions, Gibbons and Ishibashi consider *Robertson-Walker spaces*

$$M = -I \times_f F,$$

that is, M is the product manifold $I \times F$ carrying the warped product metric

$$-d\tau^2 + f^2 ds_F^2$$

where $I = (a, b)$, $a \in [-\infty, \infty)$, $b \in (-\infty, \infty]$, $f : I \to \mathbf{R}_+$, and F has constant sectional curvature. They ask when the function

$$-\, f^2/2 \tag{2.4}$$

is space-time convex (here we use f to denote both the warping function and its lift to M). For instance, various cosmological charts are considered on de-Sitter space $\mathbf{dS}^{n+1}$ and anti-de-Sitter space $\mathbf{adS}^{n+1}$. One of these yields an affirmative answer: namely, the function (2.4) is space-time convex on the region

$$(0, \pi/2) \times_{\sin} \mathbf{H}^n$$

in $\mathbf{adS}^{n+1}$.

Gibbons and Ishibashi do not consider curvature bounds when seeking examples. The perspective of space-times with curvature bounds of the form $\mathcal{R} \leq K$ suggests an alternative, namely analogs of the "square norm" ((2.3) with $\lambda = 1$). For instance, these analogs yield space-time convex functions adapted to some of the domains in de-Sitter and anti-de-Sitter space considered in [8].

Our theorems show that space-time convex functions arise naturally in all Lorentzian manifolds satisfying $\mathcal{R} \leq K$.

3 Comparisons for Lorentzian Distance

Let us mention some related works concerning the Lorentian distance functions from a point or spacelike hypersurface. All these investigations are restricted to domains containing no Lorentzian cut points of the source point or hypersurface.

(1) In [6], Erkekoglu, Garcia-Rio and Kupeli prove Hessian and Laplacian comparison theorems for level sets of the Lorentzian distance function from points or from achronal spacelike hypersurfaces, in two space-times M and $\tilde{M}$. They consider corresponding timelike, distance-realizing unit geodesics in M and $\tilde{M}$, where sectional curvatures of two-planes tangent to the geodesics at corresponding values of the time parameter are no greater in M than in $\tilde{M}$. Some space-time singularity theorems are given.

(2) In [3], Alías, Hurtado and Palmer study the restriction of Lorentzian distance from a point or achronal spacelike hypersurface to a spacelike hypersurface satisfying the Omori-Yau maximum principle. Under constant bounds either above or below on timelike sectional (or Ricci) curvatures, they obtain sharp estimates on the mean curvature of such hypersurfaces.

(3) In [12], Impera studies Hessian and Laplacian comparisons for Lorentzian distance from a point, assuming timelike sectional curvatures are bounded above or below by a function of the Lorentzian distance. Estimates are obtained on the higher order mean curvatures of spacelike hypersurfaces satisfying the Omori-Yau maximum principle.
(4) In [2], Alías, Bessa and deLira prove nonexistence results and sharp mean curvature estimates for trapped submanifolds (of arbitrary codimension), based on comparison inequalities for the Laplacian of the restriction to a spacelike submanifold of the Lorentzian distance function from a point or achronal spacelike hypersurface. They use a weak Omori-Yau maximum principle equivalent to stochastic completeness.

4 Curvature Bounds $\mathcal{R} \leq K, \mathcal{R} \geq K$

Recall that $\mathcal{R} \leq K$ means that spacelike sectional curvatures are $\leq K$ and timelike ones are $\geq K$. For $\mathcal{R} \geq K$, reverse the inequalities. (Note that $\mathcal{R} \leq K \leq K'$ does not imply $\mathcal{R} \leq K'$!)

4.1 *Geometric Meaning*

Briefly, $\mathcal{R} \leq K$ means, as in the Riemannian case, that unit geodesics radiating from a point "repel" each other at least as much as in a space of constant curvature K, assuming the same initial conditions. However, repulsion here is meant in the *signed* sense. In particular, in the Lorentzian case, if the initial direction of variation of the geodesics is timelike, we see *negative repulsion*, that is, at least as much *attraction* as in a Lorentzian space of constant curvature K. This is explained below in Sects. 4.3 and 4.4.

4.2 *GRW Spaces*

Space-times satisfying $\mathcal{R} \leq K$ and $\mathcal{R} \geq K$ are abundant. We mention as examples, *generalized Robertson-Walker (GRW) spaces*, namely warped products $M = (-I) \times_f F$ for arbitrary Riemannian manifolds F.

Lemma 4.1 *[[1], Corollary 7.2] A GRW space $M = -I \times_f F$ satisfies $\mathcal{R} \leq K$ if and only if $f : I \to \mathbf{R}_+$ satisfies*

$$f'' \geq Kf,$$

and F either is one-dimensional or has sectional curvature $\leq C$ where

$$C = \inf\,(Kf^2 - (f')^2).$$

(For $\mathcal{R} \geq K$, reverse the inequalities and substitute sup *for* inf.*)*

4.3 Comparisons Based at a Point

Let M be a semi-Riemannian manifold, and U be the diffeomorphic image under $\exp_q$ of a star-shaped region in T_qM about O. Let $\gamma_{p,q}$ be the geodesic path in U from p to q that is distinguished by this diffeomorphism.

Define the *signed energy function* $E_q : U \to \mathbf{R}$ by

$$E_q(p) = (\operatorname{sgn}\gamma_{p,q})\,(\text{length } \gamma_{p,q})^2, \tag{4.1}$$

where sgn γ take values $1, 0, -1$ according to whether $\gamma_{p,q}$ is spacelike, null or timelike, respectively.

Signing was shown in [1] to be the key to geometric understanding of the curvature bounds $\mathcal{R} \leq K$ and $\mathcal{R} \geq K$. In particular, Andersson and Howard do not consider signed distance or energy.

For a fixed choice of $K \in \mathbf{R}$ and $q \in U$, define $f_{K,q} : U \to \mathbf{R}$ by

$$f_{K,q} = \sum_{n=1}^{\infty} \frac{(-K)^{n-1}(E_q)^n}{(2n)!} = \begin{cases} E_q/2, & K = 0, \\ (1 - \cos\sqrt{KE_q})/K, & K \neq 0. \end{cases} \tag{4.2}$$

Here the argument of cos may be imaginary, yielding $\cos it = \cosh t$.

Remark 4.2 Note that on the lift of U to T_qM by $(\exp_q)^{-1}$, the lift of $f_{K,q}$ is the square norm if $K = 0$, and an analog if $K \neq 0$. The possible values of $(1 - Kf_{K,q})$ are 1, $\cos\sqrt{|KE_q|}$ and $\cosh\sqrt{|KE_q|}$.

Set $f = f_{K,q}$ as in (4.2), for a fixed choice of K and q. Define the *modified shape operator* $S = S_{K,q}$ to be the self-adjoint operator associated with the Hessian of f, namely,

$$Sv = \overline{\nabla}_v \overline{\nabla} f \tag{4.3}$$

where $\overline{\nabla}$ is the covariant derivative of M.

Note that the levels of f are the levels of E_q. The form of f was chosen for analytic convenience (following [13]), so that if M has constant curvature K then S is a scalar multiple of the identity, namely

$$S = (1 - Kf)\,I.$$

The modified shape operator S has the following further properties: along a non-null geodesic from q, its restriction to normal vectors is a scalar multiple of the second fundamental form of the level hypersurfaces of E_q; it is smoothly defined on the regular set of E_q, hence along null geodesics from q (as the second fundamental forms are not); and finally, it satisfies a matrix Riccati equation along every geodesic from q, after reparametrization as an integral curve of $\overline{\nabla} f_{K,q}$.

The proof of the following theorem is by adapting to the set up just described, a comparison theorem of Andersson and Howard [4, Theorem 3.2] that applies to exponentially embedded tubes about hypersurfaces rather than points (see Sect. 4.5).

We say two geodesic segments σ and $\tilde{\sigma}$ in semi-Riemannian manifolds (M, g) and $(\tilde{M}, \tilde{g})$ *correspond* if they are defined on the same affine parameter interval and satisfy $g(\sigma', \sigma') = \tilde{g}(\tilde{\sigma}', \tilde{\sigma}')$. Let $R_{\sigma'}$ be the self-adjoint operator $R_{\sigma'} v = R(\sigma', v)\sigma'$, and similarly for $\tilde{R}_{\tilde{\sigma}'}$.

In the special case that the geodesics σ and $\tilde{\sigma}$ are timelike, the following theorem includes comparison inequalities of Erkekoglu, Garcia-Rio and Kupeli [6, Theorem 3.1] for level hypersurfaces of the Lorentzian distance from a point. However, here we are analyzing an exponentially embedded *neighborhood of a point* rather than restricting to the chronological future.

Theorem 4.3 *[1] Let M and $\tilde{M}$ be semi-Riemannian manifolds of the same dimension and index. For $q \in M$ and $\tilde{q} \in \tilde{M}$, let U and $\tilde{U}$ be diffeomorphic images under $\exp_q$ and $\exp_{\tilde{q}}$ respectively of star-shaped regions about the origin in $T_q M$ and $T_{\tilde{q}} \tilde{M}$. Let σ and $\tilde{\sigma}$ be corresponding nonnull geodesics in U and $\tilde{U}$ respectively, radiating from q and $\tilde{q}$.*

Identify linear operators on $T_{\sigma(t)} M$ with those on $T_{\tilde{\sigma}(t)} \tilde{M}$ by parallel translation to the basepoints, together with an isometry of $T_q M$ and $T_{\tilde{q}} \tilde{M}$ that identifies $\sigma'(0)$ and $\tilde{\sigma}'(0)$.

Suppose $R_{\sigma'} \leq \tilde{R}_{\tilde{\sigma}'}$ at corresponding points of σ and $\tilde{\sigma}$. Then the modified shape operators $S = S_{K,q}$ and $\tilde{S} = \tilde{S}_{K,q}$, as in (4.3), satisfy $S \geq \tilde{S}$ (that is, $S - \tilde{S}$ is positive semidefinite) at corresponding points of σ and $\tilde{\sigma}$.

Remark 4.4 A more precise statement of Theorem 4.3 localizes at a choice of unit geodesics $\sigma : [0, a] \to M$ and $\tilde{\sigma} : [0, a] \to \tilde{M}$, where σ and $\tilde{\sigma}$ have no conjugate points. Specifically, we let $U \subset M$ and $\tilde{U} \subset \tilde{M}$ be diffeomorphic images under $\exp_q$ and $\exp_{\tilde{q}}$ of truncated cones of the form $(0, a] \times_{\mathrm{id}} D$ and $(0, a] \times_{\mathrm{id}} \tilde{D}$ with vertices at the origin, where D and $\tilde{D}$ are open disks in the unit tangent "spheres" at q and $\tilde{q}$ centered at $\sigma'(0)$ and $\tilde{\sigma}'(0)$ respectively.

The following basic lemma is verified in [1]:

Lemma 4.5 *Let M be a semi-Riemannian space of constant curvature K, and U be the diffeomorphic image under $\exp_q$ of a star-shaped region in $T_q M$ about O. Then $f_{K,q} : U \to \mathbf{R}$ satisfies*

$$\overline{\nabla}^2 f_{K,q} = (1 - K f_{K,q})\, g.$$

Combining Theorem 4.3 and Lemma 4.5, we obtain:

Theorem 4.6 *[1] Let M be a semi-Riemannian manifold satisfying $\mathcal{R} \le K$. Let U be the diffeomorphic image under $\exp_q$ of a star-shaped region in T_qM about O. Assume $E_q : U \to \mathbf{R}$ satisfies $E_q < \pi^2/K$ if $K > 0$, and $E_q > \pi^2/K$ if $K < 0$. Then $f_{K,q} : U \to \mathbf{R}$ satisfies*

$$\overline{\nabla}^2 f_{K,q} \ge (1 - K f_{K,q})\, g.$$

That is, $f_{K,q}$ is $(1 - K f_{K,q})$-convex.

4.4 Geometric Characterization of $\mathcal{R} \le K$, $\mathcal{R} \ge K$

The geometric characterization of Riemannian sectional curvature bounds Sec $\le K$ or Sec $\ge K$ is given by local triangle comparisons with Riemannian space forms of constant curvature K. This is the basis of Alexandrov geometry, which extends the theory of Riemannian manifolds with sectional curvature bounds to highly singular spaces.

It turns out that this characterization by local triangle comparisons extends to semi-Riemannian manifolds *if we take lengths of geodesics to be signed.*

Recall that in a semi-Riemannian manifold, any point q has arbitrarily small normal neighborhoods U, that is, U is the diffeomorphic exponential image of a star-shaped domain in the tangent space of each of its points. There is a unique geodesic $\gamma_{p,q}$ in U between any two points $p, q \in U$.

Theorem 4.7 ([1]) *Let M be a semi-Riemannian manifold.*

(1) If M satisfies $\mathcal{R} \le K$ ($\mathcal{R} \ge K$), and U is a normal neighborhood for K, then the signed length of the geodesic between two points on any geodesic triangle of U is at most (at least) that for the corresponding points on a model *triangle with the same signed sidelengths in a semi-Riemannian model surface M_K with constant sectional curvature K. (For a nondegenerate triangle, M_K is uniquely determined, as is the comparison model triangle up to motion.)*

(2) Conversely, if these triangle comparisons hold in some normal neighborhood of each point of M, then $\mathcal{R} \le K$ ($\mathcal{R} \ge K$).

Remark 4.8 In [10] (see also [11]), Harris proves *global* purely timelike triangle comparisons in space-times of timelike sectional curvature bounded above. Thus the theorem of Harris is a timelike version for Lorentzian manifolds of Toponogov's Globalization Theorem for Riemannian manifolds of sectional curvature bounded below [17].

4.5 Comparisons for Parallel Families of Hypersurfaces

In [4, Theorem 3.2], Andersson and Howard prove a comparison theorem for matrix Riccati equations that applies to the second fundamental forms of parallel families of hypersurfaces of any signature in semi-Riemannian manifolds, rather than only to parallel families of spacelike hypersurfaces in Lorentzian manifolds as in Sect. 3. We give an analog in Theorem 4.3.

For $\mathcal{R} \leq 0$ and $\mathcal{R} \geq 0$, Andersson and Howard prove "gap" rigidity theorems of the type first proved for Riemannian manifolds with Sec ≤ 0 by Gromov [5], and with Sec ≥ 0 by Greene and Wu [9], respectively. As applications, they obtain rigidity results for semi-Riemannian manifolds with simply connected ends of constant curvature.

We remark that while in the Riemannian case, the Ricatti comparisons of [4] reduce to one-dimensional equations (see [13]), the semi-Riemannian case seems to require matrix-valued equations. Such increased complexity is perhaps not surprising, since semi-Riemannian curvature bounds above (say) share some behavior with Riemannian curvature bounds below as well as above.

5 Results

By Theorem 4.6 we have:

Corollary 5.1 *Let M be a semi-Riemannian manifold satisfying $\mathcal{R} \leq K$. Let U be the diffeomorphic image under* $\exp_q$ *of a star-shaped region in T_qM about O. Assume $E_q : U \to \mathbf{R}$ satisfies $E_q < \pi^2/4K$ if $K > 0$, and $E_q > \pi^2/4K$ if $K < 0$. Then $f_{K,q} : U \to \mathbf{R}$ is λ-convex with $\lambda = 1 - Kf_{K,q} > 0$ (where $f_{K,q}$ is defined in (4.1) and (4.2)).*

Moreover, $f_{K,q}$ is space-time convex on a neighborhood of q.

Proof By Theorem 4.6, $f_{K,q} : U \to \mathbf{R}$ is $(1 - Kf_{K,q})$-convex. By (4.2), setting $\lambda = 1 - Kf_{K,q}$, we have

$$\lambda = \begin{cases} 1, & K = 0, \\ \cos\sqrt{KE_q}, & K \neq 0. \end{cases} \tag{5.1}$$

Suppose $K > 0$. If $E_q \leq 0$, then $\lambda = \cosh\sqrt{|KE_q|} > 0$. If $0 \leq E_q < \pi^2/4K$, then $\lambda = \cos\sqrt{|KE_q|} > 0$. Similarly, for $K < 0$.

It remains to show $\overline{\nabla}^2 f_{K,q}$ has Lorentzian signature in a neighborhood of q. This follows by continuity, since for a unit timelike geodesic γ satisfying $\gamma(0) = q$ we have $(f_{K,q} \circ \gamma)''(0) = -1$. □

In defining the second fundamental form II and mean curvature vector field H of a k-dimensional submanifold Σ of a Lorentzian manifold M, we use the convention in relativity (the opposite of that in differential geometry):

$$\overline{\nabla}_X Y = \nabla_X Y - \mathrm{II}(X, Y), \tag{5.2}$$

$$H = \frac{1}{k} \sum_i \mathrm{II}(E_i, E_i), \tag{5.3}$$

where $\overline{\nabla}$ and ∇ denote the covariant derivatives on M and Σ respectively, and $\{E_1, \ldots, E_k\}$ is a local orthonormal frame on Σ.

We are going to follow [2] in considering submanifolds Σ satisfying the *weak maximum principle* of Pigola et al. [15], according to which for any smooth function u on Σ with $u^* = \sup_\Sigma u < +\infty$, there exists a sequence of points $p_n \in \Sigma$ such that

$$u(p_n) > u^* - \frac{1}{n} \quad \text{and} \quad \Delta u(p_n) < \frac{1}{n}.$$

Pigola, Rigoli, and Setti proved that Σ satisfies the weak maximum principle if and only if Σ has the probabilistic property of stochastic completeness [15, 16].

By [8, Proposition 8], domains carrying space-time convex functions f cannot contain closed marginally inner and outer trapped surfaces. The proof extends to the following proposition, which does not depend on the behavior of $\overline{\nabla}^2 f$ on causal vectors or on the codimension, and uses the weak maximum principal to extend from closed to stochastically complete submanifolds.

Theorem 5.2 *Let M be a Lorentzian manifold and $f : M \to \mathbf{R}$ be λ-convex on spacelike vectors for some function $\lambda : M \to \mathbf{R}$. Then:*

(i) M contains no stochastically complete spacelike submanifold with vanishing mean curvature and on which f is bounded above and λ has positive infimum.
(ii) If $\lambda > 0$, then M contains no closed spacelike submanifold with vanishing mean curvature.

Proof Suppose Σ is a spacelike k-dimensional submanifold with vanishing mean curvature. Let $\overline{\nabla}$ and ∇ denote the covariant derivatives on M and Σ respectively. Let II and H denote the second fundamental form and mean curvature vector field of Σ respectively. Let $u = f|_\Sigma : \Sigma \to \mathbf{R}$ denote the restriction of f to Σ.

Then for any $x \in T_p\Sigma$,

$$(\nabla^2 u)_p(x, x) = (\overline{\nabla}^2 f)_p(x, x) - g(\mathrm{II}_p(x, x), \overline{\nabla} f_p).$$

If $\{e_i\}$ is an orthonormal basis for $T_p\Sigma$, then

$$\Delta u(p) = \sum_{i=1}^{k} (\overline{\nabla}^2 f)_p(e_i, e_i) - k\, g(H_p, \overline{\nabla} f_p). \tag{5.4}$$

Since f is λ-convex and H vanishes, u satisfies

$$\Delta u \geq k\,\lambda|_{\Sigma}.$$

Thus if the Laplacian Δu is bounded below by $k \inf_{\Sigma} \lambda > 0$, and u is bounded above, then Σ cannot be stochastically complete. This proves (i), and (ii) follows. □

Definition 5.3 In a causally orientable Lorentzian manifold, a spacelike submanifold M whose mean curvature vector field is causal and future-pointing is called a *weakly future-trapped submanifold.*

Remark 5.4 Galloway and Senovilla prove that standard singularity theorems hold in Lorentzian manifolds of arbitrary dimension with closed trapped submanifolds of arbitrary co-dimension [7]. They point out that such submanifolds appear to have many common properties independent of the codimension.

The significance of the following theorem lies in using sectional curvature bounds to examine geometric properties of a full neighborhood of a point q, rather than restricting to the chronological future of q.

If in the following theorem we restrict U and $\tilde{U}$ to the chronological future of q and assume only timelike sectional curvature $\geq K$, then taking into account Remark 4.4, we obtain a result of Alías, Bessa and deLira ([2, Corollary 4.2]).

Theorem 5.5 *Let M be a Lorentzian manifold satisfying $\mathcal{R} \leq K$. Let U be a domain in M that is the diffeomorphic image under $\exp_q$ of a star-shaped region in T_qM about O. Suppose that $E_q : U \to \mathbf{R}$ is bounded above and satisfies $E_q < \pi^2/4K$ if $K > 0$ and $E_q > \pi^2/4K$ if $K < 0$.*

(i) Then U contains no stochastically complete spacelike submanifolds Σ with vanishing mean curvature, and such that $\sup E_q|_{\Sigma} < \pi^2/4K$ if $K > 0$ and $\inf E_q|_{\Sigma} > \pi^2/4K$ if $K < 0$.

(ii) More generally, U contains no stochastically complete, weakly future-trapped submanifold whose mean curvature vector field H satisfies

$$HE_q \leq 0, \tag{5.5}$$

and such that $\sup E_q|_{\Sigma} < \pi^2/4K$ if $K > 0$ and $\inf E_q|_{\Sigma} > \pi^2/4K$ if $K < 0$.

(iii) Suppose $K \neq 0$ and $U \subset \tilde{U}$, where $\tilde{U}$ is the diffeomorphic image under $\exp_q$ of a star-shaped region in T_qM about O, and $E_q : \tilde{U} \to \mathbf{R}$ satisfies $E_q < \pi^2/K$ if $K > 0$ and $E_q > \pi^2/K$ if $K < 0$. Then no stochastically complete, weakly future-trapped submanifold in $\tilde{U}$ that satisfies $HE_q \leq 0$ enters U.

Proof By Corollary 5.1, the function $f_{K,q} : U \to \mathbf{R}$ as defined in (4.1) and (4.2) is λ-convex with $\lambda = 1 - Kf_{K,q} > 0$. Suppose Σ is a stochastically complete, weakly future-trapped k-dimensional submanifold of U whose mean curvature vector field H satisfies $HE_q \leq 0$. Let $u : \Sigma \to \mathbf{R}$ be the restriction of $f_{K,q}$ to Σ. As in Eq. (5.4),

$$\Delta u\,(p) = \sum_{i=1}^{k} (\overline{\nabla}^2 f_{K,q})_p(e_i, e_i) - k\, g(H_p, (\overline{\nabla} f_{K,q})_p)$$
$$\geq k(1 - K f_{K,q}(p)) - k\, g(H_p, (\overline{\nabla} f_{K,q})_p).$$

Simple computation yields

$$\overline{\nabla} f_{K,q} = \begin{cases} \overline{\nabla} E_q/2, & K = 0, \\ \frac{\sin \sqrt{K E_q}}{2\sqrt{K E_q}} \overline{\nabla} E_q, & K \neq 0, \end{cases}$$

where the argument of sin can be imaginary here. The function $\sin \sqrt{K E_q}/(2\sqrt{K E_q})$ is nonnegative as long as $K E_q \leq \pi^2$. Thus, $g(H_p, (\overline{\nabla} f_{K,q})_p) \leq 0$ on U since $g(H, \overline{\nabla} E_q) = H E_q \leq 0$.

Since $(1 - K f_{K,q})|_\Sigma > 0$, we conclude that u is subharmonic and satisfies the differential inequality

$$\Delta u \geq k(1 - K u) > 0. \tag{5.6}$$

By (4.2), $u^* = \sup_\Sigma u < +\infty$. Since Σ is stochastically complete, we can apply the weak maximum principle to obtain a sequence of points $p_n \in \Sigma$ such that

$$u(p_n) > u^* - \frac{1}{n} \quad \text{and} \quad \Delta u(p_n) < \frac{1}{n}.$$

Evaluating (5.6) on p_n and taking $n \to \infty$, we obtain $1 - K u^* = \cos \sqrt{K E^*} = 0$, where $E^* = \lim_{n\to\infty} E_q(p_n)$.

If $K = 0$, this is impossible. If $K > 0$ and $\sup_\Sigma E_q < \pi^2/4K$, then $K E^* < \pi^2/4$ and $\cos \sqrt{K E^*} > 0$, a contradiction. Similarly, if $K < 0$ and $\inf_\Sigma E_q > \pi^2/4K$, then $K E^* < \pi^2/4$ and $\cos \sqrt{K E^*} > 0$, a contradiction. Hence (ii) and (i).

Finally, suppose $K \neq 0$ and $U \subset \tilde{U}$, where $\tilde{U}$ is the diffeomorphic image under $\exp_q$ of a star-shaped region in $T_q M$ about O, and $E_q : \tilde{U} \to \mathbf{R}$ satisfies $E_q < \pi^2/K$ if $K > 0$ and $E_q > \pi^2/K$if $K < 0$.

Suppose Σ is a stochastically complete spacelike submanifold in $\tilde{U}$. Choose a sequence $p_n \in \Sigma$ as above and let $E^* = \lim_{n\to\infty} E_q(p_n)$. By the above calculation, we know that $K E^* \geq \pi^2/4$. If $K > 0$, then $E^* \geq \pi^2/4K$ and if $K < 0$, $E^* \leq \pi^2/4K$. If $K > 0$, then $E^* = \inf_\Sigma E_q$ and if $K < 0$, then $E^* = \sup_\Sigma E_q$. Thus, in either situation Σ does not enter U. Hence (iii). □

Note that for $K > 0$, the bounds on E_q in Theorem 5.5 affect only spacelike geodesics, and for $K < 0$, only timelike geodesics.

Remark 5.6 Where a weakly future-trapped submanifold Σ intersects the causal future of q, the condition (5.5), namely $H E_q \leq 0$, is immediate. Where Σ enters the causal past of q, (5.5) implies $H = 0$. At a point p not causally related to q, (5.5) restricts H to a subcone of the cone of future directed vectors at p: either $H \neq 0$

lies in a closed half-cone of the cone of future directed vectors at p, or H is null and future-pointing, or $H = 0$.

For example, in Minkowski space, consider points $v \in \Sigma$ where v is spacelike. If v approaches $v_0 \neq 0$ in the future null cone of the origin 0, these half-cones approach the causal future cone of 0; if v approaches $v_0 \neq 0$ in the past null cone of 0, these half-cones approach the light ray through v_0.

6 Conclusion

We have demonstrated a close connection between sectional curvature bounds of the form $\mathcal{R} \leq K$ and space-time convex and λ-convex functions ($\lambda > 0$). We have constructed new λ-convex functions. We have used these functions to find new domains that do not support trapped submanifolds.

Our goal has been to explain some viewpoints and tools, rather than to give an exhaustive treatment. We plan a more systematic treatment of results in future.

Note that the λ-convex functions considered here are based on signed energy functions. It would be interesting to identify other classes of λ-convex functions to which Theorem 5.2 can be applied.

Acknowledgements This work was partially supported by a grant from the Simons Foundation (#209053 to Stephanie Alexander). This material is partially based upon work supported by the National Science Foundation Graduate Research Fellowship to William Karr under Grant No. DGE 11-44245. Any opinion, findings, and conclusions or recommendations expressed in this material are those of the authors and do not necessarily reflect the views of the National Science Foundation.

References

1. S. Alexander, R. Bishop, Lorentz and semi-Riemannian spaces with Alexandrov curvature bounds. Comm. Anal. Geom. **16**, 251–282 (2008)
2. L. Alías, G. Bessa, J. de Lira, Geometric analysis of the Lorentzian distance function on trapped submanifolds. Class. Quantum Gravity **33**, 125007 (28 pp) (2016)
3. L. Alías, A. Hurtado, V. Palmer, Geometric analysis of Lorentzian distance function on spacelike hypersurfaces. Trans. Amer. Math. Soc. **362**, 5083–5106 (2010)
4. L. Andersson, R. Howard, Comparison and rigidity theorems in semi-Riemannian geometry. Comm. Anal. Geom. **6**, 819–877 (1998)
5. W. Ballmann, M. Gromov, V. Schroeder, *Manifolds of Nonpositive Curvature*. Progress in Mathematics, vol. 61 (Birkhauser, Boston, 1985)
6. F. Erkekoglu, E. Garcia-Rio, D.N. Kupeli, On level sets of Lorentzian distance function. Gen. Rel. Grav. **35**, 1597–1615 (2003)
7. G. Galloway, J. Senovilla, Singularity theorems based on trapped submanifolds of arbitrary curvature. Class. Quantum Gravity **27**, (10pp) (2010)
8. G. Gibbons, A. Ishibashi, Convex functions and spacetime geometry. Class. Quantum Gravity **18**(21), 4607–4627 (2001)
9. R. Greene, H. Wu, Gap theorems for noncompact Riemannian manifolds. Duke Math. J. **49**, 731–756 (1982)

10. S. Harris, A triangle comparison theorem for Lorentz manifolds. Indiana Math. J. **31**, 289–308 (1982)
11. S. Harris, Appendix A: Jacobi fields and Toponogov's theorem for Lorentzian manifolds, in *Global Lorentzian Geometry*, 2nd edn., ed. by J. Beem, P. Ehrlich, K. Easley (Dekker, New York, 1996), pp. 567–572
12. D. Impera, Comparison theorems in Lorentzian geometry and applications to spacelike hypersurfaces. J. Geom. Phys. **62**, 412–426 (2012)
13. H. Karcher, Riemannian comparison constructions, in *Global Differential Geometry*, ed. by S.S. Chern. MAA Studies in Mathematics, vol. 27 (Mathematical Association of America, 1989)
14. A. Petrunin, *Semiconcave Functions in Alexandrov's Geometry*. Surveys in Differential Geometry, vol. 11 (Int. Press, Somerville, MA, 2007), pp. 137–201
15. S. Pigola, M. Rigoli, A. Setti, *Maximum Principles on Riemannian Manifolds and Applications*, vol. 174, no. 822. (Memoirs of the American Mathematical Society, Providence, RI, 2005)
16. S. Pigola, M. Rigoli, A. Setti, *Vanishing and Finiteness Results in Geometric Analysis: A Generalization of the Bochner Technique*. Progress in Mathematics, vol. 266 (Birkhauser, Basel, 2008)
17. V. Toponogov, Riemannian spaces of curvature bounded below. Usp. Mat. Nauk. **14**, 87–130 (1959)

Recent Results on Oscillator Spacetimes

Giovanni Calvaruso

Abstract The Lorentzian oscillator group (G_μ, g_a), that is, the four-dimensional oscillator group G_μ, together with the family g_a of left-invariant metrics obtained generalizing its bi-invariant metric, 'is probably the most relevant naturally reductive Lorentzian example in the literature' Batat et al. (Differ Geometry Appl 41: 48–64, [1]) . We describe an explicit system of global coordinates for the Lorentzian oscillator group, and use it to compute its symmetries and solutions to the Ricci soliton equation.

Keywords Oscillator group · Symmetries · Collineations · Ricci solitons

2010 Mathematics Subject Classification 53C50, 53B30, 35A01

1 Introduction

Consider the $4D$ Lie algebra $\mathfrak{g}$ =span$\{H, P, Q, E\}$, described by the non-vanishing Lie brackets

$$[H, P] = -Q, \qquad [H, Q] = P, \qquad [P, Q] = E.$$

An explicit realization of this Lie algebra is the following: taking

$$\begin{array}{ll} H = \frac{1}{2}\left(x^2 - \frac{\partial^2}{\partial x^2}\right) \text{ (“}Hamiltonian\text{”)}, & P = \frac{\partial}{\partial x} \text{ (“}linear\ momentum\text{”)}, \\ Q = x \text{ (“}position\text{”)}, & E = 1, \end{array}$$

one has, for any $f = f(x)$,

$$[H, P](f) = -Q(f), \qquad [H, Q](f) = P(f), \qquad [P, Q](f) = E(f),$$

G. Calvaruso (✉)
Dipartimento di Matematica e Fisica “E. de Giorgi”, Universitá del Salento,
Prov. Lecce-Arnesano, 73100 Lecce, Italy
e-mail: giovanni.calvaruso@unisalento.it

M.A. Cañadas-Pinedo et al. (eds.), *Lorentzian Geometry and Related Topics*,
Springer Proceedings in Mathematics & Statistics 211,
DOI 10.1007/978-3-319-66290-9_3

As the above describes the harmonic oscillator problem, $\mathfrak{g}$ has been called the *oscillator algebra* [18]

The *oscillator group* is four-dimensionally connected, simply connected Lie group corresponding to the oscillator algebra. This group is given by $\mathbb{R} \times \mathbb{C} \times \mathbb{R}$, with the product

$$(x_1, z_1, y_1) \cdot (x_2, z_2, y_2) = (x_1 + x_2 + \frac{1}{2}\mathrm{Im}(\bar{z_1} e^{iy_1} z_2), z_1 + e^{iy_1} z_2, y_1 + y_2).$$

After its introduction [18], the oscillator group has been extended to a one-parameter family G_μ ($\mu > 0$), corresponding to the Lie algebra

$$[H, P] = -\mu Q, \qquad [H, Q] = \mu P, \qquad [P, Q] = E, \tag{1.1}$$

then generalized in any even dimension $2n \geq 4$, and proved several times and in very different frameworks to be an interesting object to study, both in differential geometry and in mathematical physics. Just to cite a few examples, the following properties of the oscillator group(s) have been investigated: Yang-Baxter [2] and Einstein-Yang-Mills Eq. [11], parallel hypersurfaces [7], Ricci collineations and other curvature symmetries [9], homogeneous structures [12], electromagnetic waves [15] and the Laplace–Beltrami operator [16].

Oscillator Lie groups (of any even dimension) play a very special role. In fact, apart from the trivial examples given by extensions of commutative Lie groups, they are the only solvable Lie groups carrying a bi-invariant Lorentzian metric. In dimension four, the bi-invariant metric g_0 has been generalized to a one-parameter family g_a, $-1 < a < 1$, of left-invariant Lorentzian metrics, of which g_0 is the only bi-invariant and symmetric example [12]. Equipped with these left-invariant Lorentzian metrics, the oscillator group (G_μ, g_a) is a well-known homogeneous spacetime [10], and 'one of the most celebrated examples of Lorentzian naturally reductive spaces' [1].

The success of the oscillator group as a source of interesting behaviours is due to the fact that this object has very nice and explicit descriptions, both algebraically and analytically, so that one can borrow techniques both from Analysis and Algebra to work on it. An essential tool to investigate the properties of (G_μ, g_a) is its explicit matrix description, first obtained in [18] and then adapted in [7] for any $\mu > 0$.

In Sect. 2, we shall describe this explicit matrix realization of (G_μ, g_a), and illustrate how it was applied to obtain the following results:

Symmetries. If (M, g) denotes a Lorentzian manifold and T a tensor on (M, g), codifying some either mathematical or physical quantity, a *symmetry* of T is a one-parameter group of diffeomorphisms of (M, g), leaving T invariant.

Hence, a symmetry corresponds to a vector field X satisfying $\mathcal{L}_X T = 0$, where $\mathcal{L}$ denotes the Lie derivative. Well-known examples of symmetries are: Killing vector fields ($T = g$), homotheties and conformal motions, curvature collineations (T=R is the curvature tensor), Weyl collineations (T=W is Weyl conformal curvature tensor), Ricci collineations (T=ϱ is the Ricci tensor) and matter collineation ($T = \varrho - \frac{1}{2}\tau g$ is the energy-momentum tensor). In Sect. 3, we shall report the complete classification of symmetries of homogeneous spacetimes (G_μ, g_a) obtained in [9].

Ricci solitons. A *Ricci soliton* is a pseudo-Riemannian manifold (M, g) admitting a smooth vector field X, such that

$$\mathcal{L}_X g + \varrho = \lambda g, \tag{1.2}$$

where $\mathcal{L}_X$ and ϱ, respectively, denote the Lie derivative in the direction of X and the Ricci tensor and λ is a real number. A Ricci soliton is said to be either *shrinking*, *steady* or *expanding*, according to whether $\lambda > 0$, $\lambda = 0$ or $\lambda < 0$ respectively.

In a suitable set of global coordinates on (G_μ, g_a), the above Ricci soliton Eq. (1.2) translates into a system of PDE. In Sect. 4, we shall describe this system for (G_μ, g_a) and its solutions, obtained in [4], proving that all metrics g_a are Ricci solitons.

2 The Oscillator Group

For any real number $\mu > 0$, consider the four-dimensional Lie algebra described in (1.1). Generalizing the argument used in [18] for the case $\mu = 1$, one can see that Eq. (1.1) hold for matrices H, P, Q, E given by

$$H = \begin{pmatrix} 0 & 0 & 0 & 0 \\ 0 & 0 & -\mu & 0 \\ 0 & \mu & 0 & 0 \\ 0 & 0 & 0 & 0 \end{pmatrix}, \quad P = \begin{pmatrix} 0 & 0 & 1 & 0 \\ 0 & 0 & 0 & 1 \\ 0 & 0 & 0 & 0 \\ 0 & 0 & 0 & 0 \end{pmatrix},$$

$$Q = \begin{pmatrix} 0 & -1 & 0 & 0 \\ 0 & 0 & 0 & 0 \\ 0 & 0 & 0 & 1 \\ 0 & 0 & 0 & 0 \end{pmatrix}, \quad E = \begin{pmatrix} 0 & 0 & 0 & 2 \\ 0 & 0 & 0 & 0 \\ 0 & 0 & 0 & 0 \\ 0 & 0 & 0 & 0 \end{pmatrix}.$$

Then, the oscillator group corresponds to the four-dimensional subgroup of $\mathrm{GL}(4, \mathbb{R})$

$$G_\mu = \{M_\mu(x_1, x_2, x_3, x_4) \in \mathrm{GL}(4, \mathbb{R}) \mid x_1, x_2, x_3, x_4 \in \mathbb{R}\},$$

having as typical group element

$$M_\mu(x_1, x_2, x_3, x_4) = \exp(x_1 E) \exp(x_2 P) \exp(x_3 Q) \exp(x_4 H),$$

that is,

$$M_\mu(x_1, x_2, x_3, x_4) = \begin{pmatrix} 1 & x_2 \sin(\mu x_4) - x_3 \cos(\mu x_4) & x_2 \cos(\mu x_4) + x_3 \sin(\mu x_4) & 2x_1 + x_2 x_3 \\ 0 & \cos(\mu x_4) & -\sin(\mu x_4) & x_2 \\ 0 & \sin(\mu x_4) & \cos(\mu x_4) & x_3 \\ 0 & 0 & 0 & 1 \end{pmatrix}.$$

More precisely, the above M_μ is in fact a covering, providing a diffeomorphism between G_μ and $\mathbb{R}^3 \times \mathbb{R}/\frac{2\pi}{\mu}\mathbb{Z}$.

Now denote by ∂_{x_j} the coordinate vector field corresponding to the x_j-coordinate. As a matrix in $\mathfrak{gl}(4, \mathbb{R})$, this corresponds to $\frac{\partial M_\mu}{\partial x_j}(x_1, x_2, x_3, x_4)$. We can write down explicitly a basis $\{e_1, e_2, e_3, e_4\}$ of left-invariant vector fields on G_λ such that $(e_j)_I = (\partial_{x_j})_I$, where $I = M_\mu(0, 0, 0, 2k\pi/\mu)$ for any integer k is the identity matrix:

$$\begin{aligned} e_1 &= \partial_{x_1}, \\ e_2 &= -x_3 \cos(\mu x_4)\partial_{x_1} + \cos(\mu x_4)\partial_{x_2} + \sin(\mu x_4)\partial_{x_3}, \\ e_3 &= x_3 \sin(\mu x_4)\partial_{x_1} - \sin(\mu x_4)\partial_{x_2} + \cos(\mu x_4)\partial_{x_3}, \\ e_4 &= \partial_{x_4}. \end{aligned} \tag{2.1}$$

Remark that the coordinates x_2 and x_3 do not play symmetric roles, although P and Q play similar roles in $\mathfrak{g}_\mu$, as the formula for M_μ is not symmetric in P and Q. Starting from (2.1), a direct calculation yields that the only non-vanishing Lie brackets of two of these left-invariant vector fields are given by

$$[e_2, e_3] = e_1, \qquad [e_2, e_4] = -\mu e_3, \qquad [e_3, e_4] = \mu e_2. \tag{2.2}$$

Comparing (2.2) with (1.1), we see that the Lie algebra of G_μ coincides with the oscillator Lie algebra, via the identifications $E = e_1$, $P = e_2$, $Q = e_3$ and $H = e_4$.

We now consider on G_μ the one-parameter family of left-invariant Lorentzian metrics, described by

$$\langle e_1, e_1\rangle = \langle e_4, e_4\rangle = a, \qquad \langle e_2, e_2\rangle = \langle e_3, e_3\rangle = 1, \qquad \langle e_1, e_4\rangle = \langle e_4, e_1\rangle = 1, \tag{2.3}$$

for any real constant a with $-1 < a < 1$. For $a = 0$, one has the bi-invariant metric on the oscillator group. In all other cases, g_a is only left-invariant. Using (2.1), it is easy to check that in the coordinates (x_1, x_2, x_3, x_4), these metrics are explicitly given by

$$g_a = adx_1^2 + 2ax_3dx_1dx_2 + (1 + ax_3^2)dx_2^2 + dx_3^2 + 2dx_1dx_4 + 2x_3dx_2dx_4 + adx_4^2. \tag{2.4}$$

The above explicit description of these metrics makes possible to compute their Levi-Civita connection and curvature. With respect to the basis $\{\partial_i\}$ of coordinate vector fields, the Levi-Civita connection ∇ is completely determined by the following possibly non-vanishing components:

$$\begin{aligned} &\nabla_{\partial_1}\partial_2 = -\tfrac{a}{2}\partial_3, && \nabla_{\partial_1}\partial_3 = -\tfrac{ax_3}{2}\partial_1 + \tfrac{a}{2}\partial_2, && \nabla_{\partial_2}\partial_2 = -ax_3\partial_3, \\ &\nabla_{\partial_2}\partial_3 = \tfrac{1-ax_3^2}{2}\partial_1 + \tfrac{ax_3}{2}\partial_2, && \nabla_{\partial_2}\partial_4 = -\tfrac{1}{2}\partial_3, && \nabla_{\partial_3}\partial_4 = -\tfrac{x_3}{2}\partial_1 + \tfrac{1}{2}\partial_2. \end{aligned} \tag{2.5}$$

Remark 2.1 It may be observed that the explicit description (2.4) is the same for any value of μ, since this parameter was used in (2.1). Moreover, we remark that if $a \neq a'$, then (G, g_a) is not homothetic to $(G, g_{a'})$ (in particular, they are not isometric).

In fact, for the Levi-Civita connections ∇ and ∇' of g_a and $g_{a'}$, respectively, we have $\nabla_{\partial_1}\partial_2 = -\frac{a}{2}\partial_3 \neq -\frac{a'}{2}\partial_3 = \nabla'_{\partial_1}\partial_2$.

We can then describe the Riemann-Christoffel curvature tensor R of (G_λ, g_a) with respect to $\{\partial_i\}$, computing $R(\partial_i, \partial_j)\partial_k = \nabla_{\partial_i}\nabla_{\partial_j}\partial_k - \nabla_{\partial_j}\nabla_{\partial_i}\partial_k$ for all indices i, j, k. Denoting by R_{ij}, the matrix describing $R(\partial_i, \partial_j)$ with respect to the basis of coordinate vector fields, we have

$$R_{12} = \begin{pmatrix} \frac{a^2x_3}{4} & \frac{a^2x_3^2+a}{4} & 0 & \frac{ax_3}{4} \\ -\frac{a^2}{4} & -\frac{a^2x_3}{4} & 0 & -\frac{a}{4} \\ 0 & 0 & 0 & 0 \\ 0 & 0 & 0 & 0 \end{pmatrix}, \qquad R_{13} = \begin{pmatrix} 0 & 0 & \frac{a}{4} & 0 \\ 0 & 0 & 0 & 0 \\ -\frac{a^2}{4} & -\frac{a^2x_3}{4} & 0 & -\frac{a}{4} \\ 0 & 0 & 0 & 0 \end{pmatrix},$$

$$R_{14} = 0, \qquad R_{23} = \begin{pmatrix} 0 & 0 & ax_3 & 0 \\ 0 & 0 & -\frac{3a}{4} & 0 \\ -\frac{a^2x_3}{4} & \frac{3a-a^2x_3^2}{4} & 0 & -\frac{ax_3}{4} \\ 0 & 0 & 0 & 0 \end{pmatrix},$$

$$R_{24} = \begin{pmatrix} -\frac{ax_3}{4} & -\frac{ax_3^2+1}{4} & 0 & -\frac{x_3}{4} \\ \frac{a}{4} & \frac{ax_3}{4} & 0 & \frac{1}{4} \\ 0 & 0 & 0 & 0 \\ 0 & 0 & 0 & 0 \end{pmatrix}, \quad R_{34} = \begin{pmatrix} 0 & 0 & -\frac{1}{4} & 0 \\ 0 & 0 & 0 & 0 \\ \frac{a}{4} & \frac{ax_3}{4} & 0 & \frac{1}{4} \\ 0 & 0 & 0 & 0 \end{pmatrix}.$$

Next, the Ricci tensor of (G_μ, g_a) is obtained as a contraction of the curvature tensor, by the equation $\varrho(X, Y) = \mathrm{tr}(Z \mapsto R(Z, X)Y)$. With respect to $\{\partial_i\}$, the Ricci tensor ϱ and the Ricci operator Q, defined by $g(QX, Y) := \varrho(X, Y)$, are determined by the following matrices:

$$\varrho = \begin{pmatrix} \frac{1}{2}a^2 & \frac{1}{2}a^2x_3 & 0 & \frac{1}{2}a \\ \frac{1}{2}a^2x_3 & \frac{1}{2}a(ax_3^2-1) & 0 & \frac{1}{2}ax_3 \\ 0 & 0 & -\frac{1}{2}a & 0 \\ \frac{1}{2}a & \frac{1}{2}ax_3 & 0 & \frac{1}{2} \end{pmatrix}, \quad Q = \begin{pmatrix} \frac{1}{2}a & ax_3 & 0 & \frac{1}{2} \\ 0 & -\frac{1}{2}a & 0 & 0 \\ 0 & 0 & -\frac{1}{2}a & 0 \\ 0 & 0 & 0 & 0 \end{pmatrix}. \tag{2.6}$$

Comparison between Eqs. (2.6) and (2.4) easily yields that these metrics are never Einstein (see also [17]). Moreover, the Ricci eigenvalues are 0, $\frac{1}{2}a$ and $-\frac{1}{2}a$ (twice), and so, the Ricci tensor is degenerate, for any value of a. Finally, the Weyl conformal tensor W is completely determined by the following possibly non-vanishing matrices W_{ij}, describing $W(\partial_i, \partial_j)$ with respect to the coordinate vector fields $\{\partial_i\}$:

$$
W_{12} = \begin{pmatrix} \frac{a^2 x_3}{6} & \frac{a(1+ax_3^2)}{6} & 0 & \frac{ax_3}{6} \\ -\frac{a^2}{6} & -\frac{a^2 x_3}{6} & 0 & -\frac{a}{6} \\ 0 & 0 & 0 & 0 \\ 0 & 0 & 0 & 0 \end{pmatrix}, \qquad W_{13} = \begin{pmatrix} 0 & 0 & \frac{a}{6} & 0 \\ 0 & 0 & 0 & 0 \\ -\frac{a^2}{6} & -\frac{a^2 x_3}{6} & 0 & -\frac{a}{6} \\ 0 & 0 & 0 & 0 \end{pmatrix},
$$

$$
W_{14} = \begin{pmatrix} -\frac{a}{3} & -\frac{ax_3}{3} & 0 & -\frac{a^2}{3} \\ 0 & 0 & 0 & 0 \\ 0 & 0 & 0 & 0 \\ \frac{a^2}{3} & \frac{a^2 x_3}{3} & 0 & \frac{a}{3} \end{pmatrix}, \qquad W_{23} = \begin{pmatrix} 0 & 0 & \frac{ax_3}{2} & 0 \\ 0 & 0 & -\frac{a}{3} & 0 \\ -\frac{a^2 x_3}{6} & \frac{a(2-ax_3^2)}{6} & 0 & -\frac{ax_3}{6} \\ 0 & 0 & 0 & 0 \end{pmatrix}, \tag{2.7}
$$

$$
W_{24} = \begin{pmatrix} -\frac{ax_3}{2} & -\frac{ax_3^2}{2} & 0 & -\frac{a^2 x_3}{2} \\ \frac{a}{6} & \frac{ax_3}{6} & 0 & \frac{a^2}{6} \\ 0 & 0 & 0 & 0 \\ \frac{a^2 x_3}{3} & \frac{a(2ax_3^2-1)}{6} & 0 & \frac{ax_3}{3} \end{pmatrix}, \; W_{34} = \begin{pmatrix} 0 & 0 & 0 & 0 \\ 0 & 0 & 0 & 0 \\ \frac{a}{6} & \frac{ax_3}{6} & 0 & \frac{a^2}{6} \\ 0 & 0 & -\frac{a}{6} & 0 \end{pmatrix}.
$$

In particular, by (2.7), g_a is locally conformally flat if and only if $a = 0$. Starting from the above equations, it is also easy to check the well-known fact that $\nabla R = 0$ (that is, (G_μ, g_a) is locally symmetric) if and only if $a = 0$.

Remark 2.2 Using Eq. (2.1), one can determine the components u^j of a vector field X with respect to the basis of left-invariant vector fields $\{e_1, e_2, e_3, e_4\}$, in terms of its components X^i with respect to the basis of coordinate vector fields $\{\partial_1, \partial_2, \partial_3, \partial_4\}$ (and conversely). Explicitly, if $X = X^i \partial_i = u^j e_j$, then

$$
\begin{aligned}
&(u^1, u^2, u^3, u^4) \\
&\quad = (X^1 + x_3 X^2, \cos(\mu x_4) X^2 + \sin(\mu x_4) X^3, \cos(\mu x_4) X^3 - \sin(\mu x_4) X^2, X^4).
\end{aligned} \tag{2.8}
$$

In particular, X is a left-invariant vector field if and only if the above Eq. (2.8) holds for some constants u^j, $j = 1, \ldots, 4$.

3 Symmetries of the Oscillator Group

In all the results concerning (G_μ, g_a), one constantly finds a difference in the behaviours of the bi-invariant metric g_0 and of the remaining left-invariant metrics g_a, $a \neq 0$. This already starts with the classification of Killing, homothetic and conformal vector fields, reported in the following.

Theorem 3.1 *Let $X = X^1 \partial_1 + X^2 \partial_2 + X^3 \partial_3 + X^4 \partial_4$ be an arbitrary vector field on the oscillator group (G_μ, g_a).*

(i) X is a Killing vector field if and only if one of the following cases occurs:

(a) $a = 0$ and

$$\begin{cases} X^1 = \frac{c_1}{2}(x_3^2 - x_2^2) - c_2 x_2 - x_3(c_3 \cos(x_4) - c_4 \sin(x_4)) + c_5, \\ X^2 = -c_1 x_3 + c_3 \cos(x_4) - c_4 \sin(x_4) + c_6, \\ X^3 = c_1 x_2 + c_2 + c_3 \sin(x_4) + c_4 \cos(x_4), \\ X^4 = c_7. \end{cases}$$

(b) $a \neq 0$ *and*

$$X^1 = \frac{c_1}{2}(x_3^2 - x_2^2) - c_2 x_2 + c_3, \quad X^2 = -c_1 x_3 + c_4, \quad X^3 = c_1 x_2 + c_2, \quad X^4 = c_5.$$

(ii) Proper homothetic vector fields only occur for $a = 0$. X *is a homothetic but not Killing vector field (*$\mathcal{L}_X g = \eta g$ *for some real constant* $\eta \neq 0$*) if and only if* $a = 0$ *and*

$$\begin{cases} X^1 = \eta x_1 + \frac{c_1}{2}(x_3^2 - x_2^2) - c_2 x_2 - x_3(c_3 \cos(x_4) - c_4 \sin(x_4)) + c_5, \\ X^2 = \frac{\eta}{2} x_2 - c_1 x_3 + c_3 \cos(x_4) - c_4 \sin(x_4) + c_6, \\ X^3 = \frac{\eta}{2} x_3 + c_1 x_2 + c_2 + c_3 \sin(x_4) + c_4 \cos(x_4), \\ X^4 = c_7. \end{cases}$$

(iii) X *is an affine Killing vector field if and only if one of the following cases occurs:*

(a) $a = 0$ *and*

$$\begin{cases} X^1 = \frac{c_1}{2}(x_3^2 - x_2^2) + 2c_2 x_1 - c_3 x_2 + c_4 x_4 - x_3(c_5 \cos(x_4) - c_6 \sin(x_4)) + c_7, \\ X^2 = c_2 x_2 - c_1 x_3 + c_5 \cos(x_4) - c_6 \sin(x_4) + c_8, \\ X^3 = c_2 x_3 + c_1 x_2 + c_3 + c_5 \sin(x_4) + c_6 \cos(x_4), \\ X^4 = c_9, \quad c_2^2 + c_4^2 \neq 0. \end{cases}$$

(b) $a \neq 0$ *and*

$$\begin{cases} X^1 = \frac{c_1}{2}(x_3^2 - x_2^2) - c_2 x_2 + c_3 x_4 + c_4, \\ X^2 = -c_1 x_3 + c_5, \\ X^3 = c_1 x_2 + c_2, \\ X^4 = -a c_3 x_4 + c_6, \quad c_3 \neq 0. \end{cases}$$

In all the formulas above and in the remaining part of the paper, c_i *will denote some real constants, for all indices* i.

For the proof of the above classification result, we refer to [9]. We only report below as an example the system of PDE which determines the Killing vector fields. Let $X = X^1\partial_1 + X^2\partial_2 + X^3\partial_3 + X^4\partial_4$ be an arbitrary vector field on the oscillator group (G, g_a), for some arbitrary smooth functions $X^1, \ldots, X^4$ on G. Starting from (2.4), one finds the following description of the Lie derivative of the metric tensor g_a:

$$\begin{aligned}\mathcal{L}_X g_a =& 2(\partial_1 X^4 + a x_3 \partial_1 X^2 + a \partial_1 X^1) dx_1 dx_1 \\ &+ 2(\partial_2 X^4 + x_3 \partial_1 X^4 + a x_3 \partial_2 X^2 + \partial_1 X^2 + a x_3^2 \partial_1 X^2 + a \partial_2 X^1 + a x_3 \partial_1 X^1 + a X^3) dx_1 dx_2 \\ &+ 2(\partial_3 X^4 + \partial_1 X^3 + a x_3 \partial_3 X^2 + a \partial_3 X^1) dx_1 dx_3 \\ &+ 2(\partial_4 X^4 + a \partial_1 X^4 + a x_3 \partial_4 X^2 + x_3 \partial_1 X^2 + a \partial_4 X^1 + \partial_1 X^1) dx_1 dx_4 \\ &+ 2(x_3 \partial_2 X^4 + \partial_2 X^2 + a x_3^2 \partial_2 X^2 + a x_3 \partial_2 X^1 + a x_3 X^3) dx_2 dx_2 \\ &+ 2(x_3 \partial_3 X^4 + \partial_2 X^3 + a x_3^2 \partial_3 X^2 + a x_3 \partial_3 X^1 + \partial_3 X^2) dx_2 dx_3 \\ &+ 2(a x_3 \partial_4 X^4 + a \partial_2 X^4 + \partial_4 X^2 + a x_3^2 \partial_4 X^2 + x_3 \partial_2 X^2 + a x_3 \partial_4 X^1 + \partial_2 X^1 + X^3) dx_2 dx_4 \\ &+ 2\partial_3 X^3 dx_3 dx_3 + 2(a \partial_3 X^4 + \partial_4 X^3 + x_3 \partial_3 X^2 + \partial_3 X^1) dx_3 dx_4 \\ &+ 2(a \partial_4 X^4 + x_3 \partial_4 X^2 + \partial_4 X^1) dx_4 dx_4.\end{aligned}$$

Thus, X is a Killing vector field if and only if the system of PDE obtained requiring the vanishing of all the coefficients in the above Lie derivative.

We then consider the symmetries of (G_μ, g_a) related to curvature and have the following result.

Theorem 3.2 *Let $X = X^i \partial_i$ denote an arbitrary vector field on the oscillator group (G_μ, g_a). Then:*

(i) X is a Ricci collineation if and only if one of the following cases occurs:

(a) $a = 0$ and $X^4 = c_1$.
(b) $a \neq 0$ and

$$X^2 = -c_1 x_3 + c_2, \quad X^3 = c_1 x_2 + c_3,$$
$$X^4 = -a X_1 + \frac{a}{2}(c_1 x_3^2 - c_1 x_2^2 - 2c_3 x_2) + c_4.$$

(ii) X is a curvature collineation if and only if one of the following cases occurs:

(a) $a = 0$ and X satisfies

$$\begin{cases} X^1 = \frac{c_1}{2}(x_3^2 - x_2^2) - \frac{f_1'(x_4)}{2}(x_3^2 + x_2^2) \\ \qquad +2 f_1(x_4) x_1 - (f_3'(x_4) + f_2(x_4)) x_2 - f_2'(x_4) x_3 + f_4(x_4), \\ X^2 = f_1(x_4) x_2 - c_1 x_3 + f_3(x_4), \\ X^3 = c_1 x_2 + f_1(x_4) x_3 + f_2(x_4), \\ X^4 = c_2, \end{cases}$$

for some real functions f_1, f_2, f_3, f_4 of one variable.
(b) $a \neq 0$ and

$$X^1 = \frac{c_1}{2}(x_3^2 - x_2^2) - c_2 x_2 - \frac{f(x_4)}{a} + c_3,$$
$$X^2 = -c_1 x_3 + c_4, \quad X^3 = c_1 x_2 + c_2, \quad X^4 = f(x_4),$$

for some real function f of one variable.

(iii) X is a Weyl collineation if and only if

(a) either $a = 0$ and X is arbitrary, or
(b) $a \neq 0$ and X is Killing.

For the proof, we refer again to [9]. For example, the Lie derivative of the Ricci tensor in the direction of $X = X^i \partial_i$ is given by

$$\begin{aligned}
\mathcal{L}_X \varrho = {} & a(a\partial_1 X^1 + ax_3\partial_1 X^2 + \partial_1 X^4)dx_1dx_1 \\
& +a((a\partial_2 + ax_3\partial_1)X^1 + (ax_3\partial_2 + ax_3^2\partial_1 - \partial_1)X^2 + aX^3 \\
& \quad +(\partial_2 + x_3\partial_1)X^4)dx_1dx_2 \\
& +a(a\partial_3 X^1 + ax_3\partial_3 X^2 - \partial_1 X^3 + \partial_3 X^4)dx_1dx_3 \\
& +((a^2\partial_4 + a\partial_1)X^1 + (a^2x_3\partial_4 + ax_3\partial_1)X^2 + (a\partial_4 + \partial_1)X^4)dx_1dx_4 \\
& +a(ax_3\partial_2 X^1 + (ax_3^2\partial_2 - \partial_2)X^2 + ax_3X^3 + x_3\partial_2 X^4)dx_2dx_2 \\
& +a(ax_3\partial_3 X^1 + (ax_3^2\partial_3 - \partial_3)X^2 - \partial_2 X^3 + x_3\partial_3 X^4)dx_2dx_3 \\
& +((a^2x_3\partial_4 + a\partial_2)X^1 + (a^2x_3^2\partial_4 - a\partial_4 + ax_3\partial_2)X^2 + aX^3 \\
& \quad +(ax_3\partial_4 + \partial_2)X^4)dx_2dx_4 \\
& -a\partial_3 X^3 dx_3dx_3 + (a\partial_3 X^1 + ax_3\partial_3 X^2 - a\partial_4 X^3 + \partial_3 X^4)dx_3dx_4 \\
& +(a\partial_4 X^1 + ax_3\partial_4 X^2 + \partial_4 X^4)dx_4dx_4.
\end{aligned}$$

Ricci collineations are then calculated by solving the system of PDE obtained by requiring that all the above coefficients of $\mathcal{L}_X \varrho$ vanish.

On a homogeneous space (and, more in general, whenever the scalar curvature is constant), a Killing vector field is necessarily a matter collineation. For this reason, for (G_μ, g_a), we shall consider a Killing vector field as a trivial matter collineation. The following result holds.

Theorem 3.3 *On the Lorentzian oscillator group (G_μ, g_a), nontrivial matter collineations only occur when $a = 0$ (in which case they coincide with Ricci collineations). In fact, when $a = 0$, $X = X^i \partial_i$ is a matter collineation if and only if $X^4 = c_1$ is a real constant.*

Remark 3.4 As we proved in Theorems 3.2 and 3.3, for $a = 0$, a vector field $X = X^i \partial_i$ is a Ricci (equivalently, matter) collineation if and only if X^4 is a real constant. Therefore, the homogeneous spacetime (G_μ, g_0) provides a new example where Ricci (matter) collineations form a large infinite-dimensional vector space, and the smooth ones form a large infinite-dimensional Lie algebra, since X^1, X^2, X^3 are then arbitrary smooth functions.

Again from Theorem 3.2, we see that also curvature collineations form an infinite-dimensional vector space, for any $-1 < a < 1$, since they depend on at least one arbitrary function of one variable, and the smooth ones form an infinite-dimensional Lie algebra. Finally, we observe that for $a = 0$ any curvature collineation (indeed, any smooth vector field) is a Weyl collineation, whilst in the case $a \neq 0$ the converse holds: every Weyl collineation, being a Killing vector field, is a curvature collineation.

4 Ricci Solitons of the Oscillator Group

Ricci solitons are the self-similar solutions of the *Ricci flow*. As such, they are essential in understanding its singularities. After their introduction in Riemannian settings, Ricci solitons have been intensively studied for pseudo-Riemannian metrics (see for example [3, 5, 6, 8] and references therein). Using a set of local coordinates, the Ricci soliton Eq. (1.2) translates into a system of partial differential equations, which in general is not possible to deal with. For this reason, when one considers a pseudo-Riemannian homogeneous space (in particular, a Lie group equipped with a left-invariant pseudo-Riemannian metric), the first approach in studying the Ricci soliton Eq. (1.2) is 'algebraic' in some sense. A *homogeneous Ricci soliton* is a homogeneous space $M = G/H$, together with a G-invariant metric g, for which Eq. (1.2) holds. An *invariant Ricci soliton* is a homogeneous one, such that Eq. (1.2) holds for an invariant vector field.

An *algebraic Ricci soliton* is a simply connected Lie group G, equipped with a left-invariant pseudo-Riemannian metric g, such that

$$Ric = c\,\mathrm{Id} + D,$$

where Ric denotes the Ricci operator, c is a real number and $D \in \mathrm{Der}(\mathfrak{g})$. An algebraic Ricci soliton on a solvable Lie group is called a *solvsoliton*.

Any algebraic Ricci soliton metric g is also a Ricci soliton [14, 17]. Moreover, all known examples of homogeneous *Riemannian* Ricci soliton metrics on noncompact homogeneous manifolds are isometric to some solvsolitons ([13, Remark 1.5]). But when we are considering the Ricci soliton pseudo-Riemannian metrics on a homogeneous space G, in general, neither invariant nor algebraic Ricci solitons exhaust the whole class of solutions.

Algebraic Ricci solitons on oscillator groups of every even dimension were investigated in [17], finding only g_0 as a steady algebraic Ricci soliton (nontrivial, since the metric is not Einstein). In [4], we completely solved the system of PDE which translates (1.2) in the system of coordinates for (G_μ, g_a) described in Sect. 2, proving the following result.

Theorem 4.1 *Every left-invariant metric g_a, $-1 < a < 1$ on the four-dimensional oscillator group G_μ is a Ricci soliton. More precisely,*

(a) The bi-invariant metric g_0 is a Ricci soliton (expanding, steady and shrinking, as it satisfies Eq. (1.2) for any real value of λ);
(b) The left-invariant metric g_a, for any $a \neq 0$, is a Ricci soliton, which is expanding when $a > 0$ and shrinking when $a < 0$.

With respect to the coordinate system (x_1, x_2, x_3, x_4), let $X = X^i \partial_i$ denote an arbitrary vector field on (G_μ, g_a). Using the Lie derivative $\mathcal{L}_X g_a$ already described in the previous Section, together with (2.4) and (2.6), we can write down the Ricci soliton Eq. (1.2) in this set of coordinates, obtaining that we have a Ricci soliton if and only if the following system of 10 PDE is satisfied:

$$\begin{cases} 2a\partial_1 X^1 + 2ax_3\partial_1 X^2 + 2\partial_1 X^4 + \frac{1}{2}a^2 - a\lambda = 0, \\ ax_3\partial_1 X^1 + a\partial_2 X^1 + \partial_1 X^2 + ax_3^2\partial_1 X^2 + ax_3\partial_2 X^2 + aX^3 + x_3\partial_1 X^4 + \partial_2 X^4 \\ \quad +\frac{1}{2}a^2 x_3 - a\lambda x_3 = 0, \\ a\partial_3 X^1 + ax_3\partial_3 X^2 + \partial_1 X^3 + \partial_3 X^4 = 0, \\ \partial_1 X^1 + a\partial_4 X^1 + x_3\partial_1 X^2 + ax_3\partial_4 X^2 + a\partial_1 X^4 + \partial_4 X^4 + \frac{1}{2}a - \lambda = 0, \\ 2ax_3\partial_2 X^1 + 2\partial_2 X^2 + 2ax_3^2\partial_2 X^2 + 2ax_3 X_3 + 2x_3\partial_2 X^4 + \frac{1}{2}a^2 x_3^2 - \frac{1}{2}a - \lambda - a\lambda x_3^2 = 0, \\ ax_3\partial_3 X^1 + \partial_3 X^2 + ax_3^2\partial_3 X^2 + \partial_2 X^3 + x_3\partial_3 X^4 = 0, \\ \partial_2 X^1 + ax_3\partial_4 X^1 + x_3\partial_2 X^2 + \partial_4 X^2 + ax_3^2\partial_4 X^2 + X^3 + a\partial_2 X^4 + x_3\partial_4 X^4 + \frac{1}{2}ax_3 \\ \quad -\lambda x_3 = 0, \\ 2\partial_3 X^3 - \frac{1}{2}a - \lambda = 0, \\ \partial_3 X^1 + x_3\partial_3 X^2 + \partial_4 X^3 + a\partial_3 X^4 = 0, \\ 2\partial_4 X^1 + 2x_3\partial_4 X^2 + 2a\partial_4 X^4 + \frac{1}{2} - a\lambda = 0. \end{cases} \tag{4.1}$$

The complete discussion of the above system (4.1) has been done in [4], determining the Ricci solitons of the four-dimensional oscillator group, and whether those solutions were left-invariant or gradient. Here, we only remark that integrating the eight equation of (4.1), and the first equation of (4.1) with respect to X^4, we find

$$\begin{cases} X^3 = (\frac{1}{4}a + \frac{1}{2}\lambda)x_3 + F_3(x_1, x_2, x_4), \\ X^4 = -aX^1 - ax_3X^2 + \left(\frac{1}{2}a\lambda - \frac{1}{4}a^2\right)x_1 + F_4(x_2, x_3, x_4), \end{cases}$$

for some smooth functions F_3, F_4. Replacing into the third equation of (4.1), it becomes

$$\partial_1 F_3(x_1, x_2, x_4) - aX^2 + \partial_3 F_4(x_2, x_3, x_4) = 0. \tag{4.2}$$

It is then evident that the above Eq. (4.2) (and so, the whole system (4.1)) will have different sets of solutions, depending on whether $a = 0$ or $a \neq 0$. The following Theorems give the complete description of the solutions, in the cases $a = 0$ and $a \neq 0$ respectively.

Theorem 4.2 *The bi-invariant metric g_0 is a Ricci soliton, which satisfies Eq. (1.2) for any real value of λ, where $X = X^i\partial_i$ is a smooth vector field, whose components X^i with respect to $\{\partial_i\}$ are described by*

$$\begin{cases} X^1 = \lambda x_1 + \frac{1}{2}a_3x_3{}^2 - a_3x_3\cos(x_4) + b_3x_3\sin(x_4) - \frac{1}{4}x_4 - \frac{1}{2}a_3x_2{}^2 + Kx_2 + b_2, \\ X^2 = -a_3x_3 + \frac{1}{2}\lambda x_2 + a_3\cos(x_4) - b_3\sin(x_4) + b_2, \\ X^3 = a_3x_2 + a_3\sin(x_4) + b_3\cos(x_4) - \frac{1}{2}K\lambda x_3, \\ X^4 = b_4. \end{cases} \tag{4.3}$$

This vector field X is never left-invariant, and the Ricci soliton is gradient only in the steady case (when it is also algebraic).

Theorem 4.3 *The (non-isometric) left-invariant metrics g_a, for any value of $a \in]-1, 1[$, $a \neq 0$, are Ricci solitons, which satisfy Eq. (1.2), where $X = X^i\partial_i$ is a smooth vector field, whose components X^i with respect to $\{\partial_i\}$ are described by*

$$\begin{cases} X^1 = \frac{1}{4a}\left(-4a^2 x_1 + 2 H_1 x_2{}^2 + 4 a_4 x_2 - 2 H_1 x_3{}^2 - a x_4 + 4 a s_4\right), \\ X^2 = \frac{1}{2a}\left(2 H_1 x_3 - a^2 x_2 + 2 c_4\right), \\ X^3 = -\frac{1}{2a}\left(2 H_1 x_2 + a^2 x_3 + 2 a_4\right), \\ X^4 = -\frac{3}{4} a x_4 - a s_4 + b_4 + r_4, \end{cases} \tag{4.4}$$

and $\lambda = -\frac{3}{2}a$. *In particular, this Ricci soliton is either expanding or shrinking, depending on whether* $a > 0$ *or* $a < 0$. *This vector field* X *is never left-invariant, and the Ricci soliton is not gradient.*

The above result for the case $a = 0$ also proves the existence of a nontrivial Yamabe soliton for (G_μ, g_0). A pseudo-Riemannian manifold (M, g) is said to be a *Yamabe soliton* if it admits a vector field Y, such that

$$\mathcal{L}_Y g = (\tau - \rho) g, \tag{4.5}$$

where τ denotes the scalar curvature and ρ is a real constant. Clearly, a Yamabe soliton is nontrivial when Eq. (4.5) holds with $\tau \neq \rho$, otherwise is just reduces to the equation for Killing vector fields.

We proved that g_0 satisfies Eq. (1.2) for any value of λ. As a consequence, for any distinct real constants λ_1, λ_2, let us consider two smooth vector fields $X_{\lambda_1}, X_{\lambda_2}$, with components of the form (4.3) for $\lambda = \lambda_1$ and $\lambda = \lambda_2$ respectively. Since $X_{\lambda_1}, X_{\lambda_2}$ satisfy the Ricci soliton Eq. (1.2), vector field $Y = X_{\lambda_1} - X_{\lambda_2}$ then satisfies

$$\mathcal{L}_Y g_0 = \mathcal{L}_{X_{\lambda_1}} g_0 - \mathcal{L}_{X_{\lambda_2}} g_0 = (\lambda_1 - \lambda_2) g_0.$$

Since the scalar curvature of g_0 vanishes and $\lambda_1 - \lambda_2 \neq 0$, Y is a nontrivial solution of the Yamabe soliton Eq. (4.5). Observe that the above condition satisfied by Y is also coherent with the existence of proper homothetic vector fields for the metric g_a, stated in Theorem 3.1. This proves the following.

Corollary 4.4 *The bi-invariant metric* g_0 *on the four-dimensional oscillator group* G_μ *is a Yamabe soliton.*

Acknowledgements Author partially supported by funds of the University of Salento and MIUR (PRIN).

References

1. W. Batat, M. Castrillon-Lopez, E. Rosado, Four-dimensional naturally reductive pseudo-Riemannian spaces. Diff. Geom. Appl. **41**, 48–64 (2015)
2. M. Boucetta, A. Medina, Solutions of the Yang-Baxter equations on quadratic Lie groups: the case of oscillator groups. J. Geom. Phys. **61**, 2309–2320 (2011)
3. M. Brozos-Vazquez, G. Calvaruso, E. Garcia-Rio, S. Gavino-Fernandez, Three-dimensional Lorentzian homogeneous Ricci solitons. Isr. J. Math. **188**, 385–403 (2012)
4. G. Calvaruso, Oscilator spacetimes are Ricci solitons. Nonlinear Anal. **140**, 254–269 (2016)

5. G. Calvaruso, A. Fino, Ricci solitons and geometry of four-dimensional non-reductive homogeneous spaces. Can. J. Math. **64**(4), 778–804 (2012)
6. G. Calvaruso, A. Fino, Four-dimensional pseudo-Riemannian homogeneous Ricci solitons. Int. J. Geom. Methods Mod. Phys. **12**, (2015) 1550056,21 (2015)
7. G. Calvaruso, J. van der Veken, Totally geodesic and parallel hypersurfaces of four-dimensional oscillator groups. Results Math. **64**, 135–153 (2013)
8. G. Calvaruso, A. Zaeim, A complete classification of Ricci and Yamabe solitons of non-reductive homogeneous 4-spaces. J. Geom. Phys. **80**, 15–25 (2014)
9. G. Calvaruso, A. Zaeim, On the symmetries of the Lorentzian oscillator group. Collect. Math. **68**, 51–67 (2017)
10. R. Duran Diaz, P.M. Gadea, J.A. Oubiña, The oscillator group as a homogeneous spacetime. Lib. Math. **19**, 9–18 (1999)
11. R. Duran Diaz, P.M. Gadea, J.A. Oubiña, Reductive decompositions and Einstein-Yang-Mills equations associated to the oscillator group. J. Math. Phys. **40**, 3490–3498 (1999)
12. P.M. Gadea, J.A. Oubiña, Homogeneous Lorentzian structures on the oscillator groups. Arch. Math. **73**, 311–320 (1999)
13. M. Jablonski, Homogeneous Ricci solitons. J. Reine Angew. Math. **699**, 159–182 (2015)
14. J. Lauret, Ricci solitons solvmanifolds. J. Reine Angew. Math. **650**, 1–21 (2011)
15. A.V. Levitchev, Chronogeometry of an electromagnetic wave given by a bi-invariant metric on the oscillator group. Siberian Math. J. **27**, 237–245 (1986)
16. D. Müller, F. Ricci, On the Laplace-Beltrami operator on the oscillator group. J. Reine Angew. Math. **390**, 193–207 (1988)
17. K. Onda, Examples of algebraic Ricci solitons in the pseudo-Riemannian case. Acta Math. Hung. **144**, 247–265 (2014)
18. R.F. Streater, The representations of the oscillator group. Commun. Math. Phys. **4**, 217–236 (1967)

Umbilical Spacelike Submanifolds of Arbitrary Co-dimension

Nastassja Cipriani and José M.M. Senovilla

Abstract Given a semi-Riemannian manifold, we give necessary and sufficient conditions for a Riemannian submanifold of arbitrary co-dimension to be umbilical along normal directions. We do that by using the so-called *total shear tensor*, i.e., the trace-free part of the second fundamental form. We define the *shear space* and the *umbilical space* as the spaces generated by the total shear tensor and by the umbilical vector fields, respectively. We show that the sum of their dimensions must equal the co-dimension.

Keywords Umbilical points · Umbilical submanifolds · Shear

2010 Mathematics Subject Classification 53B25 · 53B30 · 53B50

1 Introduction

The notions of umbilical point and umbilical submanifold are classical in differential geometry. They have mainly been studied in the Riemannian setting and, apart from very few exceptions, they have been applied to submanifolds of co-dimension one (hypersurfaces). In [1] the authors studied these concepts in a slightly more general

This work is partially supported by the Belgian Interuniversity Attraction Pole P07/18 (Dygest) and by the KU Leuven Research Fund project 3E160361 "Lagrangian and calibrated submanifolds". NC and JMMS are supported under grant FIS2014-57956-P (Spanish MINECO–Fondos FEDER) and project IT956-16 of the Basque Government. JMMS is also supported by EU COST action No. CA15117 "CANTATA".

N. Cipriani (✉)
Department of Mathematics, KU Leuven, Celestijnenlaan 200B,
Box 2400, 3001 Leuven, Belgium
e-mail: nastassja.cipriani@wis.kuleuven.be

N. Cipriani · J.M.M. Senovilla
Física Teórica, Universidad Del País Vasco, Apartado 644, 48080 Bilbao, Spain
e-mail: josemm.senovilla@ehu.es

M.A. Cañadas-Pinedo et al. (eds.), *Lorentzian Geometry and Related Topics*,
Springer Proceedings in Mathematics & Statistics 211,
DOI 10.1007/978-3-319-66290-9_4

framework, namely, allowing the ambient manifold to have arbitrary signature while requiring the submanifold to be *spacelike* (i.e., endowed with a Riemannian induced metric). Motivated by the applications to gravitation and general relativity, in [1] the authors focused on spacelike submanifolds of *co-dimension two*. Notice that the results presented in [1] generalized those presented in [6]. In the latter, the study of umbilical *surfaces* (*two*-dimensional) in *four*-dimensional Lorentzian manifolds was carried out.

In the present paper we generalize what has been done in [1, 6]. We consider spacelike submanifolds of *arbitrary co-dimension* and give a characterization of those which are umbilical with respect to some normal directions. Observe that when the co-dimension is one, the normal bundle is *one*-dimensional and thus the submanifold can only be umbilical along the unique normal direction. On the other hand, when the co-dimension is higher than one, there are several possibilities and the submanifold can be umbilical with respect to some normal vectors but non-umbilical with respect to others.

In order to characterize umbilical spacelike submanifolds we make use of the so-called *total shear tensor*. This is defined as the trace-free part of the second fundamental form and it appears often in the mathematical literature, especially in conformal geometry. Nevertheless, it had never been given a name prior to [1] and, surprisingly, its relationship with the umbilical properties of submanifolds seemed to be almost unknown or is not explicitly mentioned at least.

We introduce the notions of *shear space* and *umbilical space*. They are defined as the space generated by the image of the total shear tensor and the one generated by the umbilical vector fields, respectively. They both belong to the normal bundle and they happen to be mutually orthogonal. We show that the existence of umbilical directions "shrinks" the shear space reducing its dimension. More precisely, the dimensions of the shear space and the umbilical space are linked in such a way that the sum of the two must equal the co-dimension of the submanifold. Moreover, in specific situations—for instance, when the ambient manifold is Riemannian—the direct sum of the two spaces generate the whole normal space.

The plan of the paper is as follows. In Sect. 2 we recall some basic concepts of submanifold theory, we introduce the shear objects and give the definitions of umbilical point, umbilical submanifold and umbilical space. In Sect. 3 we show how the shear space and the umbilical space are related, we present necessary and sufficient conditions for the submanifold to be umbilical and give some final remarks.

2 Preliminaries

2.1 Basic Concepts of Submanifold Theory

We consider an orientable n-dimensional *spacelike* submanifold $(\mathcal{S}, g)$ of a semi-Riemannian manifold $(\mathcal{M}, \bar{g})$ with inmersion $\Phi : \mathcal{S} \longrightarrow \mathcal{M}$ and co-dimension k.

Hence, $g := \Phi^\star \bar{g}$ is positive definite everywhere on $\mathcal{S}$, so that $(\mathcal{S}, g)$ is, in particular, an oriented Riemannian manifold. Let $\mathfrak{X}(\mathcal{S})$ and $\mathfrak{X}(\mathcal{S})^\perp$ denote the set of tangent and normal vector fields, respectively, on $\mathcal{S}$. The classical formulas of Gauss and Weingarten provide the decomposition of the vector field derivatives into their tangent and normal components [3–5] as

$$\overline{\nabla}_X Y = \nabla_X Y + h(X, Y), \qquad \forall X, Y \in \mathfrak{X}(\mathcal{S})$$
$$\overline{\nabla}_X \xi = -A_\xi X + \nabla^\perp_X \xi, \qquad \forall \xi \in \mathfrak{X}(\mathcal{S})^\perp,\ \forall X \in \mathfrak{X}(\mathcal{S})$$

where $\overline{\nabla}$ and ∇ are the Levi-Civita connections of $(\mathcal{M}, \bar{g})$ and $(\mathcal{S}, g)$ respectively, h is the *second fundamental form* or *shape tensor* of the immersion and A_ξ the *Weingarten operator* relative to ξ. The derivation $\nabla^\perp$ so defined determines a connection on the normal bundle, $h(X, Y) = h(Y, X) \in \mathfrak{X}(\mathcal{S})^\perp$ for all $X, Y \in \mathfrak{X}(\mathcal{S})$ and acts linearly (as a two-covariant tensor) on its arguments while A_ξ is self-adjoint for every $\xi \in \mathfrak{X}(\mathcal{S})^\perp$. The following relation holds

$$g(A_\xi X, Y) = \bar{g}(h(X, Y), \xi), \qquad \forall X, Y \in \mathfrak{X}(\mathcal{S}),\ \forall \xi \in \mathfrak{X}(\mathcal{S})^\perp.$$

The *mean curvature vector field* $H \in \mathfrak{X}(\mathcal{S})^\perp$ is $1/n$ times the trace (with respect to g) of h, so that for instance one can write [3–5]

$$H = \frac{1}{n}\sum_{i=1}^{n} h(e_i, e_i)$$

where $\{e_1, \ldots, e_n\}$ denotes an orthonormal frame on $\mathfrak{X}(\mathcal{S})$.

2.2 The Total Shear Tensor and the Shear Operators

Using the previous notations and conventions, the following definition is taken from [1]

Definition 1 The *total shear tensor* $\widetilde{h}$ is defined as the trace-free part of the second fundamental form:

$$\tilde{h}(X, Y) = h(X, Y) - g(X, Y)H.$$

The *shear operator* associated to $\xi \in \mathfrak{X}(\mathcal{S})^\perp$ is the trace-free part of the corresponding shape operator:

$$\tilde{A}_\xi = A_\xi - \frac{1}{n}\mathrm{tr}A_\xi \mathbf{1}$$

where **1** denotes the identity operator.

The total shear tensor and shear operators are obviously related by

$$g(\widetilde{A}_\xi X, Y) = \bar{g}(\widetilde{h}(X, Y), \xi), \qquad \forall X, Y \in \mathfrak{X}(\mathcal{S}) \quad \forall \xi \in \mathfrak{X}(\mathcal{S})^\perp. \tag{1}$$

To the authors' knowledge, the trace-free part of the second fundamental form had never been given a name prior to [1]. Nevertheless, it is easy to find it in the literature, in Riemannian settings, especially in connection with the conformal properties of submanifolds. A pioneer analysis appears in [2], where an extensive exposition concerning conformal invariants was given. The total shear tensor is also on the basis of the definition of the so-called generalized Willmore functional [7].

Denote by $\{\xi_1, \ldots, \xi_k\}$ a local frame in $\mathfrak{X}(\mathcal{S})^\perp$. With respect to this frame, there exist k shear operators $\widetilde{A}_1, \ldots, \widetilde{A}_k$ such that the total shear tensor $\widetilde{h}$ decomposes as

$$\widetilde{h}(X, Y) = \sum_{i=1}^{k} g(\widetilde{A}_i X, Y)\xi_i, \qquad \forall X, Y \in \mathfrak{X}(\mathcal{S}). \tag{2}$$

If $\{\xi_1, \ldots, \xi_k\}$ is orthonormal, i.e., $\bar{g}(\xi_i, \xi_j) = \epsilon_i \delta_{ij}$ with $\epsilon_i^2 = 1$, then $\widetilde{A}_i = \epsilon_i \widetilde{A}_{\xi_i}$ for all i. However, in general, $\widetilde{A}_i$ does not need be proportional to $\widetilde{A}_{\xi_i}$, rather being a linear combination of $\widetilde{A}_{\xi_1}, \ldots, \widetilde{A}_{\xi_k}$.

Given any normal vector field $\eta \in \mathfrak{X}(\mathcal{S})^\perp$, by (2) its corresponding shear operator $\widetilde{A}_\eta$ can be expressed in terms of $\widetilde{A}_1, \ldots, \widetilde{A}_k$. Indeed, formulas (1) and (2) imply

$$\widetilde{A}_\eta = \sum_{i=1}^{k} \bar{g}(\xi_i, \eta)\widetilde{A}_i. \tag{3}$$

In the definitions that follow we introduce two useful concepts.

Definition 2 At any point $p \in \mathcal{S}$, the set

$$\mathsf{Im}\, \widetilde{h}_p := \text{span}\left\{\widetilde{h}(v, w) : v, w \in T_p\mathcal{S}\right\} \subseteq T_p\mathcal{S}^\perp$$

is called the *shear space* of $\mathcal{S}$ at p.

If $\mathcal{N}_p^1 = \text{span}\left\{h(v, w) : v, w \in T_p\mathcal{S}\right\} \subseteq T_p\mathcal{S}^\perp$ denotes the first normal space of $\mathcal{S}$ at the point $p \in \mathcal{S}$, then for every p in $\mathcal{S}$ we have $\mathsf{Im}\, \widetilde{h}_p \subseteq \mathcal{N}_p^1$, hence $\mathsf{dim}\, \mathsf{Im}\, \widetilde{h}_p \leq \mathsf{dim} \mathcal{N}_p^1 \leq k$. Furthermore, given any orthonormal basis $\{e_1, \ldots, e_n\}$ in $T_p\mathcal{S}$, the image of $\widetilde{h}_p$ is spanned by the $n(n+1)/2$ vectors $\widetilde{h}(e_i, e_j)$, for $i \leq j$. Because $\sum_{i=1}^n \widetilde{h}(e_i, e_i) = 0$, these vectors are not linearly independent. In particular, the dimension of $\mathsf{Im}\, \widetilde{h}_p$ can be at most $n(n+1)/2 - 1$. Therefore,

$$\mathsf{dim}\, \mathsf{Im}\, \widetilde{h}_p \leq \min\left\{k, \frac{n(n+1)}{2} - 1\right\}. \tag{4}$$

Formula (3) for the decomposition of any shear operator implies that if $\dim \operatorname{Im} \widetilde{h}_p = d$ then any $d+1$ shear operators must be linearly dependent in p. The converse of this is also true, so that

$$\dim \operatorname{Im} \widetilde{h}_p = \max \left\{ d \mid \exists\, \eta_1, \ldots, \eta_d \in T_p\mathcal{S}^\perp : \widetilde{A}_{\eta_1}, \ldots, \widetilde{A}_{\eta_d} \text{ are lin. ind. at } p \right\} \quad (5)$$

Notice that similar formulas to (4) and (5) also hold for the dimension of the first normal space. Indeed, for the dimension of $\mathcal{N}_p^1$ we will have $\dim \mathcal{N}_p^1 \leq \min\{k, n(n+1)/2\}$; as for (5), the Weingarten operators will just take the place of the shear operators. These two formulas for $\mathcal{N}_p^1$ imply that if $k - n(n+1)/2$ is positive, then there exist $k - n(n+1)/2$ linearly independent Weingarten operators that vanish at p.

Definition 3 Assume that the dimension of the shear spaces $\operatorname{Im} \widetilde{h}_p$ is constant on $\mathcal{S}$, i.e., there exists $d \in \mathbb{N}$ with $0 \leq d \leq k$ such that $\dim \operatorname{Im} \widetilde{h}_p = d$ for all $p \in \mathcal{S}$. The set

$$\operatorname{Im} \widetilde{h} = \operatorname{span} \left\{ \widetilde{h}(X, Y) \mid X, Y \in \mathfrak{X}(\mathcal{S}) \right\} \subseteq \mathfrak{X}(\mathcal{S})^\perp$$

is called the *shear space* of $\mathcal{S}$. Then, the shear space is a module over the ring of functions defined on $\mathcal{S}$ with dimension d.

The properties already presented relating $\dim \operatorname{Im} \widetilde{h}_p$ to the shear operators can be extended to $\dim \operatorname{Im} \widetilde{h}$ accordingly.

2.3 *Umbilical Points and Umbilical Submanifolds*

For works concerning umbilical submanifolds and some previous results in both Riemannian and semi-Riemannian settings the reader can consult, e.g., [1] and references therein.

For hypersurfaces (co-dimension 1), a point can only be umbilical along the unique normal direction. This situation changes completely for higher co-dimensions, in which case there are multiple directions along which a point can be umbilical.

Definition 4 Using the notations and conventions introduced above for the immersion $\Phi : (\mathcal{S}, g) \to (\mathcal{M}, \bar{g})$, a point $p \in \mathcal{S}$ is said to be

- *umbilical with respect to* $\xi_p \in T_p\mathcal{S}^\perp$ if A_{ξ_p} is proportional to the identity;
- *totally umbilical* if it is umbilical with respect to all $\xi_p \in T_p\mathcal{S}^\perp$.

A point $p \in \mathcal{S}$ is umbilical with respect to $\xi_p \in T_p\mathcal{S}^\perp$ if and only if $A_{\xi_p} = (\operatorname{tr} \widetilde{A}_{\xi_p}/n)\mathbf{1}$ or, equivalently, $\widetilde{A}_{a\xi_p} = 0$ for any $a \in \mathbb{R} \setminus \{0\}$. ξ_p-umbilicity is thus a property that gives information about span$\{\xi_p\}$ regardless of the length and the orientation of ξ_p. Hence, we will usually state that p is umbilical with respect to the normal *direction* spanned by ξ_p. On the other hand, p is totally umbilical if and only if $h(v, w) = g(v, w)H_p$ for all $v, w \in T_p\mathcal{S}$ or, equivalently, if and only if $\widetilde{h} = 0$ at p. This fact was already known for hypersurfaces in Riemannian settings and can be found, e.g., in [2]. However, the relationship between the total shear tensor and the umbilical properties of submanifolds, treated in [1] and in the present article, is substantially new.

Definition 5 Given any point $p \in \mathcal{S}$, the set

$$\mathcal{U}_p = \left\{\xi_p \in T_p\mathcal{S}^\perp : p \text{ is umbilical with respect to } \xi_p\right\} \subseteq T_p\mathcal{S}^\perp$$

is called the *umbilical space* of $\mathcal{S}$ at p.

Lemma 1 *The umbilical space $\mathcal{U}_p$ is a vector space for every $p \in \mathcal{S}$.*

Proof Let $\xi_p, \eta_p \in \mathcal{U}_p$, so that by definition $\widetilde{A}_{\xi_p} = \widetilde{A}_{\eta_p} = 0$. Let $a, b \in \mathbb{R}$ and consider the normal vector $a\xi_p + b\eta_p$. By linearity we have $\widetilde{A}_{a\xi_p+b\eta_p} = a\widetilde{A}_{\xi_p} + b\widetilde{A}_{\eta_p} = 0$. It follows that p is umbilical with respect to $a\xi_p + b\eta_p$, hence $a\xi_p + b\eta_p$ belongs to $\mathcal{U}_p$. □

It follows from Lemma 1 that $\dim \mathcal{U}_p$ is well defined. Notice that $\dim \mathcal{U}_p = m$ if and only if p is umbilical with respect to *exactly* m linearly independent normal directions. Moreover, by formulas (4) and (5) it follows that if $k - n(n+1)/2 + 1$ is positive, then $\dim \mathcal{U}_p \geq k - n(n+1)/2 + 1$.

Definition 6 Using the notations and conventions introduced above for the immersion $\Phi : (\mathcal{S}, g) \to (\mathcal{M}, \bar{g})$, the submanifold $(\mathcal{S}, g)$ is said to be

- *umbilical with respect to* $\xi \in \mathfrak{X}(\mathcal{S})^\perp$ if A_ξ is proportional to the identity;
- *totally umbilical* if it is umbilical with respect to all $\xi \in \mathfrak{X}(\mathcal{S})^\perp$.

The properties presented above for umbilical points can be extended to umbilical submanifolds accordingly. $\mathcal{S}$ is umbilical with respect to $\xi \in \mathfrak{X}(\mathcal{S})^\perp$ if and only if $\xi_p \in \mathcal{U}_p$ for all $p \in \mathcal{S}$. More in general, $\mathcal{S}$ is umbilical with respect to *exactly* m linearly independent nonzero normal vector fields $\xi_1, \ldots, \xi_m \in \mathfrak{X}(\mathcal{S})^\perp$ if and only if $(\xi_1)_p, \ldots, (\xi_m)_p \in \mathcal{U}_p$ for all $p \in \mathcal{S}$. Equivalently, if and only if $\dim \mathcal{U}_p = m$ for all $p \in \mathcal{S}$. This leads to the following definition.

Definition 7 Assume that the dimension of the umbilical spaces $\mathcal{U}_p$ is constant on $\mathcal{S}$, i.e. there exists $m \in \mathbb{N}$ with $0 \leq m \leq k$ such that $\dim \mathcal{U}_p = m$ for all $p \in \mathcal{S}$. Then the set

$$\mathcal{U} = \left\{\xi \in \mathfrak{X}(\mathcal{S})^\perp : \mathcal{S} \text{ is umbilical with respect to } \xi\right\} \subseteq \mathfrak{X}(\mathcal{S})^\perp$$

is called the *umbilical space* of $\mathcal{S}$.

The umbilical space of $\mathcal{S}$ is such that $\mathcal{U} = \cup_{p\in\mathcal{S}}\mathcal{U}_p$. In Lemma 1 we proved that $\mathcal{U}_p$ is a vector space for every p. Similarly, we can prove that $\mathcal{U}$ is a finitely generated module over the ring of functions defined on $\mathcal{S}$ with $\dim \mathcal{U} = m$.

3 Results

3.1 The Relationship Between $\mathcal{U}$ and $\mathsf{Im}\,\widetilde{h}$

Proposition 1 *Let $\Phi : (\mathcal{S}, g) \to (\mathcal{M}, \bar{g})$ be an isometric immersion of an n-dimensional Riemannian manifold into a semi-Riemannian manifold with co-dimension k. Let $\mathcal{U}_p$ and $\mathsf{Im}\,\widetilde{h}_p$ be the umbilical space and the shear space, respectively, of $\mathcal{S}$ at any point $p \in \mathcal{S}$. Then*

$$\mathcal{U}_p = (\mathsf{Im}\,\widetilde{h}_p)^{\perp}.$$

Moreover,

$$k - \mathit{dim}\,\mathcal{U}_p = \mathit{dim}\,\mathit{Im}\,\widetilde{h}_p.$$

Here $(\mathsf{Im}\,\widetilde{h}_p)^{\perp}$ is defined as the subspace of $T_p\mathcal{S}^{\perp}$ orthogonal to $\mathsf{Im}\,\widetilde{h}_p$, namely

$$(\mathsf{Im}\,\widetilde{h}_p)^{\perp} = \left\{\eta_p \in T_p\mathcal{S}^{\perp} \mid \forall \xi_p \in \mathsf{Im}\,\widetilde{h}_p : \bar{g}(\eta_p, \xi_p) = 0\right\}.$$

Proof By definition, a normal vector ξ_p belongs to $\mathcal{U}_p$ if $\widetilde{A}_{\xi_p} = 0$. Equivalently, if $g(\widetilde{A}_{\xi_p}(v), w) = 0$ in p, for all $v, w \in T_p\mathcal{S}$. By formula (1), this holds if and only if $\xi_p \in (\mathsf{Im}\,\widetilde{h}_p)^{\perp}$. Hence $\mathcal{U}_p = (\mathsf{Im}\,\widetilde{h}_p)^{\perp}$. Suppose $\mathsf{Im}\,\widetilde{h}_p = \{0\}$, then it is clear that $\mathcal{U}_p = (\mathsf{Im}\,\widetilde{h}_p)^{\perp} = T_p\mathcal{S}^{\perp}$ and the relation between the dimensions holds. Now assume $\mathsf{Im}\,\widetilde{h}_p \neq \{0\}$. We can choose a basis $\{(\xi_1)_p, \ldots, (\xi_k)_p\}$ of $T_p\mathcal{S}^{\perp}$ such that $\{(\xi_1)_p, \ldots, (\xi_d)_p\}$ is a basis of $\mathsf{Im}\,\widetilde{h}_p$. A normal vector η_p belongs to $\mathcal{U}_p = (\mathsf{Im}\,\widetilde{h}_p)^{\perp}$ if and only if $\bar{g}(\eta_p, (\xi_j)_p) = 0$ for $j = 1, \ldots, d$. Since $(\xi_1)_p, \ldots, (\xi_d)_p$ are linearly independent and $\bar{g}$ is nondegenerate, these are d linearly independent conditions on the components of η_p and hence $\mathsf{dim}\mathcal{U}_p = k - d$. □

By Proposition 1 we have that the umbilical space $\mathcal{U}_p$ and the shear space $\mathsf{Im}\,\widetilde{h}_p$ are such that

$$\mathcal{U}_p = (\mathsf{Im}\,\widetilde{h}_p)^{\perp} \quad \text{and} \quad \mathsf{dim}\,\mathcal{U}_p + \mathsf{dim}\,\mathsf{Im}\,\widetilde{h}_p = k.$$

However, the intersection $\mathcal{U}_p \cap \mathsf{Im}\,\widetilde{h}_p$ might be nonempty, and consequently the direct sum of the two spaces does not generate, in general, the whole normal space. For example, if p is umbilical with respect to some vector ξ_p that is null, i.e. $\bar{g}(\xi_p, \xi_p) = 0$, then ξ_p might belong to $\mathsf{Im}\,\widetilde{h}_p$. Actually, in case $\mathcal{M}$ is a *Riemannian* manifold, one easily checks that $\mathsf{Im}\,\widetilde{h}_p \cap \mathcal{U}_p = \emptyset$ and one has

$$T_p\mathcal{S}^{\perp} = \mathsf{Im}\,\widetilde{h}_p \oplus \mathcal{U}_p \qquad (\mathcal{M}\ \text{Riemannian}).$$

Remark 1 Proposition 1 implicitly shows that if the dimension of the shear spaces $\mathsf{Im}\,\widetilde{h}_p$ is constant on $\mathcal{S}$ then the dimension of the umbilical spaces $\mathcal{U}_p$ also is, and vice versa.

By Proposition 1 and Remark 1 follows the next corollary.

Corollary 1 *Assume that the dimension of the shear spaces $\mathsf{Im}\,\widetilde{h}_p$ is constant on $\mathcal{S}$ (equivalently, that the dimension of the umbilical spaces $\mathcal{U}_p$ is constant on $\mathcal{S}$). Then $\mathsf{Im}\,\widetilde{h}$ and $\mathcal{U}$ are well defined and we have*

$$\mathcal{U} = (\mathsf{Im}\,\widetilde{h})^{\perp}.$$

Moreover,

$$k - \mathsf{dim}\,\mathcal{U} = \mathsf{dim}\,\mathsf{Im}\,\widetilde{h}.$$

From now on, and for the sake of conciseness, we will assume that the dimension of the shear spaces $\mathsf{Im}\,\widetilde{h}_p$ is constant on $\mathcal{S}$. We will consider that $\mathcal{S}$ is umbilical with respect to *exactly* m linearly independent umbilical directions, that is to say, $\mathsf{dim}\,\mathcal{U} = m$. Notice, however, that all results make sense also if stated pointwise.

3.2 Characterization of Umbilical Spacelike Submanifolds

We will denote by $\wedge$ the wedge product of one-forms and by $\flat$ the musical isomorphism: if V is a vector field on $(\mathcal{M}, \bar{g})$, then its associated one-form $V^\flat$ is given by $V^\flat(Z) = \bar{g}(V, Z)$ for every vector field Z on $\mathcal{M}$.

Proposition 2 *Let $\Phi : (\mathcal{S}, g) \to (\mathcal{M}, \bar{g})$ be an isometric immersion of an n-dimensional Riemannian manifold into a semi-Riemannian manifold with co-dimension k. Let $\mathcal{U}$ be the umbilical space of $\mathcal{S}$, then $\mathsf{dim}\,\mathcal{U} = m$ if and only if the total shear tensor satisfies*

$$\bigwedge^{k-m+1} \widetilde{h}^\flat = 0$$

with

$$\bigwedge^{k-m} \widetilde{h}^\flat \neq 0.$$

Here by $\bigwedge^q \omega_r$ we mean q times the wedge product $\omega_1 \wedge \cdots \wedge \omega_q$ of q one-forms $\{\omega_r\}_{r=1}^q$. Notice that when we write, for instance, $\widetilde{h}^\flat \wedge \widetilde{h}^\flat$, we mean $\widetilde{h}(X_1, Y_1)^\flat \wedge \widetilde{h}(X_2, Y_2)^\flat$ for all $X_1, Y_1, X_2, Y_2 \in \mathfrak{X}(\mathcal{S})$.

Proof Suppose that the dimension of the umbilical space is m. By Corollary 1 it follows that $\mathsf{dim}\,\mathsf{Im}\,\widetilde{h} = k - m$ and we can then decompose the total shear tensor by means of exactly $k - m$ normal vector fields. Explicitly, there exist $k - m$ shear operators $\{\widetilde{A}_i\}_{i=1}^{k-m}$ and $k - m$ vector fields $\zeta_1, \ldots, \zeta_{k-m} \in \mathfrak{X}(\mathcal{S})^{\perp}$ such that

$$\widetilde{h}(X, Y) = \sum_{i=1}^{k-m} g(\widetilde{A}_i X, Y)\zeta_i.$$

Because these vector fields are linearly independent, their corresponding one-forms $\zeta_r^{\flat}$ also are. From this fact it easily follows that, on one hand, the wedge product $k - m$ times of $\widetilde{h}^{\flat}$ is different from zero and, on the other hand, that the wedge product $k - m + 1$ times of $\widetilde{h}^{\flat}$ must be zero.

Now suppose that the total shear tensor satisfies the conditions in the statement. By algebra's basic results, we know that l one-forms $\omega_1, \ldots, \omega_l$ are linearly independent if and only if their wedge product $\omega_1 \wedge \cdots \wedge \omega_l$ is not zero. Equivalently, they are linearly dependent if and only if their wedge product is zero. Hence, by hypothesis, among the sets of k one-forms $\{\widetilde{h}(X_1, Y_1)^{\flat}, \ldots, \widetilde{h}(X_k, Y_k)^{\flat}\}$ constructed for arbitrary $X_1, Y_1, \ldots, X_k, Y_k \in \mathfrak{X}(\mathcal{S})$, there exist exactly $k - m$ which are linearly independent. The same holds for the corresponding normal vector fields $\{\widetilde{h}(X_1, Y_1), \ldots, \widetilde{h}(X_k, Y_k)\}$. By Definition 2 of shear space, this implies that the dimension of $\mathsf{Im}\,\widetilde{h}$ is $k - m$. Finally, by Corollary 1, we obtain $\dim \mathcal{U} = m$. □

The following theorem summarizes the results presented in Corollary 1 and Proposition 2.

Theorem 1 *Let $\Phi : (\mathcal{S}, g) \to (\mathcal{M}, \bar{g})$ be an isometric immersion of an n-dimensional Riemannian manifold into a semi-Riemannian manifold with co-dimension k. Let $\mathcal{U}$ and $\mathsf{Im}\,\widetilde{h}$ be the umbilical space and the shear space, respectively, of $\mathcal{S}$. Then the following conditions are all equivalent:*

(i) *the umbilical space $\mathcal{U}$ has dimension m;*
(ii) *the shear space $\mathsf{Im}\,\widetilde{h}$ has dimension $k - m$;*
(iii) *the total shear tensor satisfies*

$$\bigwedge^{k-m+1} \widetilde{h}^{\flat} = 0$$

with

$$\bigwedge^{k-m} \widetilde{h}^{\flat} \neq 0;$$

(iv) *any $k - m + 1$ shear operators $\widetilde{A}_{\xi_1}, \ldots, \widetilde{A}_{\xi_{k-m+1}}$ are linearly dependent (and there exist precisely $k - m$ shear operators that are linearly independent).*

3.3 Special Cases

Using Theorem 1 some particular situations are worth mentioning, for example:

- If $\dim \mathcal{U} = k$ we have $\widetilde{h}(X, Y)^\flat = 0$ for every $X, Y \in \mathfrak{X}(\mathcal{S})$, equivalently $\widetilde{h} = 0$, and the submanifold $\mathcal{S}$ is totally umbilical. In particular, $\mathrm{Im}\,\widetilde{h} = \emptyset$ and $\mathfrak{X}(\mathcal{S})^\perp = \mathcal{U}$.
- If $\dim \mathcal{U} = k - 1$ then $\dim \mathrm{Im}\,\widetilde{h} = 1$ and we have $\widetilde{h}(X_1, Y_1)^\flat \wedge \widetilde{h}(X_2, Y_2)^\flat = 0$ for every $X_1, Y_1, X_2, Y_2 \in \mathfrak{X}(\mathcal{S})$. It follows that there exist a normal vector field $G \in \mathfrak{X}(\mathcal{S})^\perp$ and a properly normalized self-adjoint operator $\widetilde{A}$ such that $\widetilde{h}(X, Y) = g(\widetilde{A}X, Y)G$ for every $X, Y \in \mathfrak{X}(\mathcal{S})$. This was the case studied in [1] for $k = 2$.
- If $\dim \mathcal{U} = 0$ then there are no umbilical directions and $\mathrm{Im}\,\widetilde{h} = \mathfrak{X}(\mathcal{S})^\perp$.

Our results can be applied to all other cases too and open the door for a novel analysis of the structure of the umbilical space.

Acknowledgements We would like to thank the referee for comments and improvements.

References

1. N. Cipriani, J.M.M. Senovilla, J. Van der Veken, Umbilical properties of spacelike co-dimension two submanifolds. Results Math. **72**(1), 25–46 (2017)
2. A. Fialkow, Conformal differential geometry of a subspace. Trans. Amer. Math. Soc. **56**(2), 309–433 (1944)
3. S. Kobayashi, K. Nomizu, *Foundations of Differential Geometry, Volume II*. (Interscience Publishers, 1969)
4. M. Kriele, *Spacetime, Foundations of General Relativity and Differential Geometry* (Springer, Berlin, 1999)
5. B. O'Neill, *Semi-Riemannian Geometry with Applications to Relativity*. (Academic Press, 1983)
6. J.M.M. Senovilla, Umbilical-type surfaces in spacetime, in *Recent Trends in Lorentzian Geometry, Springer Proceedings in Mathematics and Statistics* (2013), pp. 87-109
7. F.J. Pedit, T.J. Willmore, Conformal geometry. Atti Sem. Mat. Fis. Univ. Modena **36**(2), 237–245 (1988)

A Study in Stationary: Geometric Properties of Stationary Regions and Regularity of Their Horizons

I.P. Costa e Silva

Abstract Spacetimes with stationary regions have always been among the most intensively studied ones, but by far the larger body of literature thereon has been inextricably connected with the investigation of stationary black holes arising as solutions of the Einstein field equations. Although the most important results, such as those pertaining to the issue of black hole uniqueness (Stationary black holes: uniqueness and beyond, Living Rev, Relativity, [8]), (Black hole uniqueness theorems, Cambridge University Press, Cambridge[19]) or rigidity theorems (Rigidity results in general relativity: a review [20]) do depend very sensitively on the dimension of the spacetime, assumptions of asymptotic flatness and/or the detailed analytical features of the field equations, many interesting results pertaining to stationary spacetimes and horizons therein can still be obtained by purely geometric methods. It is the purpose of this paper to give a review, without any attempt at comprehensiveness, of some global geometric consequences of the existence of a complete Killing vector field which becomes timelike at some open set, the connected components of which are referred to as *stationary regions*. If the Killing field changes causal character, *horizons* appear. I discuss their general structure and regularity under suitable assumptions on their causality and geodesic (in)completeness, but without assuming any field equations, asymptotic flatness/hyperbolicity or any dimensional restrictions. More specifically, the main focus is on presenting some old and new theorems giving descriptions of the global geometric structure of stationary regions, as well as regularity of the underlying horizons. These are meant to illustrate that a number of the extant results in the literature are not artifacts of solutions in General Relativity and/or asymptotic assumptions. Although many of the results presented are known in some form (maybe with slightly different assumptions), most are reworked in a (hopefully) didactic, unified fashion, and a number of them with new proofs.

Keywords Stationary spacetimes · Regularity of horizons · Killing horizons

I.P. Costa e Silva (✉)
Department of Mathematics, Universidade Federal de Santa Catarina,
Florianópolis-SC 88.040-900, Brazil
e-mail: pontual.ivan@ufsc.br

M.A. Cañadas-Pinedo et al. (eds.), *Lorentzian Geometry and Related Topics*,
Springer Proceedings in Mathematics & Statistics 211,
DOI 10.1007/978-3-319-66290-9_5

1 Introduction

Let (M^{n+1}, g) $(n \geq 1)$ be a *spacetime*, i.e., a smooth (C^{∞}) Lorentzian manifold with a fixed time orientation. (M, g) is said to have a *stationary region* if there exists a complete Killing vector field $X : M \to TM$ on (M, g) which becomes timelike somewhere in M. The prime examples of spacetimes with stationary regions are those in the Kerr-Newman family of electrovac exact solutions of the Einstein field equations of General Relativity (including of course Minkowski and Schwarzschild spacetimes in the vacuum case) and their generalizations with nonzero cosmological constant (including de Sitter and anti-de Sitter spacetimes) [18, 29, 34].

Of course, these spacetimes derive much of their immense importance from physical applications. Not only they are exact, analytically treatable solutions of the field equations, they are believed to more or less realistically model the final state of black holes resulting from the actual gravitational collapse of massive objects in our universe. But even on the purely mathematical side, stationary spacetimes also have a beautiful and rich geometrical structure, even when (M, g) is compact (see, e.g., [19, 29, 31, 32] for a constellation of examples).

Again, in many of the above-mentioned solutions, the underlying Killing vector field X changes causal character and is not timelike everywhere in M. Such changes are signaled by *horizons*, which are interpreted in physical applications as describing either cosmological or event horizons. In these rather special spacetimes, such horizons are *smooth* null hypersurfaces in (M, g) with a rather ubiquitous bifurcate structure, and are in fact examples of *(bifurcate) Killing horizons* [18–20].

Now, one might be tempted to think that all these nice properties of stationary regions and their horizons in the exact solutions are an artifact of their high symmetry. The main purpose of this article is to give a number of theorems which show that, on the contrary, many of these properties are fairly general, and stem from geometric and causal considerations alone, not on added symmetries nor on any field equations. Incidentally, this situation parallels what happened historically with spacetime singularities: these were present in many exact solutions and were for a long while believed to be due to the artificially imposed symmetries, but the celebrated singularity theorems [1, 18, 28] finally established that causal geodesic incompleteness is actually a fairly generic phenomenon in Lorentzian geometry.

Many of the results presented here are not new (some are fairly standard), and have appeared in one or another form in the literature. I shall, of course, refer to the original references whenever that is the case. However, in nearly all cases I give alternative, new proofs, and a streamlined, unified description of the main theorems which may be useful to researchers. The philosophy (and some of the results) pursued in this article is similar to that in Ref. [2], but I owe much of my choice of topics to [4, 6, 15, 17, 21].

I should warn the reader that this is *not* a review of the most recent results, and I make no attempt toward completeness or comprehensiveness. The literature on stationary spacetimes and horizons is prohibitively vast, and I have neither the time nor the competence to do justice to all of it. Instead, I have focused in a unified, somewhat whimsical description of the global geometry of stationary regions and horizons, with emphasis on issues of regularity in the latter case.

The rest of the paper is divided into two largely (but not entirely) independent parts. After some preliminary results in Sect. 2, including an important Lemma due to S. Harris and R. Low [17], the first part deals with the basic global principal $\mathbb{R}$-bundle structure of chronological stationary spacetimes, and comprises Sects. 3 through 5. Our discussion is embedded in the key structural theorems of S. Harris [15] (based on previous work by R. Geroch), M.A. Javaloyes and M. Sánchez [21]. In Sects. 3 and 4, we give new proofs to these now classic results, and review Geroch's reduction formalism [13, 14] for stationary spacetimes in the spirit of [2, 19], but in fully coordinate-free form (see [9] for a similar approach). As instructive illustrations of the resulting structure we use this formalism in Sect. 5 to establish two results which, albeit rather simple, to the best of my knowledge have not appeared explicitly in the literature. The first is a splitting theorem for globally hyperbolic Ricci-flat spacetimes with a compact Cauchy hypersurface and a nonempty stationary region, which is in turn related to a long-standing conjecture by R. Bartnik (see, e.g., Chap. 14 of [1]). The second is an alternative expression for the Komar energy applicable in the case of stationary Einstein manifolds. The second part of the paper spans Sects. 6, 7, and 8, and devoted to a detailed study of horizons with emphasis on regularity issues. Again, many of the stated results therein are known in more restricted settings, but are reworked here in a broader context.

2 Preliminaries

Throughout this paper, we adopt the following terminology. (M^{n+1}, g), with $n \geq 1$, denotes a spacetime with a fixed complete Killing vector field $X \in \Gamma(TM)$. The flow of X is denoted by $\phi : t \in \mathbb{R} \mapsto \phi_t \in \mathrm{Diff}(M)$. In particular, ϕ is a smooth isometric action of the abelian group $(\mathbb{R}, +)$ on M. The notation $\langle\, .\, ,\, .\, \rangle := g$ will often be used, and we denote by $\tilde{X} := \langle X, \ , \ \rangle$ the 1-form metrically associated with X.

Let S be a fixed connected component of the set

$$\{p \in M \mid g_p(X(p), X(p)) < 0\}.$$

We refer to S as a *stationary region* of (M, g). A stationary region S is said to be *static* if the orthogonal distribution $X^{\perp}|_S$ is integrable[1] in S, which by standard results [28] happens iff $\tilde{X} \wedge d\tilde{X} = 0$ iff the restriction of $d\tilde{X} \equiv curl\, X$ to $X^{\perp}$ is identically zero, where the curl of X is given by

$$(curl\, X)(V, W) = \langle \nabla_V X, W\rangle - \langle \nabla_W X, V\rangle$$

[1] In particular, every two-dimensional stationary spacetime region is static.

$\forall V, W \in \Gamma(TM)$. To avoid trivialities, I shall assume throughout that $S \neq \emptyset$. It is easy to see that S is open and (since $\langle X, X\rangle$ is constant along the orbits of ϕ) invariant by the flow ϕ. Given any $p \in M$, $\mathbb{R}p := \cup_{t\in\mathbb{R}}\phi_t(p)$ denotes the orbit of p by ϕ.

We follow throughout the standard notation (presented, for example, in [1, 28]) for objects pertaining to the causal structure of (M, g). Therefore, given $p, q \in M$, we write $p \ll q$ [resp. $p < q$] to indicate that there exists a piecewise smooth future-directed timelike [resp. causal] curve segment from p to q, and $p \leq q$ if either $p < q$ or $p = q$. Similarly, given any set $A \subset M$, we denote by

$$I^+(A)\ [\text{resp.}J^+(A)] := \{q \in M\ :\ \exists p \in A \text{ such that } p \ll q\ [\text{resp.}p \leq q]\}$$

the chronological [resp. causal] future of A. The chronological [resp. causal] past $I^-(A)$ [resp. $J^-(A)$] of A is defined time-dually. Finally, we denote $I^{\pm}(p) := I^{\pm}(\{p\})$ for any $p \in M$.

Proposition 2.1 *The vector field $X_S := X|_S$ is a complete timelike Killing vector field which is either future-directed everywhere or past-directed everywhere in the spacetime $(S, g|_S)$.*

Proof That X_S is complete and Killing timelike is immediate from the invariance of S by ϕ, and the second statement is an easy consequence of the fact that S is connected. □

In what follows, X_S is always taken to be future-directed for definiteness, although all the results will hold in the past-directed case.

We say that (M, g) is *stationary* (with respect to X) if $S \equiv M$, i.e., if M is itself a stationary region. (Of course, even if this does not happen, $(S, g|_S)$, viewed as a spacetime on its own right, is stationary. In particular, any result obtained when (M, g) remains valid for $(S, g|_S)$.) Similarly, (M, g) is *static* (with respect to X) if M is itself a static region.

The following lemma is essentially due to S. Harris and R. Low (cf. Theorem 2.3 of [17]), and will be of key importance in a number of later results. The proof we give here, however, allows us to dispense with the assumption that spacetime is *chronological* (i.e., has no closed timelike curves) which the authors of [17] adopt (although their assumption on X is weaker).

Lemma 2.2 *If (M, g) is stationary, then $\forall p, q \in M$*

$$\mathbb{R}p \cap I^{\pm}(q) \neq \emptyset,$$

so in particular, $I^{\pm}(\mathbb{R}p) = M$.

Proof Suppose, by way of contradiction, that X is everywhere timelike, but the conclusion fails. Pick then $p, q \in M$ such that, say, $\mathbb{R}p \cap I^+(q) = \emptyset$ (the proof if we assume $\mathbb{R}p \cap I^-(q) = \emptyset$ is entirely analogous). Let

$$I_p := \{r \in M \mid \mathbb{R}p \cap I^+(r) = \emptyset\}.$$

This set is nonempty, since $q \in I_p$. Also, let

$$R_p := \bigcup_{r \in I_p} I^+(r).$$

The set R_p is then manifestly open and nonempty. It is also a *future set*, i.e., $I^+(R_p) \subset R_p$. In particular, by standard results in causality theory (see, e.g., Sect. 3.2 of [1]), its boundary ∂R_p is a closed achronal Lipschitz hypersurface in (M, g). Also, it is clearly nonempty.

Claim: $\overline{R_p} = \overline{I_p}$.

Indeed, take $r \in \overline{R_p}$. If $r \notin \overline{I_p}$, then $r \in I^-(\phi_t(p))$ for some $t \in \mathbb{R}$. Since the latter set is open, we can pick $r' \in I^-(\phi_t(p)) \cap R_p$, and hence $r' \in I^+(r'')$ for some $r'' \in I_p$, whence we conclude that $r'' \in I^-(\phi_t(p))$, a contradiction. Thus, $\overline{R_p} \subset \overline{I_p}$. Now, let $r \in I_p$, and consider a neighborhood $U \ni r$. For any $r' \in I^+(r) \cap U$, $r' \in R_p$ by definition, so $U \cap R_p \neq \emptyset$. Thus, $r \in \overline{R_p}$, proving the claim.

To complete the proof, pick $r \in \partial R_p \subset \overline{R_p} = \overline{I_p}$, and any $t \in \mathbb{R}$, $t > 0$. Now, since each orbit is timelike, $\phi_t(r) \in R_p$. Then, $r' \ll \phi_t(r)$ for some $r' \in I_p$. Hence, since the isometric action ϕ preserves causal relations, $\phi_{-t}(r') \ll r$. Now, if $\phi_{-t}(r') \notin I_p$, then $\phi_{-t}(r') \ll \phi_s(p)$ for some $s \in \mathbb{R}$, which implies that $r' \ll \phi_{t+s}(p)$, which is absurd. Thus, we conclude that $\phi_{-t}(r') \in I_p$, whence $r \in R_p$. But this is impossible, since R_p is open and $R_p \cap \partial R_p = \emptyset$. This final contradiction completes the proof. □

3 Global Structure of Stationary Spacetimes I: Topological Aspects

In this section, I review some consequences for the topology of a stationary region of imposing certain causality restrictions on (M, g). These will be also be used later in this work. We shall see that relatively mild causality assumption have a dramatically simplifying effect on the global geometry of stationary spacetimes, and hence of stationary regions.

The first one is due to S. Harris (cf. [15]), and gives a key structural result for the topology of stationary chronological spacetimes.

Proposition 3.1 *If (M, g) is stationary and chronological, then ϕ is a free, proper $\mathbb{R}$-action. In particular, the space of orbits $Q := M/\mathbb{R}$ has a unique structure of a smooth n-manifold for which the standard projection $\pi : M \to Q$ is smooth; indeed, (M, π, ϕ) is a (necessarily trivial) smooth principal $\mathbb{R}$-bundle over Q, and hence M is (noncanonically) diffeomorphic to $\mathbb{R} \times Q$.*

Proof That the action is free is immediate from the fact that we have no closed (timelike) orbits. To show the action ϕ is proper, it is sufficient (cf., e.g., Proposition 9.13 in [26]) to check that given sequences $(t_k) \subset \mathbb{R}$, $(p_k), (q_k) \subset M$ and points

$p, q \in M$ such that $p_k \to p$ and $\phi_{t_k}(p_k) \to q$, then (t_k) converges up to passing to a subsequence. But suppose this is false. Then we can assume that $|t_k| \to +\infty$. Then, up to passing to a subsequence, either $t_k \to +\infty$ or $t_k \to -\infty$, and we may as well assume the first of these options holds, since the proof for the other case is entirely analogous. Fix any $t \in \mathbb{R}$, and $q' \in I^+(q)$. Eventually, $t_k > t$, and $\phi_{t_k}(p_k) \in I^-(q')$, since the latter set is open and contains q. But then $\phi_t(p_k) \ll \phi_{t_k}(p_k) \ll q'$, and since $\phi_t(p_k) \to \phi_t(p)$ we conclude that $\phi_t(p) \in \overline{I^-(q')}$. Since t was arbitrarily fixed, we conclude that $\mathbb{R}p \subset \overline{I^-(q')}$. But according to Lemma 2.2, $\mathbb{R}p \cap I^+(q') \neq \emptyset$, and hence $I^-(q') \cap I^+(q') \neq \emptyset$, violating chronology. The remaining statements follow from standard facts about principal bundles (see, e.g., p. 218, Theorem. 9.16 of [26], and Theorem 5.7, p. 58 of [24]). □

Let us see some consequences of this result for the global geometry of (M, g). *For the remainder of this section (M, g) will be assumed to be a stationary (with respect to X) chronological spacetime.*

Under our assumptions, there exists a global section $\sigma : Q \to M$. Such global sections are in one-to-one correspondence with (global) trivializations of $\pi : M \to Q$, by which I mean diffeomorphisms $\varphi : M \to \mathbb{R} \times Q$ for which the following diagrams commute:

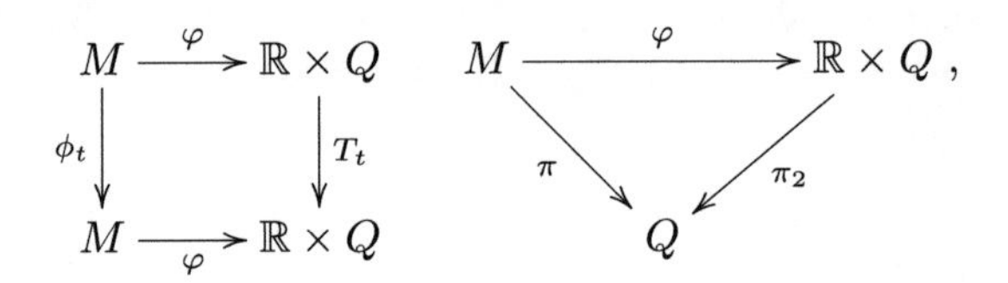

for all $t \in \mathbb{R}$. Here, π_2 denotes the projection onto the second cartesian factor, and $T_t(s, x) := (s + t, x)$, $\forall t, s \in \mathbb{R}$, $\forall x \in Q$. The relationship between the section σ and its associated trivialization φ is given by

$$\sigma(x) = \varphi^{-1}(0, x), \forall x \in Q. \tag{3.1}$$

By using (3.1) and the first of the commutative diagrams above we conclude that

$$\phi_t(\sigma(Q)) = \varphi^{-1}(\{t\} \times Q), \forall t \in \mathbb{R}. \tag{3.2}$$

Theorem 3.2 *Let $\sigma : Q \to M$ be a global section with associated trivialization $\varphi : M \to \mathbb{R} \times Q$. The following statements hold.*

(1) $\sigma(Q)$ is a smooth properly embedded hypersurface diffeomorphic to Q intersected exactly once by each integral curve of X (such a hypersurface is said to be a slice *of ϕ).*
(2) Every slice of ϕ is the image of a (unique) global section.
(3) $d\varphi_p(X(p)) = \partial_t|_{\varphi(p)}$, $\forall p \in M$, where ∂_t denotes the lift to $\mathbb{R} \times Q$ of the standard vector field d/dt on $\mathbb{R}$.

(4) *If $\sigma(Q)$ is spacelike, then the function $\tau := \pi_1 \circ \varphi : M \to \mathbb{R}$ has a past-directed timelike gradient, where $\pi_1 : \mathbb{R} \times Q \to \mathbb{R}$ is the projection onto the first cartesian factor. Moreover,*

$$\langle \nabla \tau, X \rangle = 1. \tag{3.3}$$

We shall refer to τ is the time function associated with σ.

Proof (1) follows immediately from (3.2) and from the fact that $\pi \circ \sigma \equiv Id_Q$. If $A \subset M$ is a slice of ϕ, then it is easy to check that $\pi|_A : A \to Q$ is a diffeo, and hence $\sigma := (\pi|_A)^{-1}$ is a global section whose image is A. Uniqueness is then obvious. This proves (2). To prove (3), pick $p \in M$ and write $\varphi(p) = (s_0, x_0)$. For each $f \in C^\infty(\mathbb{R} \times Q)$, we have

$$\begin{aligned} \partial_t|_{(s_0,x_0)}(f) &= \lim_{s\to 0} \frac{f(s+s_0,x_0) - f(s_0,x_0)}{s} \\ &= \lim_{s\to 0} \frac{f(T_s \circ \varphi(p)) - f \circ \varphi(p)}{s} \\ &= \lim_{s\to 0} \frac{(f \circ \varphi) \circ \phi_t(p)) - (f \circ \varphi)(p)}{s} \\ &\equiv X(p)(f \circ \varphi) \equiv d\varphi_p(X(p))(f), \end{aligned} \tag{3.4}$$

thus establishing (3).

In order to prove (4), we first claim that

$$\phi_t^* d\tau = d\tau, \forall t \in \mathbb{R}. \tag{3.5}$$

Indeed, for each $t \in \mathbb{R}$, we have

$$\phi_t^* d\tau = d(\tau \circ \phi_t) = d(\pi_1 \circ \varphi \circ \phi_t) \equiv d(\pi_1 \circ T_t \circ \varphi).$$

On the other hand, for $(s_0, x_0) \in \mathbb{R} \times Q$,

$$\pi_1 \circ T_t(s_0, x_0) = t + t_0 \equiv (\pi_1 + c_t)(s_0, x_0),$$

where $c_t : (s, x) \in \mathbb{R} \times Q \mapsto t \in \mathbb{R}$ is the constant function (for fixed t). Therefore,

$$\pi_1 \circ T_t \circ \varphi = \tau + c_t \Rightarrow d\tau = d(\pi_1 \circ T_t \circ \varphi), \tag{3.6}$$

whence (3.5) follows. In turn, from (3.5) we have

$$L_X d\tau = 0, \tag{3.7}$$

where L_X denote the Lie derivative along X. Since X is Killing, equation (3.7) is equivalent to

$$[X, \nabla \tau] = 0, \tag{3.8}$$

and therefore

$$X\langle \nabla\tau, \nabla\tau\rangle = 2\langle [X, \nabla\tau], \nabla\tau\rangle \equiv 0,$$

using again that X is Killing and (3.8). Thus $\langle \nabla\tau, \nabla\tau\rangle$ is constant along the orbits of ϕ. But from (3.2), the level hypersurfaces of τ are

$$\tau^{-1}(t) = \phi_t(\sigma(Q)), \forall t \in \mathbb{R}.$$

It follows that $\nabla\tau|_{\sigma(Q)}$ is timelike, since it is orthogonal to $\sigma(Q)$, which has been assumed to be spacelike. Thus, $\nabla\tau$ is timelike everywhere on M.

Finally, in order to establish Eq. (3.3), pick $p \in M$. Let $\gamma_p : t \in \mathbb{R} \mapsto \phi_t(p)$ be the maximal integral curve of X through p. Now,

$$(\tau \circ \gamma_p)(t) = (\pi_1 \circ \varphi \circ \phi_t)(p) = (\pi_1 \circ T_t \circ \varphi)(p) = \tau(p) + t,$$

where we have used (3.6) in the last equality. Differentiating, we get

$$1 = (\tau \circ \gamma_p)'(0) = \langle \nabla\tau(p), X(p)\rangle$$

as desired. □

Remark 3.3 In p. 05 of Ref. [16],[2] the function τ as constructed in Theorem 3.2 (4) is referred to as a "Killing time-function" even if $\sigma(Q)$ is not spacelike. However, note that according to the standard definition [1, 18, 28], a time function needs to be (continuous and) strictly increasing along causal curves, and hence its level hypersurfaces need to be acausal. Indeed, the function τ being a *bona fide* time function is *equivalent* to the assumption that $\sigma(Q)$ is acausal, since the latter is a level set of τ and the isometric flow ϕ preserves acausality. But it is well known that the existence of such a function is equivalent to the condition that (M, g) is stably causal [18]. Moreover, Bernal and Sánchez [3] have proved that stable causality is actually equivalent to the existence a *temporal function*, i.e., a smooth function $\tau \in C^\infty(M)$ with past-directed timelike gradient. Such a function is obtained right away in our context through the "stronger" assumption that $\sigma(Q)$ is spacelike. As a matter of fact, any level hypersurface of a temporal function (which is, of course, spacelike) would work as a section, by a result due to M.A. Javaloyes and M. Sánchez [21] discussed below (cf. Theorem 3.4). The upshot now is that *the existence of an acausal section implies the existence of a spacelike one*. In this sense, there is no real loss of generality in our assumption that $\sigma(Q)$ is spacelike insofar as one wishes to have a time function.

Theorem 3.4 *Suppose (M, g) is stationary, chronological and possesses an achronal edgeless set $A \subset M$.*[3] *Then each orbit of ϕ intersects A exactly once.*

---

[2] I thank the referee for bringing [16] to my attention.

[3] A is then a closed C^0 (indeed Lipschitz) hypersurface in M, see, e.g., Corollary 26, Chap. 14 in [28].

Moreover,

$$\varphi := \phi|_{\mathbb{R}\times A} : (t, p) \in \mathbb{R} \times A \mapsto \phi_t(p) \in M$$

is a homeomorphism. In particular, if A is a smooth, closed achronal spacelike hypersurface (and hence acausal), then φ is actually a diffeomorphism, and $d\varphi_{(t_0,x_0)}(\partial_t|_{(t_0,x_0)}) = X(\varphi(t_0, x_0))$, for all $(t_0, x_0) \in \mathbb{R} \times A$.

Proof The map φ is continuous (smooth if A is smooth), since A is a C^0 hypersurface, and one-to-one since A is achronal. By Invariance of Domain, φ is then a homeomorphism onto an open subset $\mathcal{O} \subset M$. The first and second claims boil down therefore to showing that $\mathcal{O} \equiv M$. Since the latter set is open and M is connected, all we need to show is that $\mathcal{O}$ is closed. To this end, consider a sequence (t_k, x_k) in $\mathbb{R} \times A$ and $p \in M$ such that $\phi_{t_k}(x_k) \to p$.

Assume first that (t_k) is unbounded. We may assume, up to passing to a subsequence that $t_k \to +\infty$, the argument if $t_k \to -\infty$ being analogous. From Lemma 2.2, we have that $\phi_s(p) \in I^-(x_1)$ for some $s \in \mathbb{R}$ (x_1 being the first term in the sequence (x_k)!). But then

$$\phi_{t_k+s}(x_k) = \phi_s(\phi_{t_k}(x_k)) \to \phi_s(p),$$

so for large enough k we have $t_k + s > 0$ and $\phi_{t_k+s}(x_k) \in I^-(x_1)$; hence,

$$x_k \ll \phi_{t_k+s}(x_k) \ll x_1,$$

which contradicts the achronality of A. Therefore, (t_k) must be bounded. But in that case, up to passing to a subsequence we may assume that it converges, say, $t_k \to t_0$. Let $x_0 := \phi_{-t_0}(p)$. Then

$$x_k = \phi(-t_k, \phi_{t_k}(x_k)) \to x_0,$$

and since A is closed we conclude that $x_0 \in A$, and $\varphi(t_0, x_0) = p$, which shows that $\mathcal{O}$ is closed, as desired. The proof of the remaining statements is straightforward, and we omit it. □

Recall that (M, g) is said to be *distinguishing* if $\forall p, q \in M$

$$I^+(p) = I^+(q) \text{ or } I^-(p) = I^-(q) \Rightarrow p = q.$$

Examining the causal ladder (cf., e.g., [1] p. 73), we see that if (M, g) is distinguishing, then it is *causal*, i.e., has no closed causal curves, and this condition is implied by strong causality, but does not usually imply it. However, M.A. Javaloyes and M. Sánchez have shown (Proposition 3.1 of [21]) the remarkable result that a distinguishing *stationary* spacetime is actually stably causal.[4] In particular,

[4]More precisely, Javaloyes and Sánchez actually show that (M, g) is *causally continuous*, a condition even stronger than stable causality.

(M, g) is strongly causal and there exists a temporal function, i.e., a smooth function $\tau : M \to \mathbb{R}$ with past-directed timelike gradient. Each level set of such a function τ is a closed, acausal spacelike hypersurface in (M, g). In view of Theorems 3.2 and 3.4 and these remarks, we then have the following immediate consequence [21].

Corollary 3.5 *If (M, g) is chronological and stationary, then the following statements are equivalent:*

(i) (M, g) is distinguishing;
(ii) there exists a smooth closed, acausal spacelike hypersurface in (M, g) which is intersected exactly once by each integral curve of X.

In the affirmative case, any one of these slices is diffeomorphic to Q (and hence to one another), and (M, g) is causally continuous.[5] □

4 Global Structure of Stationary Spacetimes I: Metrical Aspects

In the previous section, a topological splitting for chronological stationary spacetimes was obtained. We now proceed to review how the metric responds to this splitting. Most facts stated here are fairly well known [2, 9, 13, 14, 19], so the goal of repeating them here is twofold: to establish terminology and set the stage for the results of the next section.

We assume again, throughout this section, that (M, g) is stationary and chronological, with the single exception of Theorem 4.5 *below.*

Since X is Killing, the orthogonal distribution $X^{\perp} \subset TM$ is covariant, in the sense that $(d\phi_t)_p(X_p^{\perp}) = X_{\phi_t(p)}^{\perp}$ for all $t \in \mathbb{R}$ and all $p \in M$, and of course $T_pM = X_p^{\perp} \oplus \mathbb{R}X_p$. In other words, $X^{\perp}$ is a principal (Ehresmann) $\mathbb{R}$-connection on the principal bundle $\pi : M \to Q$, for which the orthogonal complements of X at each point are the horizontal spaces, and having an associated ϕ-invariant 1-form $\theta \in \Omega^1(M)$. Indeed, θ is uniquely defined by the conditions $\theta(X) \equiv 1$ and $\theta(X^{\perp}) \equiv 0$, and hence

$$\theta \equiv -\tilde{X}/\beta^2, \tag{4.1}$$

where

$$\beta := \sqrt{|\langle X, X\rangle|}.$$

Thus, its curvature $\tilde{\Omega} := d\theta|_{X^{\perp}\wedge X^{\perp}}$ measures the integrability of the orthogonal distribution, i.e., $\tilde{\Omega} = 0$ iff $X^{\perp}$ is integrable, that is, if (M, g) is static. Moreover, β is constant along the orbits of ϕ, and $\tilde{\Omega}$ is a horizontal ϕ-invariant 2-form, and hence there exists a unique function $u \in C^{\infty}(Q)$ and a unique 2-form $\Omega \in \Omega^2(Q)$ such that

[5] See also [16]. I am grateful to the referee for pointing out this reference.

$$\beta = \pi^* u = u \circ \pi \text{ and } \tilde{\Omega} = \pi^* \Omega. \tag{4.2}$$

Now, for each $p \in M$, $\ker d\pi_p$ is the span of X_p, and since π is a submersion $d\pi_p$ has maximal rank, i.e., $\dim X_p^{\perp} = \dim T_{\pi(p)} S$, so $d\pi_p|_{X_p^{\perp}} : X_p^{\perp} \to T_{\pi(p)} Q$ is an isomorphism. Since X is Killing, the induced inner product on $T_{\pi(p)} Q$ is the same for all points along the orbit of p, so there exists a unique Riemannian metric h on Q defined by requiring that this pointwise isomorphisms are linear isometries. (Hence, when Q is endowed with this metric, π becomes a semi-Riemannian submersion.) The splitting of vectors into vertical and horizontal parts implies that the metric g can then be written as

$$g = -\beta^2 \theta \otimes \theta + \pi^* h. \tag{4.3}$$

Equation (4.3) implies that the geometrical information of (M, g) is entirely encoded in the 1-form θ, the function u (or equivalently β) and the Riemannian metric h.

When $n \geq 2$, another geometric quantity of interest is the *twist* of X which is the $(n-2)$-form ω given[6] by

$$\omega := \frac{1}{2} \star_g (\tilde{X} \wedge d\tilde{X}). \tag{4.4}$$

Of course $\omega = 0$ iff (M, g) is static with respect to X. Thus, one imagines this is not independent of the other geometric quantities, especially of θ. Indeed, we have:

Proposition 4.1 *The twist ω is a horizontal, invariant form. In fact,*

$$\omega = \pi^* (\frac{1}{2} u^3 \star_h \Omega).$$

Proof Fix $p \in M$, and let $x = \pi(p)$. Let $U \subset Q$ be some neighborhood of x where is defined an h-orthonormal frame $\{(E_1)_*, \dots, (E_n)_*\} \subset \Gamma(TU)$ with dual frame $\{(E^1)_*, \dots, (E^n)_*\} \subset \Omega^1(U)$. Then, there exists a unique g-orthonormal frame $\{E_0, E_1, \dots, E_n\} \subset \Gamma(T\pi^{-1}(U))$ of vector fields defined around p for which

$$\pi_*(E_i) \equiv (E_i)_* \ (i = 1, \dots, n)$$

and $E_0 := X/\beta$. Let $\{E^0, E^1, \dots, E^n\} \subset \Omega^1(\pi^{-1}(U))$ denote the corresponding dual frame. Using the Killing equation, we have, for each $i \in \{1, \dots, n\}$,

$$d\tilde{X}(E_0, E_i) = -2\langle \nabla_{E_i} X, \frac{X}{\beta} \rangle \equiv 2E_i(\beta). \tag{4.5}$$

However, $\forall q \in \pi^{-1}(U)$ we have

$$E_i(\beta)(q) = d\beta_q(E_i(q)) = (\pi^* du)_q(E_i(q)) \equiv du_{\pi(q)}((E_i)_*(\pi(q))),$$

[6]The symbols $\star_g$ and $\star_h$ will be used for the Hodge star operators of g and h, respectively.

and so we conclude that

$$d\tilde{X}(E_0, E_i) = 2[du((E_i)_*)] \circ \pi. \tag{4.6}$$

On the other hand, from (4.1) we have

$$d\tilde{X} = -2\beta d\beta \wedge \theta - \beta^2 d\theta, \tag{4.7}$$

and hence, since the E_i's annihilate θ, we have, for $i < j$,

$$d\tilde{X}(E_i, E_j) = -\beta^2 d\theta(E_i, E_j) \equiv -\beta^2 \tilde{\Omega}_{ij}. \tag{4.8}$$

But since $\tilde{\Omega} = \pi^*\Omega$, $\tilde{\Omega}_{ij} = \Omega_{ij} \circ \pi$. Gathering Eqs. (4.6) and (4.8), we get

$$d\tilde{X} = 2\sum_{i=1}^{n}\{[du((E_i)_*)] \circ \pi\}(E^0 \wedge E^i) - \sum_{i<j}\{[u^2\Omega_{ij}] \circ \pi\}(E^i \wedge E^j). \tag{4.9}$$

Taking the exterior product with $\tilde{X}$, and noting that $E^0 = -\tilde{X}/\beta$,

$$\tilde{X} \wedge d\tilde{X} = \sum_{i<j}\{[u^3\Omega_{ij}] \circ \pi\}(E^0 \wedge E^i \wedge E^j). \tag{4.10}$$

Now, by just noting that

$$\star_g(E^0 \wedge E^i \wedge E^j) = \pi^*(\star_h((E^i)_*, (E^j)_*)),$$

whence the result follows. □

Example 4.2 (Standard stationary spacetimes)
Let (M_0, g_0) be any smooth Riemannian n-manifold. On M_0, pick a smooth, real-valued, strictly positive function β_0, a smooth 1-form $\delta_0 \in \Omega^1(M_0)$. The *standard stationary spacetime* associated with the data $(M_0, g_0, \beta_0, \delta_0)$ is (M, g), where $M := \mathbb{R} \times M_0$,

$$g = -\beta^2 d\pi_1 \otimes d\pi_1 + \delta \otimes d\pi_1 + d\pi_1 \otimes \delta + \pi_2^* g_0, \tag{4.11}$$

$\beta := \beta_0 \circ \pi_2$, $\delta := \pi_2^*\delta_0$, and π_1 [resp. π_2] is the projection of M onto the $\mathbb{R}$ [resp. M_0] factor. The time orientation is chosen such that ∂_t, the lift to M of the standard vector field d/dt on $\mathbb{R}$, is future-directed. If $\delta_0 \equiv 0$, then (M, g) thus defined is said to be *standard static*. The following facts are very easily verified.

(a) $X = \partial_t$ is a complete Killing timelike vector field on (M, g), with flow given by $\phi_t(s, x) = (t + s, x)$, $\forall t, s \in \mathbb{R}$, $\forall x \in M_0$. The quotient Q of M by this action can be canonically identified with M_0 with projection $\pi = \pi_2$.
(b) Each hypersurface $t \times M_0$ is a spacelike slice for ϕ; thus (cf. Corollary 3.5) (M, g) is distinguishing, and hence stably causal.

(c) π_1 is a temporal function on (M, g), and indeed this is precisely the time function associated with the section $\sigma : x \in M_0 \mapsto (0, x) \in M$ (cf. Theorem 3.2 (4)).
(d) g can be cast in the form (4.3), with

$$\theta \equiv d\pi_1 - \frac{\delta}{\beta^2} \text{ and } h \equiv \frac{\delta_0 \otimes \delta_0}{\beta_0^2} + g_0, \tag{4.12}$$

from which we deduce that

$$\Omega = -d(\delta_0/\beta_0^2). \tag{4.13}$$

This example is of course quite illustrative, because stationary regions in the Kerr-Newman family, for example, assume the general form (4.11) with Boyer-Lindquist coordinates. Indeed, the authors of [21] have proven the following (cf. their Lemma 3.3).

Theorem 4.3 *Suppose (M, g) is distinguishing, and let $\sigma : Q \to M$ be a section such that $\sigma(Q)$ is spacelike. Let $\varphi : M \to \mathbb{R} \times Q$ and be the associated trivialization. Then $(\mathbb{R} \times Q, (\varphi^{-1})^* g)$ is standard stationary.* □

It is important to realize that *although every distinguishing stationary spacetime is isometric to a standard stationary spacetime, this isometry is highly noncanonical, and depends on the choice of a spacelike slice*. Therefore, each such isometry behaves much like a "coordinate system", and so one must exercise care when extracting geometric information from it. The following simple example illustrates this.

Example 4.4 Pick any smooth function $f : \mathbb{R} \to \mathbb{R}$ with $f(0) = 0$, and let $\epsilon > 0$ for which

$$-\epsilon < x < \epsilon \Rightarrow |f(x)| < 1.$$

Consider the open strip $M = \{(t, x) \in \mathbb{R}^2 \mid -\epsilon < x < \epsilon\}$ with flat metric

$$g = -dt^2 + dx^2.$$

This is static with respect to the complete Killing vector field $X = \partial_t$, which again we assume future-directed. If we introduce a new coordinate in M given by

$$\xi = t + \int_0^x f(\lambda) d\lambda,$$

the metric becomes

$$g = -d\xi^2 + 2f(x)d\xi dx + (1 + f(x)^2)dx^2. \quad (4.14)$$

The metric g in the form (4.14) is in standard stationary form, and corresponds to the choice of section as $\sigma : x \in (-\epsilon, \epsilon) \mapsto (x, f(x))$, i.e., the graph of f is the spacelike slice and ξ is the associated time function. But as f is quite arbitrary, it gives no interesting geometric information for this spacetime: the standard stationary form is unduly mystifying.

Of course, with the appropriate choices it can be highly illuminating to deal with the standard stationary form, as the coordinate treatment of Kerr itself shows. Moreover, the standard stationary form can be used to perform a detailed analysis of general stationary spacetimes using an associated Randers type Finsler structure (see, e.g., [5]), which has intrinsic interest.

The situation changes when (M, g) is *static* with respect to X. In this case, we have the following result, due to M. Sánchez [32]:

Theorem 4.5 *The universal covering of any spacetime (chronological or not) (M, g), static with respect to the Killing vector field X is (canonically isometric to) a standard static spacetime with respect to the lift of X, and is, in particular, causally continuous.*

Comment on the proof. Adopting our previous notation, the 1-form

$$\theta = -\tilde{X}/\beta^2$$

is well-defined independently of its status as a connection 1-form (which applies only in the chronological case discussed above) and $\theta(X) = 1$. It is closed when (M, g) is static. Since we are interested in the universal covering, one can assume without loss of generality that M is simply connected and $\theta = d\tau$ for a smooth function $\tau \in C^\infty(M)$ unique up to an additive constant. The level sets of τ are of course embedded hypersurfaces, and they coincide precisely with the leaves of the integral foliation of $X^\perp$. Moreover, the condition $d\tau(X) = 1$ implies that each such leaf is a section for the action ϕ. Taking $\mathcal{N} = \tau^{-1}(0)$, the map $(t, x) \in \mathbb{R} \times \mathcal{N} \mapsto \phi_t(x) \in M$ is a diffeomorphism, and can be used to pullback the metric, thus yielding the desired isometry. □

Example 4.6 For an example of a static spacetime which is not standard static, take (M, g) as the 2-dimensional cylinder with the flat metric

$$g = -dt^2 + dx^2.$$

obtained by the isometric identification $(t, x) \sim (t, x + 1)$ in the Minkowski plane. Pick any number $0 < a < 1$ and

$$X = \partial_t + a\partial_x.$$

The leaves of $X^{\perp}$ are embedded spacelike hypersurfaces, but they are *not* sections, because each orbit will intersect each leaf infinitely many times. This spacetime is globally hyperbolic, and hence does possess the general bundle structure described above, with $Q \simeq S^1$. But although the bundle is flat, its holonomy group is $\mathbb{Z}$, i.e., it coincides with the fundamental group of M, and this translates into the infinitely many intersections mentioned above.

5 Static Splitting and Komar Energy of Stationary Spacetimes

We apply the ideas in the previous sections in the so-called reduction formalism [2, 13, 14, 19]. The first result we quote is crucial for that. Its proof is a long calculation using O'Neill's equations of a semi-Riemannian submersion [30], and is carried out in [9], so we omit it here.

Theorem 5.1 *Pick, as in the proof of Proposition 4.1, an h-orthonormal frame* $\{(E_1)_*, \ldots, (E_n)_*\}$ *on* Q *and the unique g-orthonormal frame* $\{E_0, E_1, \ldots, E_n\}$ *on* M *such that*

$$\pi_*(E_i) \equiv (E_i)_* \ (i = 1, \ldots, n)$$

and $E_0 := X/\beta$. *The g-Ricci tensor components of in this frame are given in terms of the data* (θ, u, h) *described above by*

$$Ric(E_0, E_0) = \left[\frac{\Delta_h u}{u} + \frac{u^2}{4}\|\Omega\|_h^2\right] \circ \pi, \tag{5.1}$$

$$Ric(E_0, E_i) = \left\{\frac{u}{2}[(div\,\Omega_h)((E_i)_*) + 3\Omega_h(\nabla^h(\log u), (E_i)_*)]\right\} \circ \pi \tag{5.2}$$

$$Ric(E_i, E_j) = \left[Ric_h((E_i)_*, (E_j)_*) - \frac{1}{u}Hess_u^h((E_i)_*(E_j)_*) + \frac{u^2}{2}\langle i_{(E_i)_*}\Omega, i_{(E_i)_*}\Omega\rangle_h\right] \circ \pi, \tag{5.3}$$

where Δ_h *is the Laplacian operator,* $Hess_u^h$ *is the Hessian of* u *and* $\langle .\,, .\rangle_h$ *is the inner product on forms, all relative to* h. □

Using these equations we can prove the following Bochner-like splitting theorem (see Sect. 5 of [31] and references therein for related results).

Theorem 5.2 *Suppose that:*

(i) (M, g) *is globally hyperbolic with compact Cauchy hypersurfaces,*
(ii) the complete Killing vector field X *is timelike everywhere, and*
(iii) $Ric(v, v) \leq 0$ *for any* $v \in TM$ *timelike.*

Then X *is parallel.*
As a consequence, the universal cover of (M, g) *splits isometrically as* $(\mathbb{R} \times N, -dt^2 \oplus g_0)$*, where* (N, g_0) *is a Riemannian manifold.*

Proof Clearly Q is homeomorphic to any given Cauchy hypersurface, and hence compact. Equation (5.1) together with condition (iii) implies that u is superharmonic on the compact set Q and hence u is constant by the maximum principle. Rescaling if necessary we can assume $u \equiv 1$. Substituting this information back in Eq. (5.1) allows us to conclude that $\Omega = 0$. But then $d\tilde{X} = 0$, and since X is Killing, it is parallel. This also shows that (M, g) is actually static with respect to X. By Theorem 4.5, the universal covering $(\hat{M}, \hat{g})$ of (M, g) is standard static with respect to the lift $\hat{X}$ of X, Since the corresponding covering is a local isometry, $\hat{X}$ is also unit and parallel. We conclude that $(\hat{M}, \hat{g}) = (\mathbb{R} \times \hat{N}, -dt^2 \oplus \hat{g}_0)$ and is also globally hyperbolic. □

Remark 5.3 A couple of comments about Theorem 5.2 are in order:

(a) Condition (iii) in Theorem 5.2 holds, in particular, if (M, g) is Einstein, i.e., if $Ric = c \cdot g$, with $c \geq 0$, as it will happen with a solution of the vacuum Einstein field equations with a positive or zero cosmological constant.
(b) Eq. (5.3) implies that if (M, g) is Ricci-flat, then so is (N, g_0). Thus, if $n \leq 4$ this implies (M, g) is flat, and hence (N, g_0) is an n-torus.

Let us now turn to the problem of obtaining a simplified expression for the so-called Komar energy [25, 34]. The generalized Komar integral expression used here has been deduced (apart from an overall normalization factor which need not concern us here) by A. Magnon [27][7] For the rest of this section, I shall make the following two assumptions:

(1) (M, g) is *distinguishing* and stationary, so, in particular, it has the bundle structure $\pi : M \to Q$ described in the previous section, and we fix a spacelike slice $\sigma : Q \to M$ for the action ϕ.
(2) (M, g) is Einstein: $Ric = c \cdot g$ with a fixed $c \in \mathbb{R}$. We make no assumptions about c.

Definition 5.4 A open set $R \subset Q$ is said to be an *admissible region* if

(i) the closure $\overline{R}$ of R is a submanifold of dimension n with smooth compact boundary ∂R, and
(ii) $|c| \int_R u\, vol_h < +\infty$, i.e., $c\, u$ is integrable in R. (This is, of course, automatic if $c = 0$.)

For such an admissible region R, the *Komar energy* associated with the data (σ, R) is

$$E_{\sigma,R} := \frac{1}{2} \int_{\sigma(\partial R)(\equiv \partial\sigma(R))} \star_g d\tilde{X} - c \int_{\sigma(R)} \star_g \tilde{X}. \tag{5.4}$$

Interesting choices of admissible regions involve noncompactness, as the next result shows. The notation and conventions on the operators d and $d^\dagger$ on forms are as in [19].

[7] Although the definition (5.4) has been obtained in [27] when analyzing asymptotically anti-de Sitter metrics of a very specific nature, the way the author obtains the expression is fairly general and uses only the Einstein character of the metric and the Ricci identity.

Proposition 5.5 *If $R \subset Q$ is precompact, then $E_{\sigma,R} = 0$.*

Proof If $\overline{R}$ is compact, we may use Stokes' theorem to get

$$\int_{\partial\sigma(R)} \star_g d\tilde{X} = \int_{\sigma(R)} d \star_g d\tilde{X}.$$

But

$$d \star_g d\tilde{X} = (-1)^{n+1} \star_g^2 d \star_g d\tilde{X} = -(-1)^{n+1+3(n+1)+1} \star_g d^\dagger d\tilde{X} \equiv - \star_g (\triangle_H \tilde{X}), \tag{5.5}$$

where $\triangle_H := -(dd^\dagger + d^\dagger d)$ is the Hodge Laplacian. (Since X is Killing, $d^\dagger \tilde{X} \equiv 0$, cf. Equation (2.2) of [19].) Now the Ricci identity (see discussion around Eq. (2.7) in [19]) for a Killing vector field is

$$\triangle_H \tilde{X} = -2\, Ric(X) \equiv -2c\, \tilde{X},$$

due to our assumption (2). From (5.5) we deduce that

$$\int_{\partial\sigma(R)} \star_g d\tilde{X} = 2c \int_{\sigma(R)} \star_g \tilde{X},$$

which proves the claim. □

Remark 5.6 Note that Proposition 5.5 implies, in particular, that if $R, R' \subset Q$ are two admissible regions with $\overline{R} \subset R'$, then

$$E_{\sigma,R} = E_{\sigma,R'}.$$

This fact is of importance in certain applications, in which, given suitable assumptions on the topology of Q (and hence of M) one may define a notion of energy which depends only on σ. But I shall not pursue this matter any further here.

Theorem 5.7 *If $R \subset Q$ is an admissible region, then*

$$E_{\sigma,R} = \int_{\partial R} \star_h du + c \cdot \int_R u\, vol_h - \int_{\partial\sigma(R)} \theta \wedge \omega, \tag{5.6}$$

where ω is the twist of X, given by (4.4).

Proof Using the same conventions and notation as in the proof of Proposition 4.1, we may use Eq. (4.9): just note that $\forall i, j \in \{1, \dots, n\}, i < j$,

$$\star_g (E^0 \wedge E^i) = \pi^*(\star_h E_*^i), \tag{5.7}$$
$$\star_g (E^i \wedge E^j) = E^0 \wedge \pi^*(\star_h (E_*^i \wedge E_*^j)). \tag{5.8}$$

Therefore, Eq. (4.9) gives

$$\star_g \, d\tilde{X} = 2\,\pi^*(\star_h du) - E^0 \wedge \pi^*(u^2 \star_h \Omega). \tag{5.9}$$

Then, using Proposition 4.1 and Eq. (4.1),

$$\frac{1}{2} \star_g d\tilde{X} = \pi^*(\star_h du) - \theta \wedge \omega. \tag{5.10}$$

But

$$\begin{aligned}\int_{\sigma(\partial R)} \pi^*(\star_h du) &= \int_{\partial R} \sigma^* \pi^*(\star_h du) \\ &= \int_{\partial R} (\pi \circ \sigma)^*(\star_h du) \equiv \int_{\partial R} \star_h du,\end{aligned} \tag{5.11}$$

and hence

$$\frac{1}{2} \int_{\partial\sigma(R)} \star_g d\tilde{X} = \int_{\partial R} \star_h du - \int_{\partial\sigma(R)} \theta \wedge \omega. \tag{5.12}$$

We also have

$$\star_g \, \tilde{X} = -\beta \star_g E^0 = -\beta \pi^*(E^1_* \wedge \ldots \wedge E^n_*) \tag{5.13}$$

$$= -\pi^*(u \, vol_h), \tag{5.14}$$

whence we deduce that

$$\int_{\sigma(R)} \star_g \tilde{X} = -\int_R u \, vol_h. \tag{5.15}$$

Now, collecting Eqs. (5.12) and (5.15) we obtain the desired result. □

Remark 5.8 Note that we can add a total derivative term to the last integrand in (5.6) and write it as

$$E_{\sigma,R} = \int_{\partial R} \star_h du + c \cdot \int_R u \, vol_h - \int_{\partial\sigma(R)} d_\theta \omega, \tag{5.16}$$

where

$$d_\theta \omega = d\omega + \theta \wedge \omega$$

is the covariant derivative with respect to the connection 1-form θ.

6 General Properties of Stationary Regions

From now on, our focus will shift to the situation where we have a stationary region $S \neq M$, and start the study of features of its horizons. I shall also drop, for the time being, the assumption that (M, g) is chronological, adopted in the previous sections. Instead, I will invoke any causal assumptions insofar as they are needed for a given proof.

The following definitions will play an important role for the remainder of this paper.

Definition 6.1 In (M, g):

(a) the *future* [resp. *past*] *S-horizon* is

$$\mathcal{H}^{+} := \partial I^{-}(S) \cap I^{+}(S) \text{ [resp.} \mathcal{H}^{-} := \partial I^{+}(S) \cap I^{-}(S)],$$

(b) the *S-domain of outer communication* is

$$D := I^{+}(S) \cap I^{-}(S), \text{ and}$$

(c) the *S-bifurcation set* is

$$\Sigma := \overline{\mathcal{H}^{+}} \cap \overline{\mathcal{H}^{-}}.$$

(In particular D is open, connected, causally convex and contains S; moreover, $\mathcal{H}^{+} \cap \mathcal{H}^{-} = \emptyset$.)

The following proposition summarizes some general structural facts about the S-horizons and the S-domain of outer communication.

Proposition 6.2 *In* (M, g)*:*

(i) $\forall t \in \mathbb{R}$, $\phi_t(D) = D$, $\phi_t(\Sigma) = \Sigma$ *and* $\phi_t(\mathcal{H}^{\pm}) = \mathcal{H}^{\pm}$*;*
(ii) $\mathcal{H}^{\pm} \subset I^{\pm}(D)$ *and* $J^{\pm}(\mathcal{H}^{\pm}) \cap D = \emptyset$*;*
(iii) if (M, g) *is globally hyperbolic, then so is* $(D, g|_D)$*. (In this case,* $(S, g|_S)$ *is stably causal.)*

Proof (i)
Given $t \in \mathbb{R}$, we have

$$\begin{aligned} \phi_t(\mathcal{H}^{\pm}) &= \phi_t(\partial I^{\mp}(S)) \cap \phi_t(I^{\pm}(S)) \\ &= \partial[\phi_t(I^{\mp}(S))] \cap \phi_t(I^{\pm}(S)) \\ &= \partial[I^{\mp}(\phi_t(S))] \cap I^{\pm}(\phi_t(S)) \\ &= \partial I^{\mp}(S) \cap I^{\pm}(S) \equiv \mathcal{H}^{\pm}, \end{aligned} \tag{6.1}$$

where the first and second equalities follow from the fact that ϕ_t is a diffeomorphism, the third equality follows because ϕ_t is an isometry and the fourth equality holds because S is ϕ-invariant. Moreover, again because ϕ_t is a diffeomorphism,

$$\phi_t(\Sigma) = \overline{\phi_t(\mathcal{H}^+)} \cap \overline{\phi_t(\mathcal{H}^-)} \equiv \Sigma$$

from the previous part. The proof that $\phi_t(D) = D$ is entirely analogous.
(ii)
We show that $\mathcal{H}^+ \subset I^+(D)$ and $J^+(\mathcal{H}^+) \cap D = \emptyset$, since the time-dual cases are analogous. The first of these claims is immediate from the definition of $\mathcal{H}^+$ and the easy-to-check fact that $I^{\pm}(D) = I^{\pm}(S)$. To establish the second statement, suppose, by way of contradiction, there exists $p \in J^+(\mathcal{H}^+) \cap D$, and pick $r \in \mathcal{H}^+ = \partial I^-(S) \cap I^+(S)$ with $r \in J^-(p)$. Now, $p \in D \Rightarrow p \in I^-(q)$ for some $q \in S$, and hence $r \in I^-(q) \subset I^-(S)$, which is impossible since $r \in \partial I^-(S)$ and $I^-(S)$ is open.
(iii)
Suppose (M, g) is globally hyperbolic. Then, since D is an open subset of a strongly causal spacetime it is also strongly causal. It is straightforward to check that for all $p, q \in D$

$$J^+(p, D) \cap J^-(q, D) = J^+(p) \cap J^-(q),$$

and since the latter set is compact in M, so is the former compact in D. Hence $(D, g|_D)$ is globally hyperbolic, and, in particular, it is stably causal. Pick a *temporal function* on D, i.e., a smooth function $f : D \to \mathbb{R}$ with timelike gradient. Clearly, $f|_S$ is a time function in S, so $(S, g|_S)$ is also stably causal. □

The first key consequence of Lemma 2.2 is a structural theorem for stationary regions.

Theorem 6.3 *Suppose $(S, g|_S)$ is chronological. Then the following statements hold.*

(i) $\forall p \in S$,

$$D = I^+(\mathbb{R}p) \cap I^-(\mathbb{R}p). \tag{6.2}$$

(ii) If $\mathcal{H}^-$ is non empty, then $\mathcal{H}^+ \subset I^+(\mathcal{H}^-)$.
(iii) The future [resp. past] S-horizon $\mathcal{H}^+$ [resp. $\mathcal{H}^-$], if non empty, is an achronal Lipschitz null hypersurface with future [resp. past] inextendible null generators in (M, g).
(iv) The $\mathbb{R}$-action induced by the flow of X_S on S is free and proper, in particular S is diffeomorphic to $\mathbb{R} \times Q$, where Q is the smooth n-dimensional quotient manifold by this action. If $(S, g|_S)$ is distinguishing, then it is stably causal, and given any smooth function $f \in C^\infty(M)$ with timelike gradient, each maximal integral curve of X_S intersects the (acausal) level hypersurfaces of f exactly once.

Proof (i)
Let $p \in S$. Since $\mathbb{R}p \subset S$, $I^+(\mathbb{R}p) \cap I^-(\mathbb{R}p) \subset D$ by the definition of D. Conversely, given $q \in D$, then $q \in I^+(r) \cap I^-(s)$ for $r, s \in S$. From Lemma 2.2 there exist $t, t' \in \mathbb{R}$ such that $\phi_t(p) \in I^+(s) \Rightarrow q \in I^-(\phi_t(p))$ and $\phi_{t'}(p) \in I^-(r) \Rightarrow q \in I^+(\phi_{t'}(p))$. Hence $q \in I^+(\mathbb{R}p) \cap I^-(\mathbb{R}p)$ and the other inclusion follows.

(ii)
Let $p \in \mathcal{H}^+$. From Proposition 6.2(ii), there exists $q \in D$ with $q \ll p$. From the definition of D we can assume, without loss of generality, that $q \in S$. Pick $r \in \mathcal{H}^-$. Again from Proposition 6.2(ii), there exists $q' \in S$ with $r \ll q'$. Thus, $\forall t \in \mathbb{R}$, $\phi_t(r) \ll \phi_t(q')$. But from Lemma 2.2, $\mathbb{R}q' \cap I^-(q) \neq \emptyset$, so for some $t_0 \in \mathbb{R}$, $\phi_{t_0}(r) \ll \phi_{t_0}(q') \ll q \ll p$. Now, $\mathcal{H}^-$ is ϕ-invariant by Proposition 6.2(i), and hence $\phi_{t_0}(r) \in \mathcal{H}^-$, and we conclude that $p \in I^+(\mathcal{H}^-)$, which establishes our claim.
(iii)
Again, we focus on the future S-horizon $\mathcal{H}^+$, as an analogous proof applies for the past one. Pick any $p \in S$. Arguing just as in (i), we may conclude that $I^-(S) = I^-(\mathbb{R}p)$, and, in particular, $\partial I^-(S) = \partial I^-(\mathbb{R}p)$. By standard properties of boundaries of past sets of inextendible timelike curves, which is, in particular, an achronal boundary, $\partial I^-(\mathbb{R}p)$ is a Lipschitz null hypersurface with future-inextendible null generators in (M, g), and so these properties also hold for $\partial I^-(S)$ and hence for $\mathcal{H}^+$.
(iv)
This result is just a restatement of the main theorems in Refs. [15, 21], and the discussion in Sects. 3 and 4. □

Another important consequence of Lemma 2.2 concerns the zeros of the Killing vector field X.

Theorem 6.4 *Suppose $(D, g|_D)$ is chronological. Then $\forall t \in \mathbb{R} \setminus \{0\}$, and $\forall p \in D$, $\phi_t(p) \neq p$. In particular, $X(p) \neq 0\ \forall p \in D$, i.e., X has no zeros in D. In addition, if strong causality holds on $\mathcal{H}^+ \cup D$ [resp. $\mathcal{H}^- \cup D$], then $\forall t \in \mathbb{R} \setminus \{0\}$, and $\forall p \in \mathcal{H}^+ \cup D$ [resp. $\mathcal{H}^- \cup D$], $\phi_t(p) \neq p$. In particular, $X(p) \neq 0\ \forall p \in \mathcal{H}^+ \cup D$ [resp. $\mathcal{H}^- \cup D$], i.e., X has no zeros in $\mathcal{H}^+ \cup D$ [resp. $\mathcal{H}^- \cup D$].*

Proof We shall adapt an idea set forth in the proof of Lemma 2.1 of Ref. [11]. Suppose $\phi_{t_0}(p) = p$ for some $t_0 \in \mathbb{R} \setminus \{0\}$, and some $p \in D$. Then $\forall n \in \mathbb{Z}$, $\phi_{nt_0}(p) = p$. Pick any $q \in D$. For some $r \in S$, $q \ll r$, and for some $r' \in S$, $r' \ll p$. Now, from Lemma 2.2, $r \ll \phi_s(r')$ for some $s \in \mathbb{R}$, and since the orbit of r' is timelike there exists $n_0 \in \mathbb{Z}$ such that $\phi_s(r') \ll \phi_{n_0t_0}(r') \ll \phi_{n_0t_0}(p) = p$. We conclude that $q \ll p$. Analogously, we can show that $p \ll q$, and hence $p \ll p$, violating chronology at p. This contradictions establishes the first claim.

If strong causality holds on $\mathcal{H}^+ \cup D$, since $(D, g|_D)$ is, in particular, chronological, X has no zeros in D from the previous part. Pick $p \in \mathcal{H}^+$, and again assume by contradiction that $\phi_{t_0}(p) = p$ for some $t_0 \in \mathbb{R} \setminus \{0\}$. Arguing just as before we conclude that $D \subset I^-(p)$. Let $\alpha : [0, a) \rightarrow M$ be the future-inextendible (by (iv)) null geodesic generator of $\mathcal{H}^+$ starting at p, and pick any number $0 < s_0 < a$. From causality, we have that $\alpha(s_0) \neq p$. Using the Hausdorff property of M, pick $U \ni \alpha(s_0)$ and open set such that $p \notin U$. Now, the strong causality at $\alpha(s_0)$ allows us to pick a neighborhood $V \subset U$ of $\alpha(s_0)$ such that any causal curve segment with endpoints in V is entirely contained in U. Clearly the set $I^-(p) \cap V \cap I^+(S)$ is nonempty, so $V \cap D \neq \emptyset$. Let $r \in V \cap D$. We have seen that $r \ll p$, so choose a future-directed timelike curve $\beta : [0, 1] \rightarrow M$ from r to p. The curve $\gamma : [0, s_0 + 1] \rightarrow M$ given by

$$\gamma(s) = \begin{cases} \beta(s) & \text{if } s \in [0,1] \\ \alpha(t-1) & \text{if } s \in [1, s_0+1] \end{cases}$$

is future-directed causal from $\gamma(0) = r$ to $\gamma(s_0+1) = \alpha(s_0)$, and so has endpoints in V, but $\gamma(1) = p \notin U$, a contradiction. The proof for $\mathcal{H}^-$ goes by time duality. □

The following consequence on $\mathcal{H}^+$ shows how the orbits behave on the horizon.

Corollary 6.5 *If strong causality holds on $\mathcal{H}^+ \cup D$, then $\forall p \in \mathcal{H}^+$, $\mathbb{R}p \subset \mathcal{H}^+$ and the curve $\alpha_p : t \in \mathbb{R} \mapsto \phi_t(p) \in \mathcal{H}^+$ is either spacelike or a null pregeodesic whose image coincides with the geodesic null generators of $\mathcal{H}^+$.*

Proof The fact that the range of α_p is contained in $\mathcal{H}^+$ is immediate from Proposition 6.2(i). From Theorem 6.3 (v), $X(q) \neq 0$, $\forall q \in \mathcal{H}^+$, and hence α_p has everywhere nonzero tangent vector. Since X is Killing, α_p has a definite causal character. If it were timelike, this would violate the achronality of $\mathcal{H}^+$ (cf. Theorem 6.3(iv)), and thus α_p must be either spacelike or null. In the latter case, the achronality of $\mathcal{H}^+$ again implies that α_p is a null pregeodesic whose image coincides with the geodesic null generators of $\mathcal{H}^+$. □

(Of course, a time-dual version of this result holds for $\mathcal{H}^-$.)

Remark 6.6 The situation when the orbits are null is realized, e.g., on the event horizon in Schwarzschild spacetime or on any cosmological horizon in de Sitter spacetime, while the spacelike situation occurs on "rotating" horizons in spacetimes such as slow ($a < m$) Kerr's.

The next result also establishes that S-bifurcation sets have similar properties to the ones in the usual black hole solutions.

Theorem 6.7 *Regarding the S-bifurcation set Σ, the following statements hold.*

(i) $\Sigma \cap D = \emptyset$.
(ii) $\Sigma = edge(\mathcal{H}^+) = edge(\mathcal{H}^-)$.
(iii) $\partial D = \Sigma \dot{\sqcup} \mathcal{H}^+ \dot{\sqcup} \mathcal{H}^+$.
(iv) *If $(S, g|_S)$ is chronological, then Σ is an acausal set, and for each $p \in \Sigma$, there exists a future-directed, future-inextendible null geodesic ray $\eta : [0, a) \to M$ with $\eta(0) = p$ such that $\eta|_{(0,a)}$ is a null geodesic generator of $\mathcal{H}^+$. If in addition Σ is a (necessarily spacelike) codimension two submanifold of M, then η is a Σ-ray. (A time-dual statement holds for $\mathcal{H}^-$.)*

Proof (i)
$p \in D \Rightarrow p \notin \partial I^+(S) \cup \partial I^-(S) \supset \overline{\mathcal{H}^+} \cap \overline{\mathcal{H}^-} \equiv \Sigma$. Hence, $\Sigma \subset M \setminus D$.
(ii)
We give the proof only for $\mathcal{H}^+$, the past case being analogous. Let $p \in \Sigma \subset \overline{\mathcal{H}^+}$. First, note that if $p \in \mathcal{H}^+$, then $p \in I^+(S) \Rightarrow p \notin \partial I^+(S) \supset \overline{\mathcal{H}^-}$, a contradiction. Therefore, $p \notin \mathcal{H}^+$. We wish to establish that $p \in$ edge$(\mathcal{H}^+)$. To do that, pick any

open set $U \ni p$ and $p_\pm \in U$ such that $p \in I^+(p_-, U) \cap I^-(p_+, U)$. Choose a future-directed timelike curve $\alpha : [0, 1] \to U$ such that $\alpha(0) = p_-, \alpha(1) = p_+$ and $\alpha(t_0) = p$ for some $t_0 \in [0, 1]$. Now, α already crosses $\overline{\mathcal{H}^+}$ (at p) and it cannot do that more than once because of the achronality of the latter set. We conclude that α does not intersect $\mathcal{H}^+$, and hence $p \in \text{edge}(\mathcal{H}^+)$ as claimed. This establishes that $\Sigma \subseteq \text{edge}(\mathcal{H}^+)$.

Conversely, let $q \in \text{edge}(\mathcal{H}^+)$. In particular, $q \in \overline{\mathcal{H}^+}$. Pick an open set $V \ni q$ and a future-directed timelike curve $\beta : [0, 1] \to V$ not intersecting $\mathcal{H}^+$, such that $q_- := \beta(0) \in I^-(q, V)$ and $q_+ := \beta(1) \in I^+(q, V)$. Since $q \in \overline{\mathcal{H}^+} \subset \partial I^-(S)$ and the latter set is edgeless, we may assume that β *does* intersect $\partial I^-(S)$ at a point r, say. In other words, $r \in \partial I^-(S) \setminus I^+(S)$. However, $I^-(q_+, V) \cap I^+(q_-, V) \cap \mathcal{H}^+ \neq \emptyset$, and, in particular, $I^-(q_+, V) \cap I^+(q_-, V) \cap I^+(S) \neq \emptyset$, which in turn implies that $q_+ \in I^+(S)$. Therefore, $V \cap I^+(S) \neq \emptyset$, and $V \cap (M \setminus I^+(S)) \neq \emptyset$. We conclude that $q \in \partial I^+(S)$. In particular, $q \notin I^+(S)$, and hence $q_- \notin I^+(S)$. Now, pick any $s \in I^-(q_+, V) \cap I^+(q_-, V) \cap \mathcal{H}^+$, and choose a future-directed timelike curve $\gamma : [0, 1] \to V$ such that $q_- = \gamma(0), q_+ = \gamma(1)$, and $\gamma(t_0) = s$ for some $t_0 \in [0, 1]$. We have chosen $q_\pm \notin \mathcal{H}^+$, and hence $0 < t_0 < 1$. For some $0 < t_0' < t_0, \gamma(t_0') \in \partial I^+(S)$, since $\gamma(t_0) = s \in \mathcal{H}^+ \subset I^+(S)$ but $q_- \notin I^+(S)$. Since $\gamma(t_0') \ll s \in \partial I^-(S)$, we conclude that $\gamma(t_0') \in \partial I^+(S) \cap I^-(S) \equiv \mathcal{H}^-$. In other words, $V \cap \mathcal{H}^- \neq \emptyset$. This last fact now implies that $q \in \overline{\mathcal{H}^-}$, and hence that $q \in \Sigma$, thus concluding the proof that $\Sigma = \text{edge}(\mathcal{H}^+)$.

(iii)
Clearly $\mathcal{H}^\pm \subset \partial D$, whence $\Sigma \subset \partial D$ and hence the inclusion $\Sigma \,\dot\cup\, \mathcal{H}^+ \,\dot\cup\, \mathcal{H}^+ \subset \partial D$ easily follows. Let $p \in \partial D \subset \overline{D} \subset \overline{I^+(S)} \cap \overline{I^-(S)}$. Assume that $p \notin \mathcal{H}^+ \cup \mathcal{H}^-$. We wish to show that $p \in \Sigma$.

Since D is open, $p \notin D$, and hence either $p \notin I^+(S)$ or $p \notin I^-(S)$. However, if $p \in I^+(S)$, then $p \in \partial I^-(S) \cap I^+(S) \equiv \mathcal{H}^+$, a contradiction. It follows that $p \in \partial I^+(S)$. By an analogous argument, $p \notin I^-(S) \Rightarrow p \in \partial I^+(S) \cap \partial I^-(S)$.

As in the proof of (ii), we pick an open set $U \ni p$, $p_\pm \in U$ such that $p \in I^+(p_-, U) \cap I^-(p_+, U)$ and a timelike curve $\alpha : [0, 1] \to U$ such that $\alpha(0) = p_-$, $\alpha(1) = p_+$. From the fact that $I^+(p_-, U) \cap I^-(p_+, U) \cap D \neq \emptyset$ and by standard facts about achronal boundaries, we may assume that there exist $t_\pm, t_0 \in [0, 1]$ for which $\alpha(t_0) \in D$ and $\alpha(t_\pm) \in \partial I^\pm(S)$. Clearly $p_\pm \in I^\pm(S) \setminus I^\mp(S)$, so $0 < t_0 < 1$, and $t_\pm \neq t_0$. If $t_0 < t_+$, then $\alpha(t_0) \ll \alpha(t_+) \Rightarrow \alpha(t_+) \in I^+(S)$, an impossibility. We conclude that $t_+ < t_0$. But then $\alpha(t_+) \in \partial I^+(S) \cap I^-(S) \equiv \mathcal{H}^-$, from which we conclude that $U \cap \mathcal{H}^- \neq \emptyset$, and thus that $p \in \overline{\mathcal{H}^-}$. Similarly, we show that $t_0 < t_-$ and hence $p \in \overline{\mathcal{H}^+}$, i.e., $p \in \Sigma$ as desired.

(iv)
If Σ is empty, it is vacuously acausal, so we assume $\Sigma \neq \emptyset$. In particular, both $\mathcal{H}^\pm$ are then nonempty. Assume that $(S, g|_S)$ is chronological, and let $\alpha : [0, 1] \to M$ be a future-directed causal curve such that $\alpha(0), \alpha(1) \in \Sigma$. Now, $\Sigma \subset \partial I^+(S) \cap \partial I^-(S)$, and hence by standard properties of achronal boundaries α is, up to reparametrization, a segment of a common null generator η of both $\partial I^+(S)$ and $\partial I^-(S)$. The (proof of) Theorem 6.3(iii) shows that η can be taken to be both past and future inextendible, and

indeed a globally achronal null geodesic line. A simple extension argument shows that the connected components of $\partial I^+(S)$ and $\partial I^-(S)$ containing η coincide. Let us denote this common connected component as Λ. Obviously, Λ can intersect neither $I^+(S)$ nor $I^-(S)$, and, in particular, cannot intersect either S-horizon. Since $\partial I^+(S)$ and $\partial I^-(S)$ are closed C^0 hypersurfaces (cf. Theorem 6.3(iii)) and M is a normal topological space, there exists an open set U containing Λ which does not intersect either $\mathcal{H}^+$ or $\mathcal{H}^-$, which is impossible since $\alpha(0), \alpha(1) \in \Lambda$ are in the closure of $\mathcal{H}^\pm$. This contradiction establishes the first claim.

Finally, let $p \in \Sigma$. Pick a background complete Riemannian metric h and a sequence $(\gamma_k : [0, +\infty) \to M)_{k\in\mathbb{N}}$ of future-directed, future-inextendible null pregeodesics such that:

(a) $\gamma_k[0, +\infty) \subseteq \mathcal{H}^+$,
(b) $\gamma_k(0) \to p$ as $k \to +\infty$,
(c) for each $k \in \mathbb{N}$, γ_k is the h-arc length reparametrized future portion of the (future-inextendible) null geodesic generator of $\mathcal{H}^+$ passing through $\gamma_k(0)$ (cf. Theorem 6.3(iii)).

Then the final claim follows from a standard argument using the Limit Curve Lemma applied to this sequence, using the achronality of $\overline{\mathcal{H}^+}$. □

Corollary 6.8 *Assume that strong causality holds on $\mathcal{H}^+ \cup D$, and that $D \equiv S$, i.e., that X is timelike everywhere on D. Then $\forall p \in \mathcal{H}^+$, the p-orbit $\alpha_p : t \in \mathbb{R} \mapsto \phi_t(p) \in \mathcal{H}^+$ is a null pregeodesic whose image coincides with the geodesic null generators of $\mathcal{H}^+$. (An analogous statement holds for $\mathcal{H}^-$.). Moreover, for all $q \in \Sigma$, $X(q) = 0$.*

Proof The strong causality on $\mathcal{H}^+ \cup D$ implies, via Theorem 6.4, that X has no zeros on $\mathcal{H}^+$. Moreover, given $p \in \mathcal{H}^+$, according to Corollary 6.5 $X(p)$ is either null or spacelike. If $X(p)$ were spacelike, then for some open set $U \ni p$, $X(q)$ would be spacelike for each $q \in U$. However, $U \cap D \neq \emptyset$ by Theorem 6.7(iii), in contradiction with our hypothesis that X is timelike in D. Then $X(p)$ is null, and the first claim follows. Now, since $\langle X, X\rangle = 0$ on $\mathcal{H}^+$, we will have $\langle X, X\rangle|_\Sigma \equiv 0$ by continuity. Let $q \in \Sigma$. If $X(q) \neq 0$, the q-orbit would be null. However, by Proposition 6.2(i) this orbit would be contained in Σ, contradicting Theorem 6.7(iv). Hence $X(q) = 0$. □

Theorem 6.9 *Assume that $(D, g|_D)$ is globally hyperbolic, and let $\mathcal{P} \subset D$ be a Cauchy hypersurface thereof. Then $\overline{\mathcal{H}^\pm} = H^\pm(\mathcal{P})$.*[8] *In particular, any null geodesic generators of $\mathcal{H}^+$ [resp. $\mathcal{H}^-$] is either future and past inextendible or else has a past [resp. future] endpoint on Σ. Moreover, $\Sigma = edge(\mathcal{P})$, so that if $\Sigma = \emptyset$, then $\mathcal{P}$ is a partial Cauchy hypersurface in (M, g). Finally, each point $p \in \Sigma$ is the past [resp. future] endpoint of a null geodesic generator of $\mathcal{H}^+$ [resp. $\mathcal{H}^-$].*

[8] Here and hereafter we will use the following standard notation: if $A \subseteq M$ is an achronal set, then $D^+(A)$ [resp. $D^-(A)$] denotes its future [resp. past] Cauchy development, and $H^+(A)$ [resp. $H^-(A)$] its future [resp. past] Cauchy horizon.

Proof As usual, we shall establish this result for $\mathcal{H}^+$, since the other case is analogous. Let $p \in \overline{\mathcal{H}^+}$.
Claim: $p \in \overline{D^+(\mathcal{P})}$.
To see this, pick an open set $U \ni p$ and $q \in U \cap \mathcal{H}^+$. If $p \in \overline{\mathcal{P}}$ we are done, hence assume $p \notin \overline{\mathcal{P}}$ and $U \cap \mathcal{P} = \emptyset$. Now let $q_\pm \in U$ be such that $q \in I^-(q_+, U) \cap I^+(q_-, U)$. From Theorem 6.7(iii), $\mathcal{H}^+ \subset \partial D$, so we can choose $r \in I^-(q_+, U) \cap I^+(q_-, U) \cap D$. In particular, $r \in D^+(\mathcal{P}) \cup D^-(\mathcal{P})$. But if $r \in D^-(\mathcal{P})$, then $q_+ \in D^-(\mathcal{P})$, and hence $q \in \mathcal{H}^+ \cap J^-(D)$, contradicting Proposition 6.2(ii). Thus $r \in D^+(\mathcal{P})$, whence $U \cap D^+(\mathcal{P}) \neq \emptyset$, which establishes the claim. Using Proposition 6.2(ii) again, we conclude that $I^+(p) \cap D^+(\mathcal{P}) = \emptyset$, and therefore $p \in H^+(\mathcal{P})$. In other words, $\overline{\mathcal{H}^+} \subset H^+(\mathcal{P})$.

Conversely, let $p \in H^+(\mathcal{P})$, and again pick an open set $U \ni p$.

First, suppose $p \in H^+(\mathcal{P}) \cap \overline{\mathcal{P}}$. In this case, we may again choose $p_\pm \in U$ be such that $p \in I^-(p_+, U) \cap I^+(p_-, U)$ and $q \in I^-(p_+, U) \cap I^+(p_-, U) \cap \mathcal{P}$. Let $\alpha : [0, 1] \to U$ be a future-directed timelike curve such that $\alpha(0) = p_-$, $\alpha(1) = p_+$ and $\alpha(t_0) = q$ for some $t_0 \in [0, 1]$. By construction, $p_+ \in I^+(\mathcal{P}) \subset I^+(S)$. If $p_+ \in I^+(S)$, then $p_+ \in D$, and since $I^+(p) \cap D^+(\mathcal{P})$ from the definition of a future Cauchy horizon, p_+ cannot be in $D^+(\mathcal{P})$ and must be in $D^-(\mathcal{P}) \setminus \mathcal{P} \subset I^-(\mathcal{P})$, i.e., $p \in I^+(\mathcal{P}) \cap I^-(\mathcal{P})$, in violation of the achronality of $\mathcal{P}$. We conclude that $p_+ \notin I^-(S)$. In particular, $q \neq p_+$, and hence $t_0 < 1$. Therefore, there exists $t_0 < s_0 < 1$ such that $\alpha(s_0) \in \partial I^-(S) \cap I^+(S) \equiv \mathcal{H}^+$. Therefore, $U \cap \mathcal{H}^+ \neq \emptyset$.

Finally, assume that $p \in H^+(\mathcal{P}) \setminus \overline{\mathcal{P}}$, and shrinking U if necessary we can take $U \cap \overline{\mathcal{P}} = \emptyset$ and U connected. If we again pick $p_\pm \in U$ be such that $p \in I^-(p_+, U) \cap I^+(p_-, U)$, it is easy to check that $p \in I^+(S)$, and hence $p_+ \in I^+(S)$. Since $I^+(S)$ is open, there is no loss of generality if we assume that $U \subset I^+(S)$. If $p_+ \in I^-(S)$ an argument just like the one in the previous paragraph would give a contradiction with the achronality of $\mathcal{P}$, hence $p_+ \notin I^-(S)$. Clearly, $p \in D \subset I^-(S)$, and since we assumed U connected we conclude that $U \cap \partial I^-(S) \neq \emptyset$, and therefore $U \cap \mathcal{H}^+ \neq \emptyset$.

Thus, in any case $U \cap \mathcal{H}^+ \neq \emptyset$, so $p \in \overline{\mathcal{H}^+}$, and hence $\overline{\mathcal{H}^+} = H^+(\mathcal{P})$. The other statements follow from standard properties of Cauchy horizons and from the fact (cf. Theorem 6.7(ii)) that $\Sigma = \text{edge}(\mathcal{H}^+) = \text{edge}(\mathcal{H}^-)$. This completes the proof. □

7 Regularity of *S*-Horizons

After studying the general geometric structure of horizons, we turn to their regularity. A first regularity result for S-horizons arises when the S-bifurcation set Σ is empty.

We shall often need to impose a convergence condition on null geodesics. Recall that a spacetime is said to satisfy the *null convergence condition* if

$$Ric(v, v) \geq 0, \text{ for any null vector } v \in TM.$$

(This condition is standard in a number of singularity theorems, and has a natural physical motivation as manifesting the "attractive" character of gravity.)

Theorem 7.1 *Suppose $(D, g|_D)$ is globally hyperbolic, $\Sigma = \emptyset$, $\mathcal{H}^+ \neq \emptyset$ and assume that the null convergence condition holds in (M, g). Then either one inextendible null geodesic generator of $\mathcal{H}^+$ is incomplete, or else $\mathcal{H}^+$ is a closed smooth null totally geodesic hypersurface in (M, g). (An analogous statement holds for $\mathcal{H}^-$.)*

Proof If $\Sigma = \emptyset$, then, in particular, (cf. Theorem 6.7(ii)) edge$(\mathcal{H}^+) = \emptyset$, and hence $\mathcal{H}^+ = \overline{\mathcal{H}^+}$ is a closed achronal C^0 hypersurface with future-inextendible null generators (by Theorem 6.3(iii)). Using the result and notation from Theorem 6.9, we have $\mathcal{H}^+ = H^+(\mathcal{P})$, where $\mathcal{P}$ is a Cauchy hypersurface for $(D, g|_D)$. The standard properties of the latter set reveal (again because edge$(\mathcal{H}^+) = \emptyset$) that its null geodesic generators are also past inextendible, and hence inextendible. If each one of these is complete, they will therefore be null geodesic lines. The conclusion now follows from Theorem 4.1 of [12]. □

We shall now seek more regularity results for S-horizons when $\Sigma \neq \emptyset$. We will see that this problem is related, at least for globally hyperbolic spacetimes, with the regularity of Σ.

A useful result toward our end is the following, which is a simple consequence of Theorem 3.2.31 of [33].

Lemma 7.2 *Suppose (M, g) is globally hyperbolic and let $S \subset M$ be an acausal, future causally complete*[9] *submanifold with codimension ≥ 2. Then $\mathcal{H} = \partial I^+(S) \setminus (S \cup Cut(S))$ is a smooth null hypersurface, where $Cut(S)$ is the set of null cut points to S along future-directed null geodesics starting at and normal to S.* □

(Again, a time-dual of this result for past causally complete submanifolds is understood.)

To consider connected components, the following simple technical lemma is useful.

Lemma 7.3 *Assume that $(D, g|_D)$ is chronological, and let $\Sigma_0 \neq \emptyset$ be a connected component of Σ. Then there exist (unique) connected components $\mathcal{H}_0^+$ of $\mathcal{H}^+$ and $\mathcal{H}_0^-$ of $\mathcal{H}^-$, respectively, such that $\Sigma_0 = \overline{\mathcal{H}_0^+} \cap \overline{\mathcal{H}_0^-}$.*

Proof Let $p \in \Sigma_0$. By Theorem 6.7(iv) and its time-dual, there exists a future-directed [resp. past-directed], future-inextendible [resp. past-inextendible] null geodesic ray $\eta_+ : [0, a_+) \to M$ [resp. $\eta_- : [0, a_-) \to M$] with $\eta_\pm(0) = p$ such that $\eta_\pm|_{(0,a_\pm)}$ is a null geodesic generator of $\mathcal{H}^\pm$. Let $\mathcal{H}_0^+$ [resp. $\mathcal{H}_0^-$] be the connected component of $\mathcal{H}^+$ [resp. $\mathcal{H}^-$] containing $\eta_+(0, a_+)$ [resp. $\eta_-(0, a_-)$]. Clearly $p \in \overline{\mathcal{H}_0^+} \cap \overline{\mathcal{H}_0^-}$, which together with connectivity of the sets involved imply that $\Sigma_0 = \overline{\mathcal{H}_0^+} \cap \overline{\mathcal{H}_0^-}$ as desired. Uniqueness is obvious. □

[9] Recall that a subset $C \subseteq M$ is *future causally complete* (FCC) if for all $p \in J^+(C)$, the set $J^-(p) \cap C$ has compact closure in C. A time-dual notion of *past causally complete* (PCC) is evident. Obviously, every compact set is both FCC and PCC. It is easy to check that every FCC(PCC) subset of M is closed.

Our second regularity result is now as follows.

Theorem 7.4 *Suppose* (M, g) *is globally hyperbolic. Let* $\Sigma_0 \neq \emptyset$ *be a connected component of the* S*-bifurcation set* Σ. *If* Σ_0 *is a smooth and future [resp. past] causally complete submanifold of* M*, then the connected component* $\mathcal{H}_0^+$ *[resp.* $\mathcal{H}_0^-$*] of* $\mathcal{H}^+$ *[resp.* $\mathcal{H}^-$*] such that* $\Sigma_0 = \overline{\mathcal{H}_0^+} \cap \overline{\mathcal{H}_0^-}$ *according to Lemma 7.3 is a smooth null hypersurface. In particular, if* Σ_0 *is smooth and compact, then both* $\mathcal{H}_0^\pm$ *are smooth.*

Proof As usual, we develop the future case, and the other case follows from time-dual arguments. Clearly, codim$(\Sigma_0)\geq 2$. We have to show that $\mathcal{H}_0^+ \subseteq \partial I^+(\Sigma_0) \setminus (\Sigma_0 \cup \text{Cut}(\Sigma_0))$ in order to apply Lemma 7.3. From Proposition 6.2(iii), $(D, g|_D)$ is globally hyperbolic, so in particular Theorem 6.7(iv) Σ_0 is acausal. Let $\mathcal{P} \subset D$ be a Cauchy hypersurface for $(D, g|_D)$. If we had $p \in \mathcal{H}^+ \cap \text{Cut}(\Sigma_0)$, then by standard properties of Cut(Σ_0) we would have in particular (cf. e.g., Proposition 3.2.28 of [33]) have either more than one maximal null geodesic from Σ_0 to p or else p would be the first Σ_0-focal point along a maximal null geodesic from Σ_0 to p. However, the achronality of $\partial I^-(S)$ together with the fact that all null geodesic generators of $\mathcal{H}^+$ are future-inextendible in (M, g) (cf. Theorem 6.3(iii)) preclude this. We conclude that $\mathcal{H}^+ \cap \text{Cut}(\Sigma_0) = \emptyset$. From Theorem 6.9, $\mathcal{H}_0^+ \subset \overline{\mathcal{H}^+} = H^+(\mathcal{P}) \subset \partial I^+(\text{edge}(\mathcal{P})) \equiv \partial I^+(\Sigma)$. But then the proof of Lemma 7.3 actually implies that $\mathcal{H}_0^+ \subset \partial^+(\Sigma_0)$, which in turn establishes the desired conclusion. □

Although Theorem 7.4 gives a regularity criterion for *smoothness*, unlike Theorem 7.1 its says nothing about the *(mean) convexity* properties of the S-horizons. We will now refine this result in terms of the convexity properties of the S-bifurcation set Σ. Recall that a submanifold $\mathcal{S} \subset M$ of codimension $<$ dim(M) is *extremal* if its mean curvature vector is identically zero. Of course, this will happen, in particular, if $\mathcal{S}$ is totally geodesic.

Theorem 7.5 *Suppose* (M, g) *is globally hyperbolic, and satisfies the null convergence condition. Assume also that the null geodesic generators of* $\mathcal{H}^+$ *[resp.* $\mathcal{H}^-$*] are future-[resp. past-]complete. Let* $\Sigma_0 \neq \emptyset$ *be a connected component of the* S*-bifurcation set* Σ. *If* Σ_0 *is smooth, extremal, and compact, and* $2 \leq \mathit{codim}(\Sigma_0) < \mathit{dim}(M)$*, then the connected component* $\mathcal{H}_0^+$ *and* $\mathcal{H}_0^-$ *of* $\mathcal{H}^\pm$ *such that* $\Sigma_0 = \overline{\mathcal{H}_0^+} \cap \overline{\mathcal{H}_0^-}$ *according to Lemma 7.3 are smooth and totally geodesic null hypersurfaces.*

Proof We again argue only for $\mathcal{H}_0^+$. Theorem 7.4 implies that $\mathcal{H}_0^+$ is a smooth null hypersurface. Pick any null geodesic generator γ of $\mathcal{H}_0^+$, which by assumption is future complete. Since Σ_0 is compact, there exists a Cauchy hypersurface $\mathcal{C}$ of (M, g) such that $\Sigma_0 \subset I^+(\mathcal{C})$. Since from Theorem 6.9 every null generator of $\mathcal{H}_0^+$ when maximally extended has to intersect $\mathcal{C}$, then every null generator of $\mathcal{H}_0^+$ must have a past endpoint on Σ_0. Thus, apart from an affine reparametrization, there exists a null future-complete geodesic Σ_0-ray $\overline{\gamma} : [0, +\infty) \to M$ such that $\overline{\gamma}|_{(0,+\infty)} = \gamma$. We conclude that for some smooth future-directed null section ℓ of the normal bundle of Σ_0,

$$\mathcal{H}_0^+ \cup \Sigma_0 = \{\exp^\perp(t\ell(p)) \,:\, p \in \Sigma_0,\, t \in [0, +\infty)\},$$

where $\exp^\perp$ denotes the normal exponential map. One can now argue just as in Proposition 2.2(ii) in [10], using the Raychaudhuri equation, the null convergence condition and the extremality of Σ_0 to conclude that the null mean curvature θ of $\mathcal{H}_0^+$ is non-positive. But again the fact that its generators are future-complete and the null convergence condition, Lemma IV.2 in [12] applied to a smooth null hypersurface implies that $\theta \geq 0$, so $\theta \equiv 0$. Basic comparison theory for null hypersurfaces, using again the the null convergence condition, now yields that $\mathcal{H}_0^+$ is totally geodesic. □

Another approach to the study of the regularity of the S-horizons is to explore a little closer the symmetries of (M, g). We proceed to do this in the next section.

8 Killing Horizons

Systematic studies of various properties of Killing horizons, especially bifurcate one, has been carried out by many authors. We adapt the approach in [4, 22].

Definition 8.1 A connected smooth null hypersurface $\mathcal{N} \subset M$ is a *Killing horizon* if there exists a complete Killing vector field $K \in \Gamma(TM)$ with flow $\psi : t \in \mathbb{R} \mapsto \psi_t \in \mathrm{Diff}(M)$ such that

KH1) $\psi_t(\mathcal{N}) \subset \mathcal{N}\ \forall t \in \mathbb{R}$,
KH2) $\langle K, K\rangle|_{\mathcal{N}} \equiv 0$, and
KH3) $K(p) \neq 0\ \forall p \in \mathcal{N}$.

If the complete Killing vector field K satisfies conditions (KH1)-(KH3), we say K is *adapted* to the Killing horizon $\mathcal{N}$.

Remark 8.2 Given a Killing horizon $\mathcal{N} \subset M$, there may very well exist more than one Killing field adapted to $\mathcal{N}$. Consider the following example, adapted from Ref. [7], pp. 580-581. Take the 3-dimensional Minkowski spacetime, i.e., let $M = \mathbb{R}^3$, with the flat metric given in standard cartesian coordinates (t, x, y) as

$$g = -dt^2 + dx^2 + dy^2.$$

As a null hypersurface take

$$\mathcal{N} := \{t + x = 0 \,:\, y > 0, t > 0\}.$$

Then the Killing vector fields

$$K_1 = x\partial_t + t\partial_x \text{ and } K_2 = y(\partial_t - \partial_x) + (t + x)\partial_y$$

are both adapted to $\mathcal{N}$.

Proposition 8.3 *Every Killing horizon is totally geodesic.*

Proof Let $\mathcal{N} \subset M$ be a Killing horizon, and let $K \in \Gamma(TM)$ be a complete Killing adapted to $\mathcal{N}$. Fix a smooth future-directed null tangent vector field $Z \in \Gamma(T\mathcal{N})$ on $\mathcal{N}$. Since $T_p\mathcal{N} = Z(p)^{\perp}$ for each $p \in \mathcal{N}$, we have that

$$\langle \nabla_V Z, W \rangle = \langle \nabla_W Z, V \rangle \tag{8.1}$$

for all $V, W \in \Gamma(T\mathcal{N})$. Let $K_{\mathcal{N}} := K|_{\mathcal{N}}$. The conditions on K imply that there exists $\lambda \in C^{\infty}(\mathcal{N})$ such that $K_{\mathcal{N}} = \lambda Z$, and from (8.1) we conclude that

$$\langle \nabla_V K_{\mathcal{N}}, W \rangle = \langle \nabla_W K_{\mathcal{N}}, V \rangle \tag{8.2}$$

for all $V, W \in \Gamma(T\mathcal{N})$. But since K is Killing we also have

$$\langle \nabla_V K_{\mathcal{N}}, W \rangle + \langle \nabla_W K_{\mathcal{N}}, V \rangle = 0. \tag{8.3}$$

Equations (8.2) and (8.3) together now give

$$\langle \nabla_V K_{\mathcal{N}}, W \rangle = 0,$$

and therefore

$$\langle \nabla_V Z, W \rangle = 0$$

for all $V, W \in \Gamma(T\mathcal{N})$, from which the claim follows. □

Proposition 8.4 *Let $\mathcal{N} \subset M$ be a Killing horizon, and let $K \in \Gamma(TM)$ be a Killing field adapted to $\mathcal{N}$. Then there exists a unique smooth function $\kappa : \mathcal{N} \to \mathbb{R}$ such that*

$$\nabla_K K(p) = \kappa(p) K(p) \tag{8.4}$$

$$\nabla f_K(p) = -2\kappa(p) K(p), \tag{8.5}$$

$\forall p \in \mathcal{N}$, where $f_K := \langle K, K \rangle$. κ is called the surface gravity *associated with K and $\mathcal{N}$. In particular, the orbits of K contained in $\mathcal{N}$ are null pregeodesics whose images coincide with the null geodesic generators of $\mathcal{N}$. Moreover, κ is constant along each one of these orbits. Finally, if $\kappa \neq 0$ along a given orbit of K contained in $\mathcal{N}$, then the corresponding null generator η is incomplete in $\mathcal{N}$, and if η is in addition extendible as a geodesic in (M, g), then it has an endpoint on a zero of K.*

Proof Note that $K_{\mathcal{N}} := K|_{\mathcal{N}}$ must be an everywhere nonzero, null vector field tangent to $\mathcal{N}$. For every $V \in \Gamma(TM)$, we have

$$\langle \nabla f_K, V \rangle = V\langle K, K \rangle = 2\langle \nabla_V K, K \rangle = -2\langle \nabla_K K, V \rangle,$$

where we have used the fact that K is Killing in the last equality. We then conclude that

$$-2\nabla_K K = \nabla f_K. \tag{8.6}$$

Now, given $p \in \mathcal{N}$, and $v \in T_p\mathcal{N}$, we must have

$$g_p(v, \nabla f_K(p)) = v(f_K) \equiv 0$$

since $f_K|_{\mathcal{N}} = 0$. But $T_p\mathcal{N} \equiv K(p)^{\perp}$, from which we deduce that there exists a unique number $\kappa(p)$ for which $\nabla f_K(p) = -2\kappa(p)K(p)$. From this and from Eqs. (8.6), (8.4), and (8.5) follow. To establish the smoothness of κ, pick any Riemannian metric h on M. We have, recalling that K is everywhere nonzero in $\mathcal{N}$ that

$$\kappa = \frac{h(\nabla_K K, K)}{h(K, K)}$$

from which smoothness immediately follows. Equation (8.4) clearly means that the orbits of K contained in $\mathcal{N}$ are null pregeodesics whose images coincide with the null geodesic generators of $\mathcal{N}$.

The remaining statements are an immediate consequence of the proof of Lemmas 1 and 2 in Ref. [4]. However, those proofs are given in index notation and for the convenience of the reader we essentially repeat the arguments here in coordinate-free form. Let $V \in \Gamma(TM)$. We have

$$\begin{aligned}\langle \nabla_K \nabla f_K, V\rangle &= K\langle \nabla f_K, V\rangle - \langle \nabla f_K, \nabla_K V\rangle \\ &= KV(f_K) - \langle \nabla f_K, \nabla_V K\rangle - \langle \nabla f_K, [K, V]\rangle \\ &= KV(f_K) - \langle \nabla f_K, \nabla_V K\rangle - [K, V](f_K) \\ &= V(K(f_K)) - \langle \nabla f_K, \nabla_V K\rangle.\end{aligned} \tag{8.7}$$

But since K is Killing, $K(f_K) \equiv 0$. Thus

$$\langle \nabla_K \nabla f_K, V\rangle \equiv -\langle \nabla f_K, \nabla_V K\rangle. \tag{8.8}$$

Now, let $p \in \mathcal{N}$. Using (8.8) and (8.5) we get

$$\begin{aligned}\langle \nabla_K \nabla f_K(p), V(p)\rangle &= 2\kappa(p)\langle K(p), \nabla_V K(p)\rangle \\ &= \kappa(p)V(f_K)(p) \equiv \kappa(p)\langle \nabla f_K(p), V(p)\rangle \\ &= -2\kappa(p)^2\langle K(p), V(p)\rangle,\end{aligned} \tag{8.9}$$

and since V is arbitrary we conclude that

$$\nabla_K \nabla f_K(q) = -2\kappa(q)^2 K(q) \tag{8.10}$$

for all $q \in \mathcal{N}$. Let $\alpha : \mathbb{R} \to M$ be an integral curve of K contained in $\mathcal{N}$. Denote by D/dt the covariant derivative of vector fields along α. Then on the one hand, (8.10) yields

$$\frac{D(\nabla f_K \circ \alpha)}{dt} = (\nabla_K \nabla f_K) \circ \alpha = -2(\kappa \circ \alpha)^2 \alpha', \tag{8.11}$$

and on the other hand, using Eqs. (8.4) and (8.5) we get

$$\frac{D(\nabla f_K \circ \alpha)}{dt} = -2\frac{d(\kappa \circ \alpha)}{dt}\alpha' - 2(\kappa \circ \alpha)\alpha'' \overset{\text{Eq.(8.5)}}{=} -2\left(\frac{d(\kappa \circ \alpha)}{dt} + (\kappa \circ \alpha)^2\right)\alpha'. \tag{8.12}$$

Putting (8.11) and (8.12) together we conclude that

$$\frac{d(\kappa \circ \alpha)}{dt} \equiv 0,$$

and hence that κ is constant along α. Thus $c := (\kappa \circ \alpha)(t)$, $\forall t \in \mathbb{R}$ is a constant. Suppose $c \neq 0$. The curve $\eta : (0, +\infty) \to M$ given by

$$\eta(s) := \alpha(\frac{1}{c} \log s)$$

$\forall s \in (0, +\infty)$ is an incomplete null geodesic whose image coincides with that of α. Hence, it is the corresponding null geodesic generator of $\mathcal{N}$, and is in particular inextendible in $\mathcal{N}$. Suppose η is extendible as a geodesic in (M, g), and let $\overline{\eta} : (A, +\infty) \to M$ denote the maximal extension, with $-\infty \leq A < 0$. Let h be a Riemannian metric on M. For every $s > 0$ a direct calculation reveals that

$$h(\overline{\eta}'(s), \overline{\eta}'(s)) = \frac{1}{c^2 s^2} h(K(\overline{\eta}(s)), K(\overline{\eta}(s))),$$

and the continuity of the left-hand side at $s = 0$ therefore requires that

$$h(K(\overline{\eta}(0)), K(\overline{\eta}(0))) = 0,$$

i.e., $K(\overline{\eta}(0)) = 0$. □

We shall call a Killing horizon $\mathcal{N} \subset M$ *degenerate* with respect to an adapted Killing vector field K if the surface gravity κ associated with $\mathcal{N}$ and K is identically zero throughout $\mathcal{N}$. We then have the immediate consequence of Proposition 8.4.

Corollary 8.5 *A Killing horizon $\mathcal{N} \subset M$ is degenerate with respect to an adapted Killing field $K \in \Gamma(TM)$ if and only if all null geodesic generators are complete geodesics entirely contained in $\mathcal{N}$ coinciding with the orbits of K in $\mathcal{N}$.* □

Corollary 8.6 *Suppose (M, g) is a globally hyperbolic spacetime. Assume that each connected component $\mathcal{N}_\lambda$ of $\mathcal{H}^+$ is a Killing horizon with an adapted complete Killing vector field $K_\lambda \in \Gamma(TM)$. Then the following statements hold.*

(i) If for each λ, the associated connected component $\mathcal{N}_\lambda$ is degenerate with respect to K_λ, then $\Sigma \equiv \emptyset$.

(ii) If $\Sigma = \emptyset$, then either some null geodesic generator of $\mathcal{H}^+$ is incomplete in (M, g) or else $\mathcal{N}_\lambda$ is degenerate with respect to K_λ, $\forall\lambda$.

Proof (i)
Assume, by way of contradiction, that for each λ, the associated component $\mathcal{N}_\lambda$ is degenerate with respect to K_λ, and yet there exists $p \in \Sigma$. By Proposition 6.2(iii), $(D, g|_D)$ is globally hyperbolic, and using Theorem 6.7(iv), we conclude that there exists a future-inextendible null geodesic ray $\eta : [0, a) \to M$ starting at p such that $\eta|_{(0,a)}$ is a *past-incomplete* null geodesic generator of $\mathcal{H}^+$, contradicting Corollary 8.5.
(ii)
Suppose $\Sigma = \emptyset$. Then, by Theorem 6.9(i), each null geodesic generator of $\mathcal{H}^+$ is both past and future inextendible in (M, g). Suppose any such generator is complete. Then, given any index λ, and any null generator α in $\mathcal{N}_\lambda$, the final claims in Proposition 8.4 imply that the surface gravity κ_λ associated with $\mathcal{N}_\lambda$ and K_λ must vanish identically along α, and the proof is complete. □

The following result is an important consequence of Theorem 7.1 of Ref. [6], and ensures regularity of the S-horizons provided one has adapted isometries.

Theorem 8.7 *Suppose (M, g) is a stably causal spacetime satisfying the null convergence condition. Let $\mathcal{N} \subset M$ be a connected C^0 achronal hypersurface with the following properties:*

(i) $\forall p \in \mathcal{N}$, there exists a future-complete null geodesic $\gamma : [0, +\infty) \to M$ such that $p = \gamma(0)$ and $\gamma[0, +\infty) \subseteq \mathcal{N}$;
(ii) there exists a complete Killing vector field $K \in \Gamma(TM)$ with flow $\psi : t \in \mathbb{R} \mapsto \psi_t \in \mathrm{Diff}(M)$ such that $\psi_t(\mathcal{N}) \subset \mathcal{N}$ $\forall t \in \mathbb{R}$, $\langle K, K\rangle|_{\mathcal{N}} \equiv 0$ and $K(p) \neq 0$, $\forall p \in \mathcal{N}$.

Then $\mathcal{N}$ is a Killing horizon. In particular, $\mathcal{N}$ is a smooth totally geodesic null hypersurface in (M, g), and K is adapted to $\mathcal{N}$.

Proof All we need to show is that $\mathcal{N}$ is smooth. Let $\tilde{\mathcal{N}} := \partial I^-(\mathcal{N})$. Since $\mathcal{N}$ is achronal, $\mathcal{N} \subseteq \tilde{\mathcal{N}}$. Condition (i) is clearly satisfied for $\tilde{\mathcal{N}}$ as well, so in particular $\tilde{\mathcal{N}}$ is a *future horizon* in the sense of Ref. [6].

We claim first that $\psi_t(\tilde{\mathcal{N}}) \subset \tilde{\mathcal{N}}$ $\forall t \in \mathbb{R}$. Note that $\forall p \in \mathcal{N}$, $p = \psi_t(\psi_{-t}(p)) \in \psi_t(\mathcal{N})$, so in fact

$$\psi_t(\mathcal{N}) \equiv \mathcal{N}$$

$\forall t \in \mathbb{R}$. Thus, for each $t \in \mathbb{R}$

$$\psi_t(\tilde{\mathcal{N}}) = \partial I^-(\psi_t(\mathcal{N})) = \partial I^-(\mathcal{N}) \equiv \tilde{\mathcal{N}},$$

which proves the claim.

Condition (ii) implies in particular that $K|_{\mathcal{N}}$ is a null, everywhere nonzero vector field. Since we can clearly work with either K or $-K$, from the connectedness of $\mathcal{N}$ there is no loss of generality in assuming that $K|_{\mathcal{N}}$ is future-directed. Let $p \in \mathcal{N}$. Condition (ii) now implies that the orbit $\mathbb{R}p \subset \mathcal{N}$ is a future-directed null pregeodesic. Condition (i) together with the achronality of $\mathcal{N}$ then imply that there exists a number $\epsilon > 0$ and a future-directed null (affinely parameterized) future-complete geodesic $\gamma : [-\epsilon, +\infty) \to M$ contained in $\mathcal{N}$ such that $\gamma(0) = p$. Let $t_\epsilon > 0$ be the unique number such that $\psi_{t_\epsilon}(p) = \gamma(\epsilon)$. Since $\mathcal{N}$ is an open submanifold of $\tilde{\mathcal{N}}$, there exists a connected open set $U \ni p$ such that $U \cap \tilde{\mathcal{N}} = U \cap \mathcal{N}$. We can assume that $\gamma[-\epsilon, \epsilon] \subset U \cap \mathcal{N}$.

Using stable causality, we choose a smooth time function $\mathcal{T} : M \to \mathbb{R}$ such that $\nabla\mathcal{T}$ is past-directed timelike. The set

$$\mathcal{S} := U \cap \mathcal{T}^{-1}(\mathcal{T}(p))$$

is a smooth spacelike acausal hypersurface in (M, g), and $\mathcal{S} \cap \tilde{\mathcal{N}} = \mathcal{S} \cap \mathcal{N}$. Then

$$\psi_{t_\epsilon}(\mathcal{S} \cap \tilde{\mathcal{N}}) \subset J^+(\mathcal{S} \cap \tilde{\mathcal{N}}).$$

The latter inclusion means that the conditions of Theorem 7.1 of [6] are all satisfied and therefore

$$V := (J^-(\psi_{t_\epsilon}(\mathcal{S} \cap \tilde{\mathcal{N}})) \setminus \psi_{t_\epsilon}(\mathcal{S} \cap \tilde{\mathcal{N}})) \cap (J^+(\mathcal{S} \cap \tilde{\mathcal{N}}) \setminus \mathcal{S} \cap \tilde{\mathcal{N}}) \subset \mathcal{N}$$

is a smooth totally geodesic null hypersurface in (M, g). But clearly $p \in V$, which proves that $\mathcal{N}$ is smooth, as desired. □

Joining Theorems 6.3(iii) and 8.7, together with Corollary 6.8 yields the following regularity result which is independent of the S-bifurcation set.

Corollary 8.8 *Assume that (M, g) is a stably causal spacetime satisfying the null convergence condition, and that X is timelike everywhere on D. Then either some null generator of $\mathcal{H}^+$ is future-incomplete or else any connected component $\mathcal{N}_\lambda$ of $\mathcal{H}^+$ is a Killing horizon, and hence a smooth totally geodesic null hypersurface in (M, g), with X adapted to $\mathcal{N}_\lambda$.*

Remark 8.9 Both Theorem 8.7 and Corollary 6.8 have time-dual versions, from which we deduce that in the conditions of Corollary 8.8, each connected component of $\mathcal{H}^-$ is also a Killing horizon, and hence a smooth totally geodesic null hypersurface in (M, g) having X as an adapted Killing vector field.

We shall henceforth adopt the following terminology. Given a Killing vector field $K \in \Gamma(TM)$, we denote by

$$\mathcal{Z}(K) := \{p \in M \,:\, K(p) = 0\}$$

the set of its zeros.

The following lemma is a well-know result by S. Kobayashi [23]. (Although it is proved therein for Riemannian manifolds, the proof remains valid in the semi-Riemannian setting, see also [29], Theorem 1.7.12.)

Lemma 8.10 *Let $K \in \Gamma(TM)$ be a complete Killing vector field. Then each connected component of $\mathcal{Z}(K)$ is a smooth, closed totally geodesic submanifold of M.* □

Lemma 8.10, together with Theorems 7.4 and 7.5 now yield the following result.

Corollary 8.11 *Let $K \in \Gamma(TM)$ be a complete, not identically zero Killing vector field, and let Σ_0 be a nonempty connected component of Σ. Assume the following.*

(1) (M, g) is globally hyperbolic and satisfies the null convergence condition.
(2) $\mathcal{H}^+$ [resp. $\mathcal{H}^-$] has future-complete [resp. past-complete] null generators.
(3) Σ_0 coincides with a compact connected component $\mathcal{Z}_0$ of $\mathcal{Z}(K)$.

Then $\mathcal{H}^{\pm}$ such that $\Sigma_0 = \overline{\mathcal{H}_0^+} \cap \overline{\mathcal{H}_0^-}$ according to Lemma 7.3 are Killing horizons and K is adapted to both.

Comment on the proof. The fact that K is adapted to the smooth surfaces $\mathcal{H}^{\pm}$ follows from a discussion entirely analogous to that in pp. 57–58 in Ref. [22]. □

Acknowledgements This work has been partially supported by a research scholarship from CNPq, Brazil, Programa Ciências sem Fronteira, process number 200428/2015-2. I also gratefully acknowledge support from the Spanish MICINN-FEDER Grant MTM2013-47828-C2-2-P, with FEDER funds. I wish to thank José Luis Flores (University of Málaga) and Steven Harris (Saint Louis University) for stimulating discussions, and the staff and faculty members of the Department of Mathematics of the University of the Miami, where this work was carried out. My special thanks go to Gregory Galloway, for his kind hospitality throughout the year of 2016.

References

1. J.K. Beem, P.E. Ehrlich and K.L. Easley *Global Lorentzian Geometry*, 2nd edn. (Marcel Dekker, New York, 1996)
2. R. Beig, B.G. Schmidt (eds.), *Time-independent gravitational fields*, in *Einsteins Field Equations and Their Physical Implications*, Lecture Notes in Physics, vol. 540 (Springer, Heidelberg, 2000), pp 325–372
3. A.N. Bernal, M. Sánchez, Smoothness of time function and the metric splitting of globally hyperbolic spacetimes. Commun. Math. Phys. **257**, 43–50 (2005)
4. R.H. Boyer, Geodesic killing orbits and bifurcate killing horizons. Proc. R. Soc. A **311**, 245–52 (1969)
5. E. Caponio, M.A. Javaloyes, A. Masiello, On the energy functional on Finsler manifolds and applications to stationary spacetimes. Math. Ann. **351**, 365–392 (2011)
6. P.T. Chruściel, E. Delay, G.J. Galloway, R. Howard, Regularity of horizons and the area theorem. Ann. Henri Poincaré **2**, 109–178 (2001)
7. P.T. Chruściel, G.J. Galloway, D. Pollack, Mathematical general relativity: a sampler. Bull. Amer. Math. Soc. **47**, 567–638 (2010)
8. P.T. Chruściel, J.L. Costa, M. Heusler, *Stationary Black Holes: Uniqueness and Beyond* (Living Rev, Relativity, 2012)

9. J. Cortier, V. Minerbe, On complete stationary vacuum initial data. J. Geom. Phys. **99**, 20–27 (2016)
10. I.P. Costa e Silva J.L. Flores, *Extremal surfaces and the rigidity of null geodesic incompleteness*. Class. Quantum Grav. **32**, 055010 (2015)
11. H. Friedrich, I. Rácz, R.M. Wald, On the rigidity theorem for spacetimes with a stationary event horizon or a compact Cauchy horizon. Commun. Math. Phys. **204**, 691–707 (1999)
12. G.J. Galloway, Maximum principles for null hypersurfaces and null splitting theorems. Ann. Henri Poincaré **1**, 543–567 (2000)
13. R.P. Geroch, A method for generating solution of Einstein's equations. J. Math. Phys. **12**, 918–924 (1971)
14. R.P. Geroch, A method for generating solution of Einstein's equations 2. J. Math. Phys. **13**, 394–404 (1972)
15. S.G. Harris, Conformally stationary spacetimes. Class. Quantum Grav. **9**, 1823–1827 (1992)
16. S.G. Harris, Static- and stationary-complete spacetimes: algebraic and causal structures. Class. Quantum Gravity **32**, 135026 (2015)
17. S.G. Harris, R.J. Low, Causal monotonicity, omniscient foliations and the shape of space. Class. Quantum Grav. **18**, 27–43 (2001)
18. S.W. Hawking, G.F.R. Ellis, *The Large Scale Structure of Space-time* (Cambridge University Press, Cambridge, 1973)
19. M. Heusler, *Black Hole Uniqueness Theorems* (Cambridge University Press, Cambridge, 1996)
20. A. Ionescu, S. Klainerman, *Rigidity Results in General Relativity: A Review*, arXiv:1501.01587v1 [gr-qc]
21. M.A. Javaloyes, M. Sánchez, A note on the existence of standard splittings for conformally stationary spacetimes. Class. Quantum Grav. **25**, 168001 (2008)
22. B.S. Kay, R.M. Wald, R M, Theorems on the uniqueness and thermal properties of stationary, nonsingular, quasifree states on spacetimes with a bifurcate Killing Horizon. Phys. Rep. **201**, 49–136 (1991)
23. S. Kobayashi, Fixed points of isometries. Nagoya Math. J. **13**, 63–68 (1958)
24. S. Kobayashi, K. Nomizu, *Foundations of Differential Geometry*, vol. 1 (Wiley, New York, 1962)
25. A. Komar, Covariant conservation laws in general relativity. Phys. Rev. **113**, 934 (1959)
26. J.M. Lee, *Introduction to smooth manifolds* (Springer, New York, 2003)
27. A. Magnon, On Komar integrals in asymptotically anti-de Sitter space-times. J. of Math. Phys. **26**, 3112–3117 (1985)
28. B. O'Neill, *Semi-Riemannian Geometry with Applications to Relativity* (Academic Press, New York, 1983)
29. B. O'Neill, *The Geometry of Kerr Black Holes* (Academic Press, New York, 1995)
30. B. O'Neill, The fundamental equations of a submersion. Michigan Math. J. **13**, 459–469 (1966)
31. M. Sánchez, Lorentzian Manifolds Admitting a Killing Vector Field. Nonlinear Anal. **30**, 643–654 (1997)
32. M. Sánchez, On causality and closed geodesics of compact Lorentzian manifolds and static spacetimes. Differ. Geom. Appl. **24**, 21–32 (2006)
33. J.H. Treude, *Ricci curvature comparison in Riemannian and Lorentzian geometry*, diplomarbeit, Albert-Ludwigs-Universität, Freiburg (2011). http://www.freidok.uni-freiburg.de/volltexte/8405/pdf/diplomarbeit_-treude.pdf
34. R.M. Wald, *General Relativity* (University of Chicago Press, Chicago, IL, 1984)

Anti-De Sitter Spacetimes and Isoparametric Hypersurfaces in Complex Hyperbolic Spaces

J. Carlos Díaz-Ramos, Miguel Domínguez-Vázquez and Víctor Sanmartín-López

Abstract By lifting hypersurfaces in complex hyperbolic spaces to anti-De Sitter spacetimes, we prove that an isoparametric hypersurface in the complex hyperbolic space has the same principal curvatures as a homogeneous one.

Keywords Complex hyperbolic space · Isoparametric hypersurface · Kähler angle · Anti-De Sitter spacetime

2010 Mathematics Subject Classification 53C40 · 53B25

1 Introduction

One of the final aims of the research of the authors of this paper is guided by the following question: to what extent do the symmetries of an object determine its shape? It is intuitively clear that the existence of symmetries reduces the number of degrees of freedom in the description of a geometric object and imposes constraints on how the different parameters defining it are related. The more symmetries an object has, the more likely it is that this object is uniquely determined. Somewhat more complicated is to address the following converse problem: if the shape of an object A is the same as that of an object B with symmetries, can we assert that object A is the same as B?

In order to make this broad question more concrete, we introduce the mathematical context into which we will tackle the problem. Our area of research is submanifold

---

J.C. Díaz-Ramos (✉) · V. Sanmartín-López
Department of Mathematics, Universidade de Santiago de Compostela, Santiago, Spain
e-mail: josecarlos.diaz@usc.es

V. Sanmartín-López
e-mail: victor.sanmartin@usc.es

M. Domínguez-Vázquez
ICMAT - Instituto de Ciencias Matemáticas (CSIC-UAM-UC3M-UCM), Madrid, Spain
e-mail: miguel.dominguez@icmat.es

M.A. Cañadas-Pinedo et al. (eds.), *Lorentzian Geometry and Related Topics*,
Springer Proceedings in Mathematics & Statistics 211,
DOI 10.1007/978-3-319-66290-9_6

geometry of Riemannian manifolds. Our "symmetric objects" will be the so-called homogeneous submanifolds. Let $\bar{M}$ be a Riemannian manifold, and M a submanifold of $\bar{M}$. We say that M is extrinsically homogeneous, henceforth simply homogeneous, if for any two points $p, q \in M$ there exists an isometry g of $\bar{M}$ such that $g(M) = M$ and $g(p) = q$. Equivalently, M is homogeneous if it is an orbit of a subgroup G of the isometry group of $\bar{M}$, that is, $M = G \cdot p$, for some $p \in \bar{M}$. In this paper we will actually be interested in homogeneous hypersurfaces, that is, homogeneous submanifolds of codimension one.

A homogeneous hypersurface has a great deal of symmetries, namely, the isometries of G. It is thus conceivable that homogeneous hypersurfaces can be classified in a broad class of Riemannian manifolds with large isometry groups. (If isometry groups are small, homogeneous hypersurfaces might not exist at all.) This is true for example in Euclidean spaces [17], spheres [11], real hyperbolic spaces [4], complex projective spaces [18], irreducible symmetric spaces of compact type [12], complex hyperbolic spaces, and the Cayley hyperbolic plane [3]. Remarkably, no such classification is known for quaternionic hyperbolic spaces.

Homogeneous hypersurfaces have two interesting properties related to their shape: they are isoparametric and have constant principal curvatures. A hypersurface is called isoparametric if its nearby parallel hypersurfaces have constant mean curvature. A hypersurface has constant principal curvatures if the eigenvalues of its shape operator are constant. These two concepts are equivalent for spaces of constant curvature. Indeed, this property of their shape is characteristic of homogeneous hypersurfaces in Euclidean and real hyperbolic spaces: Segre, for Euclidean spaces [17], and Cartan, for real hyperbolic spaces [4], proved that isoparametric hypersurfaces are homogeneous and derived their classification. Surprisingly, this is not true of spheres, as the examples in [9] show. The classification of isoparametric hypersurfaces in spheres has been the aim of several recent and important works (see for example [5] and [14]).

In complex space forms, isoparametric hypersurfaces do not necessarily have constant principal curvatures. A classification of isoparametric hypersurfaces in complex projective spaces $\mathbb{C}P^n$, $n \neq 15$, has been obtained by the second author in [8] as a consequence of the available classifications in spheres. As a consequence, there exist inhomogeneous isoparametric hypersurfaces in complex projective spaces.

In this paper we are interested in isoparametric hypersurfaces in complex hyperbolic spaces. Their classification has recently been obtained by the authors in [7]:

Theorem 1.1 *Let M be a connected real hypersurface in the complex hyperbolic space $\mathbb{C}H^n$, $n \geq 2$. Then, M is isoparametric if and only if M is congruent to an open part of:*

(i) *a tube around a totally geodesic complex hyperbolic space $\mathbb{C}H^k$, $k \in \{0, \dots, n-1\}$, or*
(ii) *a tube around a totally geodesic real hyperbolic space $\mathbb{R}H^n$, or*
(iii) *a horosphere, or*
(iv) *a ruled homogeneous minimal Lohnherr hypersurface W^{2n-1}, or some of its equidistant hypersurfaces, or*

(v) *a tube around a ruled homogeneous minimal Berndt-Brück submanifold* W_{φ}^{2n-k}, *for* $k \in \{2, \ldots, n-1\}$, $\varphi \in (0, \pi/2]$, *where* k *is even if* $\varphi \neq \pi/2$, *or*
(vi) *a tube around a ruled homogeneous minimal submanifold* $W_{\mathfrak{w}}$, *for some proper real subspace* $\mathfrak{w}$ *of* $\mathfrak{g}_\alpha \cong \mathbb{C}^{n-1}$ *such that* $\mathfrak{w}^\perp$, *the orthogonal complement of* $\mathfrak{w}$ *in* $\mathfrak{g}_\alpha$, *has nonconstant Kähler angle.*

We give a brief description of the examples (iv) through (vi); see [6] or [7] for more details. Let $\mathfrak{g}$ denote the Lie algebra of $SU(1, n)$, the isometry group of $\mathbb{C}H^n$, and let $\mathfrak{g} = \mathfrak{g}_{-2\alpha} \oplus \mathfrak{g}_{-\alpha} \oplus \mathfrak{g}_0 \oplus \mathfrak{g}_\alpha \oplus \mathfrak{g}_{2\alpha}$ be a restricted root space decomposition of $\mathfrak{g}$ with respect to some point $o \in \mathbb{C}H^n$ and some point at infinity $x \in \mathbb{C}H^n(\infty)$. The point x determines a maximal flat $\mathfrak{a} \subset \mathfrak{g}_0$. It turns out that $\mathfrak{a}$ and $\mathfrak{g}_{2\alpha}$ are 1-dimensional, and $\mathfrak{g}_\alpha$ is a complex vector space of complex dimension $n-1$, whose complex structure we denote by J. If $\mathfrak{w}$ is a real subspace of $\mathfrak{g}_\alpha$, we define $\mathfrak{s}_{\mathfrak{w}} = \mathfrak{a} \oplus \mathfrak{w} \oplus \mathfrak{g}_{2\alpha}$. Then, $\mathfrak{s}_{\mathfrak{w}}$ is a Lie subalgebra of $\mathfrak{g}$, and the connected subgroup $S_{\mathfrak{w}}$ of $SU(1, n)$ whose Lie algebra is $\mathfrak{s}_{\mathfrak{w}}$ acts isometrically on $\mathbb{C}H^n$. We define $W_{\mathfrak{w}} = S_{\mathfrak{w}} \cdot o$. Then, the tubes around $W_{\mathfrak{w}}$ are isoparametric hypersurfaces of $\mathbb{C}H^n$. If $\mathfrak{w}$ is a hyperplane, then $W_{\mathfrak{w}}$ is denoted by W^{2n-1} and we obtain the examples in (iv) (see also [13]). If $\mathfrak{w}^\perp$, the orthogonal complement of $\mathfrak{w}$ in $\mathfrak{g}_\alpha$, has constant Kähler angle $\varphi \in (0, \pi/2]$ and codimension k, then $W_{\mathfrak{w}}$ is denoted by W_{φ}^{2n-k} and we get (v) (see [1]). Recall that $\mathfrak{w}^\perp$ has constant Kähler angle φ if for any nonzero $\xi \in \mathfrak{w}^\perp$ the angle between $J\xi$ and $\mathfrak{w}^\perp$ is always φ. If $\mathfrak{w}^\perp$ does not have constant Kähler angle, then we obtain the examples in (vi).

Theorem 1.1 implies the classification of homogeneous hypersurfaces in complex hyperbolic spaces. In fact,

Corollary 1.2 [3, 7] *A real hypersurface in* $\mathbb{C}H^n$ *is homogeneous if and only if it belongs to one of the families (i)–(v) in* Theorem 1.1.

Thus, if $n \geq 3$, there are uncountably many families of inhomogeneous isoparametric hypersurfaces in complex hyperbolic spaces.

The aim of this paper is to prove the following result.

Theorem 1.3 *Let* M *be an isoparametric hypersurface in* $\mathbb{C}H^n$. *Then, the principal curvatures of* M *are pointwise the same as the principal curvatures of a homogeneous hypersurface of* $\mathbb{C}H^n$.

It is clear that, by working out the principal curvatures of the examples appearing in Theorem 1.1, the conclusion of Theorem 1.3 follows from the classification of isoparametric hypersurfaces in $\mathbb{C}H^n$. The purpose of this paper is to prove Theorem 1.3 by a more direct approach, which avoids several intricate arguments needed for the proof of Theorem 1.1.

The complex hyperbolic space $\mathbb{C}H^n$ is the quotient of the anti-De Sitter spacetime $H_1^{2n+1} \subset \mathbb{C}^{1,n}$ by S^1. Let us call $\pi\colon H_1^{2n+1} \to \mathbb{C}H^n$ the projection map, the so-called Hopf map. See Sect. 2. Recall that $\mathbb{C}H^n$ is a Kähler manifold of constant holomorphic sectional curvature. We denote its Kähler structure by J. The anti-De Sitter space is, in turn, a Lorentzian manifold of constant negative curvature. It can be shown that, if a hypersurface M of $\mathbb{C}H^n$ is isoparametric, then $\pi^{-1}(M)$ has constant principal curvatures. A generalization of a result by Cartan [4] implies that the number of real principal curvatures of $\pi^{-1}(M)$ is bounded by two. This allows us to deduce many interesting properties of M just by using the fundamental equations of a submersion and some algebraic calculations. It is remarkable, for example, that the principal curvatures of M at a point coincide with the principal curvatures of some homogeneous hypersurface in $\mathbb{C}H^n$. In particular, if $g(p)$ denotes the number of principal curvatures of M at p, and $h(p)$ denotes the number of nontrivial projections of $J\xi_p$ onto the principal curvature spaces, where ξ_p is a normal vector of M at p, we have

Proposition 1.4 *If M is an isoparametric hypersurface of $\mathbb{C}H^n$, then $h \leq 3$ and $g \leq 5$.*

The results obtained by the authors in this paper precede those of [7]. Although the principal curvatures of isoparametric hypersurfaces in $\mathbb{C}H^n$ are the same as in the homogeneous examples, we found inhomogeneous examples of isoparametric hypersurfaces in $\mathbb{C}H^n$. These examples, corresponding to case (vi) of Theorem 1.1, were first constructed in [6]. It was surprising at the moment to notice that, pointwise, the principal curvatures of these examples are the same as those of a homogeneous hypersurface. Nonetheless, the principal curvatures of these examples are nonconstant, so they are not homogeneous.

A major disadvantage of working with $\pi^{-1}(M)$ instead of M is that the shape operator of the former is not necessarily diagonalizable. There are exactly four different types of Jordan canonical forms for this shape operator, described in Sect. 3. Using the algebraic approach that we describe in this paper we can get Theorem 1.3. However, we will only deal with Type III points. There are two reasons for this. Firstly, Type III is considerably more involved than the other types. Once Type III is sorted out, the other types can be handled with similar arguments. Secondly, Types I, II and IV are tackled in [7] by this very same method. The arguments in [7] diverge considerably from our approach here for Type III points. In this paper we get a weaker result, but the argument is much shorter. The core of this paper is the proof of Theorem 3.4 from where Theorem 1.3 and Proposition 1.4 follow.

2 Anti-De Sitter Spacetime and Complex Hyperbolic Space

In $\mathbb{C}^{n+1}$ we define the flat semi-Riemannian metric

$$\langle z, w \rangle = \mathrm{Re}\Big(-z_0\bar{w}_0 + \sum_{k=1}^{n} z_k\bar{w}_k\Big).$$

It is customary to denote by $\mathbb{C}^{1,n}$ the vector space $\mathbb{C}^{n+1}$ endowed with the previous scalar product. The anti-De Sitter spacetime of radius $r > 0$ is defined as

$$H_1^{2n+1}(r) = \left\{ z \in \mathbb{C}^{1,n} : \langle z, z \rangle = -r^2 \right\}.$$

This hypersurface of $\mathbb{C}^{1,n}$ is a Lorentzian manifold of constant negative curvature $c = -4/r^2$ and dimension $2n+1$. The map $S^1 \times H_1^{2n+1}(r) \to H_1^{2n+1}(r)$, $(\lambda, z) \mapsto \lambda z$ defines an S^1-action on $H_1^{2n+1}(r)$. The quotient space $\mathbb{C}H^n(c) = H_1^{2n+1}(r)/S^1$ turns out to be a Kähler manifold of real dimension $2n$ with constant holomorphic sectional curvature c. The natural projection map $\pi\colon H_1^{2n+1}(r) \to \mathbb{C}H^n(c)$ is called the Hopf map. The complex hyperbolic space $\mathbb{C}H^n(c)$ inherits its metric by requiring the Hopf map to be a semi-Riemannian submersion with timelike totally geodesic fibers. We denote by $\tilde{\nabla}$ and $\bar{\nabla}$ the Levi-Civita connections of $H_1^{2n+1}(r)$ and $\mathbb{C}H^n(c)$, respectively. From now on, we will drop r and c in the notations of the anti-De Sitter spacetime and the complex hyperbolic space.

Let V denote the vector field on H_1^{2n+1} defined by $V_z = i\sqrt{-c}\, z/2$ for each $z \in H_1^{2n+1}$. This is a unit timelike vector field that is tangent to the S^1-flow. Now, we have the isomorphism

$$T_z H_1^{2n+1} \cong T_{\pi(z)}\mathbb{C}H^n \oplus \mathbb{R}V_z,$$

and $\ker \pi_{*z} = \mathbb{R}V_z$. Vectors in $\ker \pi_*$ are called vertical, and vectors orthogonal to $\ker \pi_*$ are called horizontal. The subspace of horizontal vectors of $T_z H_1^{2n+1}$ is a spacelike complex vector subspace of $\mathbb{C}^{1,n}$; this induces a Kähler structure on $\mathbb{C}H^n$ which we denote by J. Note that, for a vector field $X \in \Gamma(T\mathbb{C}H^n)$ there is a unique horizontal vector field $X^L \in \Gamma(TH_1^{2n+1})$, the horizontal lift of X, such that $\pi_* X^L = X$.

Since π is a semi-Riemannian submersion, the fundamental equations of a semi-Riemannian submersion [15] relate the Levi-Civita connections of H_1^{2n+1} and $\mathbb{C}H^n$ as

$$\begin{aligned} \tilde{\nabla}_{X^L} Y^L &= (\bar{\nabla}_X Y)^L + \frac{\sqrt{-c}}{2} \langle JX^L, Y^L \rangle V, \\ \tilde{\nabla}_V X^L &= \tilde{\nabla}_{X^L} V = \frac{\sqrt{-c}}{2} (JX)^L = \frac{\sqrt{-c}}{2} JX^L, \end{aligned} \tag{1}$$

for all $X, Y \in \Gamma(T\mathbb{C}H^n)$.

Now let M be a real hypersurface in $\mathbb{C}H^n$ and denote by ξ a (local) unit normal vector field to M. Then, $\tilde{M} = \pi^{-1}(M)$ is a hypersurface in H_1^{2n+1} that is invariant under the S^1-action, and ξ^L is a (local) spacelike normal unit vector field to $\tilde{M}$. We denote by ∇ the Levi-Civita connection on M or $\tilde{M}$, as there will not be a chance for confusion. We also denote by S and $\tilde{S}$ the shape operators of M and $\tilde{M}$, respectively. Since M is Riemannian, S_p is diagonalizable for each p. The eigenvalues of S_p are called the principal curvatures of M at p. The sum of the eigenvalues of S_p, which is also the trace of S_p, is called the mean curvature at p. We denote by $g(p)$ the number of principal curvatures of M at p.

Recall that M is said to be *Hopf* at $p \in M$ if $J\xi_p$ is an eigenvector of S_p; M is said to be Hopf if it is Hopf at all points $p \in M$. We also denote by $h(p)$ the number of nontrivial projections of $J\xi_p$ onto the principal curvature spaces. Thus, M is Hopf at p if and only if $h(p) = 1$.

The Gauss and Weingarten formulas for $\tilde{M}$ are

$$\tilde{\nabla}_X Y = \nabla_X Y + \langle \tilde{S}X, Y \rangle \xi^L, \qquad \tilde{\nabla}_X \xi^L = -\tilde{S}X.$$

Thus, (1) implies

$$\tilde{S}X^L = (SX)^L + \frac{\sqrt{-c}}{2}\langle J\xi^L, X^L \rangle V, \qquad \tilde{S}V = -\frac{\sqrt{-c}}{2} J\xi^L. \tag{2}$$

Hence, if $\lambda_1, \dots, \lambda_{2n-1}$ are the principal curvatures of M, then (2) implies that, with respect to a suitable basis of TH_1^{2n+1}, the endomorphism $\tilde{S}$ can be represented by the matrix

$$\begin{pmatrix} \lambda_1 & & 0 & -\frac{b_1\sqrt{-c}}{2} \\ & \ddots & & \vdots \\ 0 & & \lambda_{2n-1} & -\frac{b_{2n-1}\sqrt{-c}}{2} \\ \frac{b_1\sqrt{-c}}{2} & \dots & \frac{b_{2n-1}\sqrt{-c}}{2} & 0 \end{pmatrix}, \tag{3}$$

where $b_i = \langle J\xi, X_i \rangle \circ \pi$, $i \in \{1, \dots, 2n-1\}$, are S^1-invariant functions on (an open set of) $\tilde{M}$. In particular, it follows that M and $\tilde{M}$ have the same mean curvature.

3 Isoparametric Hypersurfaces

We say that a hypersurface M of $\mathbb{C}H^n$ is *isoparametric* if all sufficiently close parallel hypersurfaces have constant mean curvature. We take $\tilde{M} = \pi^{-1}(M)$, which is a Lorentzian hypersurface of anti-De Sitter spacetime, and note that, since it is a semi-Riemannian submersion, π maps parallel hypersurfaces of $\tilde{M}$ to parallel hypersurfaces of M. As we have seen above, a hypersurface and its lift have the same mean curvature, and thus, if M is isoparametric, parallel hypersurfaces to $\tilde{M}$ have constant mean curvature. It follows from the work of Hahn [10] that $\tilde{M}$ has constant principal curvatures with constant algebraic multiplicities. However, it is important to point out that M does not necessarily have constant principal curvatures. Even more, the functions g and h do not have to be constant.

The rest of this paper is devoted to proving Theorem 1.3.

The shape operator $\tilde{S}_q$ at a point $q \in \tilde{M}$ is a self-adjoint endomorphism of $T_q\tilde{M}$. Since $\tilde{M}$ is Lorentzian, $\tilde{S}$ is not necessarily diagonalizable, but it is known to have one of the following Jordan canonical forms (see for example [16, Chapter 9]):

$$\text{I.}\begin{pmatrix}\lambda_1 & & 0\\ & \ddots & \\ 0 & & \lambda_{2n}\end{pmatrix} \qquad \text{II.}\begin{pmatrix}\lambda_1 & 0 & & & \\ \varepsilon & \lambda_1 & & & \\ & & \lambda_2 & & \\ & & & \ddots & \\ & & & & \lambda_{2n-1}\end{pmatrix},\ \varepsilon = \pm 1$$

$$\text{III.}\begin{pmatrix}\lambda_1 & 0 & 1 & & & \\ 0 & \lambda_1 & 0 & & & \\ 0 & 1 & \lambda_1 & & & \\ & & & \lambda_2 & & \\ & & & & \ddots & \\ & & & & & \lambda_{2n-2}\end{pmatrix} \qquad \text{IV.}\begin{pmatrix}a & -b & & & \\ b & a & & & \\ & & \lambda_3 & & \\ & & & \ddots & \\ & & & & \lambda_{2n}\end{pmatrix}$$

The eigenvalues $\lambda_i \in \mathbb{R}$ can be repeated and, in case IV we have $\lambda_1 = a + ib, \lambda_2 = a - ib$ $(b \neq 0)$. In cases I and IV, the basis with respect to which $\tilde{S}_q$ is represented is orthonormal, and the first vector of this basis is timelike. In cases II and III the basis is semi-null. A basis $\{u, v, e_1, \ldots, e_{m-2}\}$ is semi-null if all inner products are zero except $\langle u, v\rangle = \langle e_i, e_i\rangle = 1$, for each $i \in \{1, \ldots, m-2\}$. A point $q \in \tilde{M}$ is said to be of type I, II, III or IV according to the type of the Jordan canonical form of $\tilde{S}_q$.

In his work on isoparametric hypersurfaces in spaces of constant curvature [4], Cartan proved a fundamental formula relating the curvature of the ambient manifold and the principal curvatures. A similar argument works for the anti-De Sitter spacetime. In particular, the following consequence can be derived from this fundamental formula [7, Lemma 3.4]:

Lemma 3.1 *Let $q \in \tilde{M}$ be a point of type I, II or III. Then the number $\tilde{g}(q)$ of constant principal curvatures at q satisfies $\tilde{g}(q) \in \{1, 2\}$. Moreover, if $\tilde{g}(q) = 2$ and the principal curvatures are λ and μ, then $c + 4\lambda\mu = 0$.*

The objective of this paper is to analyze the eigenvalue structure of the shape operator of an isoparametric hypersurface in $\mathbb{C}H^n$ and obtain, as a consequence of this study, Theorem 1.3. As a corollary, we derive a bound for h and g (Proposition 1.4). The proof of these facts will be mostly algebraic, and is carried out by analyzing the possible Jordan canonical forms for the shape operator of $\tilde{M}$ at a point q as described above.

As stated in the introduction, we only deal with Type III points. For points of Types I, II and IV the proof is very similar and can be found in [7, Section 3].

Proposition 3.2 *Let $\tilde{M}$ be the lift of an isoparametric hypersurface in $\mathbb{C}H^n$ to the anti-De Sitter spacetime, and let $q \in \tilde{M}$ and $p = \pi(q)$. Then:*

(i) *If q is of type I, then M is Hopf at p, and $g(p) \in \{2, 3\}$. The principal curvatures of M at p are:*

$$\lambda \in \Big(-\frac{\sqrt{-c}}{2}, \frac{\sqrt{-c}}{2}\Big), \lambda \neq 0, \quad \mu = -\frac{c}{4\lambda} \in \Big(-\infty, -\frac{\sqrt{-c}}{2}\Big) \cup \Big(\frac{\sqrt{-c}}{2}, \infty\Big), \quad \lambda + \mu.$$

The last principal curvature has multiplicity one and corresponds to the Hopf vector.

(ii) *If q is of type II, then M is Hopf at p, and $g(p) = 2$. Moreover, $\tilde{M}$ has one principal curvature $\lambda = \pm\sqrt{-c}/2$, and the principal curvatures of M at p are λ and 2λ. The second one has multiplicity one and corresponds to the Hopf vector.*

(iii) *If q is of type IV, then M is Hopf at p. Let λ and $\mu = -c/(4\lambda)$ be the real principal curvatures of $\tilde{M}$ at q (μ might not exist). Then the principal curvatures of M at p are*

$$\lambda,\ \mu,\ \text{and}\ 2a = \frac{4c\lambda}{c - 4\lambda^2} \in \left(-\sqrt{-c}, \sqrt{-c}\right),$$

where $2a$ is the principal curvature associated with the Hopf vector.

Remark 3.3 Proposition 3.2 implies Theorem 1.3 for types I, II and IV.

Indeed, the values given in part (i) correspond to the principal curvatures of a tube of radius r around a totally geodesic $\mathbb{C}H^k$ in $\mathbb{C}H^n$, where

$$\lambda = \frac{\sqrt{-c}}{2}\tanh\left(\frac{r\sqrt{-c}}{2}\right), \tag{4}$$

and $2(n-k)$ is the multiplicity of μ.

The values obtained in (ii) correspond to the principal curvatures of a horosphere in $\mathbb{C}H^n$.

Finally, the values obtained in (iii) correspond to the principal curvatures of a tube of radius r around a totally geodesic $\mathbb{R}H^n$ in $\mathbb{C}H^n$, where r is given by the same formula as in (4).

In the rest of the paper we deal with Type III points. The arguments that follow are not contained in [7]. Thus, let M be an isoparametric hypersurface of $\mathbb{C}H^n$, whose lift to the anti-De Sitter spacetime is denoted by $\tilde{M}$. We fix a point $q \in \tilde{M}$ and assume that q is of Type III. We analyze the possible principal curvatures of M at the point $p = \pi(q)$.

Theorem 3.4 *Let λ be the principal curvature of $\tilde{M}$ at q whose algebraic and geometric multiplicities do not coincide. Then $h(p) \in \{2, 3\}$ and $\lambda \in \left(-\sqrt{-c}/2, \sqrt{-c}/2\right)$.*

There exists a number $\varphi \in (0, \pi/2]$ such that the zeroes of the polynomial

$$f_{\lambda,\varphi}(x) = -x^3 + \left(-\frac{c}{4\lambda} + 3\lambda\right)x^2 + \frac{1}{2}\left(c - 6\lambda^2\right)x + \frac{-c^2 - 16c\lambda^2 + 16\lambda^4 + (c + 4\lambda^2)^2\cos(2\varphi)}{32\lambda},$$

are principal curvatures of M at p. If $\varphi = \pi/2$, then $h(p) = 2$ and $g(p) \in \{2, 3, 4\}$. Moreover, we have the following possibilities:

(i) *If $\varphi = \pi/2$ and $g = 4$, then $0 \neq \lambda \neq \pm\sqrt{-c}/(2\sqrt{3})$, and the principal curvatures of M at p are:*

$$\frac{1}{2}\left(3\lambda \pm \sqrt{-c-3\lambda^2}\right), \qquad \lambda, \qquad \mu = -\frac{c}{4\lambda}.$$

The principal curvature spaces corresponding to the first two principal curvatures are one dimensional and the Hopf vector has nontrivial projection onto both of them.

(ii) *If* $\varphi = \pi/2$ *and* $g \in \{2, 3\}$ *then we have two cases:*

(a) *If* $\lambda = \pm\sqrt{-c}/(2\sqrt{3})$ *then the principal curvatures of* M *at* p *are*

$$0, \qquad \mu = -\frac{c}{4\lambda} = \pm\frac{\sqrt{-3c}}{2}, \qquad \lambda = \pm\frac{\sqrt{-c}}{2\sqrt{3}}.$$

The principal curvature space associated with 0 *is one dimensional, and the Hopf vector has nontrivial projection onto the principal curvature spaces corresponding to the first two principal curvatures. The value* λ *might not appear as a principal curvature.*

(b) *If* $0 \neq \lambda \neq \pm\sqrt{-c}/(2\sqrt{3})$, *then the principal curvatures of* M *at* p *are*

$$\frac{1}{2}\left(3\lambda \pm \sqrt{-c-3\lambda^2}\right), \qquad \lambda \text{ or } \mu = -\frac{c}{4\lambda}.$$

The principal curvature spaces corresponding to the first two principal curvatures are one dimensional and the Hopf vector has nontrivial projection onto both of them.

(iii) *If* $\varphi \in (0, \pi/2)$, *then* $\lambda \neq 0$ *and the three zeros of the polynomial* $f_{\lambda,\varphi}$ *are different, and also different from* λ *and* $-c/(4\lambda)$. *Therefore,* M *has* $g(p) \in \{3, 4, 5\}$ *principal curvatures at* p:

$$\text{the zeroes of } f_{\lambda,\varphi}, \qquad \lambda, \qquad \mu = -\frac{c}{4\lambda}.$$

The principal curvature spaces corresponding to the first three principal curvatures are one dimensional and the Hopf vector has nontrivial projection onto all of them. The values λ *and/or* μ *might not appear as principal curvatures.*

Proof For the sake of readability we will shorten the notation and write $v = V_q$. We also write $J\xi$ instead of $J\xi_p$, $\tilde{S}$ instead of $\tilde{S}_q$ and so on.

Assume that the shape operator $\tilde{S}$ has a type III matrix expression at q with respect to a semi-null basis $\{e_1, e_2, e_3 \ldots, e_{2n}\}$, where

$$\begin{gathered} \langle e_1, e_1\rangle = \langle e_2, e_2\rangle = \langle e_1, e_3\rangle = \langle e_2, e_3\rangle = 0, \qquad \langle e_1, e_2\rangle = \langle e_3, e_3\rangle = 1, \\ \tilde{S}e_1 = \lambda e_1, \qquad \tilde{S}e_2 = \lambda e_2 + e_3, \qquad \tilde{S}e_3 = e_1 + \lambda e_3. \end{gathered} \tag{5}$$

We denote by $T_\lambda(q)$ and $T_\mu(q)$ the eigenspaces of λ and μ at q. Then $T_\lambda(q) \ominus \mathbb{R}e_2$ and $T_\mu(q)$ are spacelike. As a matter of caution, $e_2 \notin T_\lambda(q)$, and $T_\lambda(q) \ominus \mathbb{R}e_2$ denotes the vectors of $T_\lambda(q)$ that are orthogonal to e_2. For example, $e_1 \notin T_\lambda(q) \ominus \mathbb{R}e_2$ because e_1 and e_2 are not orthogonal.

Assume first that there are two distinct principal curvatures λ, μ. By Lemma 3.1 we have $c + 4\lambda\mu = 0$ and thus, $\lambda, \mu \neq 0$. We can write $v = r_1 e_1 + r_2 e_2 + r_3 e_3 + u + w$, where $u \in T_\lambda \ominus \mathbb{R}e_2$, and $w \in T_\mu(q)$. Changing the orientation of $\{e_1, e_2, e_3\}$ if necessary, we can also assume $r_2 \geq 0$. We have

$$-1 = \langle v, v\rangle = 2r_1 r_2 + r_3^2 + \langle u, u\rangle + \langle w, w\rangle.$$

Thus, $r_2 > 0$ and $r_1 < 0$. If $u \neq 0$ we define

$$e_1' = e_1, \qquad e_2' = -\frac{\langle u, u\rangle}{2r_2^2} e_1 + e_2 + \frac{1}{r_2} u, \qquad e_3' = e_3.$$

Then, the vectors in $\{e_1', e_2', e_3'\}$ satisfy the same equations as in (5), and $v = (r_1 + \langle u, u\rangle/(2r_2))e_1' + r_2 e_2' + r_3 e_3' + w$. This shows that we can assume, swapping to $\{e_1', e_2', e_3'\}$ if necessary, that $u = 0$.

Thus, we have

$$\begin{aligned} -1 &= \langle v, v\rangle = 2r_1 r_2 + r_3^2 + \langle w, w\rangle, \\ \tilde{S}v &= (r_1\lambda + r_3)e_1 + r_2\lambda e_2 + (r_2 + r_3\lambda)e_3 + \mu w. \end{aligned}$$

Using (2) we get

$$J\xi^L = -\frac{2}{\sqrt{-c}}\tilde{S}v = -\frac{2}{\sqrt{-c}}\Big((r_1\lambda + r_3)e_1 + r_2\lambda e_2 + (r_2 + r_3\lambda)e_3 + \mu w\Big),$$

and since $2r_1 r_2 = -1 - r_3^2 - \langle w, w\rangle$ we obtain

$$\begin{aligned} 1 = \langle J\xi^L, J\xi^L\rangle &= -\frac{4}{c}\left(2r_1 r_2\lambda^2 + 4r_2 r_3\lambda + r_2^2 + r_3^2\lambda^2 + \langle w, w\rangle\mu^2\right) \\ &= -\frac{4}{c}\left(4r_2 r_3\lambda + r_2^2 - \lambda^2 + (\mu^2 - \lambda^2)\langle w, w\rangle\right), \end{aligned}$$

$$0 = \langle \tilde{S}v, v\rangle = 2r_1 r_2\lambda + 2r_2 r_3 + r_3^2\lambda + \mu\langle w, w\rangle = 2r_2 r_3 - \lambda + (\mu - \lambda)\langle w, w\rangle.$$

Hence, we get

$$r_2^2 + (\mu - \lambda)^2\langle w, w\rangle = -\frac{c}{4} - \lambda^2,$$

or equivalently,

$$\left(\frac{2r_2}{\sqrt{-c - 4\lambda^2}}\right)^2 + \left(\frac{2(\mu - \lambda)\|w\|}{\sqrt{-c - 4\lambda^2}}\right)^2 = 1.$$

Since $r_2 > 0$ we obtain $\lambda \in \left(-\sqrt{-c}/2, \sqrt{-c}/2\right) \setminus \{0\}$. Note that, since $c + 4\lambda\mu = 0$ we have $-c - 4\lambda^2 = 4\lambda(\mu - \lambda)$. Solving the previous equations yields

$$r_2 = \sin(\varphi)\frac{\sqrt{-c-4\lambda^2}}{2}, \quad \|w\| = \cos(\varphi)\frac{2\lambda}{\sqrt{-c-4\lambda^2}}, \quad r_3 = \frac{\lambda}{\sqrt{-c-4\lambda^2}}\sin(\varphi), \quad (6)$$

for a suitable $\varphi \in \left(0, \pi/2\right]$. The proof now diverges from the one that can be found in [7].

Assume $\varphi \neq \pi/2$, that is, $w \neq 0$. We have that the vectors in $T_\lambda(q) \ominus \mathbb{R}e_2$ and in $T_\mu(q) \ominus \mathbb{R}w$ are orthogonal to v and $J\xi^L$. These vectors project bijectively, via the Hopf map π_{*q}, to eigenvectors of the principal curvatures λ and μ respectively, and they are all orthogonal to $J\xi$. Let $L = T_q\tilde{M} \ominus \left((T_\lambda(q) \ominus \mathbb{R}e_2) \oplus (T_\mu(q) \ominus \mathbb{R}w) \oplus \mathbb{R}v\right)$. Then, L is a 3-dimensional space, and thus, $h(p) \leq 3$. Furthermore, by (3) we see that $h(p) \neq 1$; otherwise $\tilde{S}$ would contain at most a 2×2 nondiagonal block, and so q would not be of type III. In fact, L is spanned by the following basis: $l_1 = r_1e_1 - r_2e_2$, $l_2 = r_3e_1 - r_2e_3$ and $l_3 = -\langle w, w\rangle e_1 + r_2w$. We have $\operatorname{span}\{e_1, e_2, e_3, w\} = L \oplus \mathbb{R}v$. After some long calculations, and using (2) and $\pi_{*q}v = 0$, we get that the matrix expression of the shape operator of M at p restricted to $\pi_{*q}L$, with respect to the basis $\{\pi_*l_1, \pi_*l_2, \pi_*l_3\}$ is

$$\begin{pmatrix} \lambda + r_2r_3 & r_2^2 & r_2(\lambda-\mu)\langle w, w\rangle \\ 1 + r_3^2 & \lambda + r_2r_3 & r_3(\lambda-\mu)\langle w, w\rangle \\ -r_3 & -r_2 & \mu - (\lambda-\mu)\langle w, w\rangle \end{pmatrix}.$$

Using the expressions we got for r_2, r_3, and $\langle w, w\rangle$, together with $4\lambda\mu + c = 0$, we can calculate the characteristic polynomial of the previous matrix. This polynomial turns out to be precisely $f_{\lambda,\varphi}$, as defined in the statement of Theorem 3.4. This is the same characteristic polynomial as that of the nontrivial part of the shape operator of a tube around the submanifolds $W_{\mathfrak{w}}$ in Theorem 1.1 (vi) (see also [6]). We have

$$f_{\lambda,\varphi}(\lambda) = -\frac{(c+4\lambda^2)^2\sin^2(\varphi)}{16\lambda} > 0, \quad f_{\lambda,\varphi}(\mu) = \frac{(c+4\lambda^2)^2\cos^2(\varphi)}{16\lambda} > 0.$$

Therefore, neither λ nor μ are eigenvalues of the matrix above. Moreover, the same argument as in [2, p. 146] proves that the three zeroes of $f_{\lambda,\varphi}$ are different. Hence, if $\varphi \in (0, \pi/2)$, M has $g(p) \in \{3, 4, 5\}$ principal curvatures at p: the zeroes of $f_{\lambda,\varphi}$, possibly λ, and possibly μ. Indeed, $g(p) = 3$ if $T_\lambda(q) \ominus \mathbb{R}e_2 = T_\mu(q) \ominus \mathbb{R}w = 0$, $g(p) = 4$ if either $T_\lambda(q) = \mathbb{R}e_1$ or $T_\mu(q) = \mathbb{R}w$, and $g(p) = 5$ otherwise.

We now prove that, in this case ($\varphi \neq \pi/2$), we have $h(p) = 3$. The characteristic polynomial of the shape operator $\tilde{S}$ restricted to $L \oplus \mathbb{R}v$ is $(x-\lambda)^3(x-\mu)$. Define x_1, x_2, x_3 to be unit eigenvectors of S_p whose corresponding eigenvalues are the three different zeroes λ_1, λ_2, λ_3 of the polynomial $f_{\lambda,\varphi}$, respectively. Set $b_i = \langle J\xi, x_i\rangle$, for $i = 1, 2, 3$. Then, according to (3), the shape operator $\tilde{S}$ of $\tilde{M}$ at q restricted to $L \oplus \mathbb{R}v$ with respect to the basis $\{x_1^L, x_2^L, x_3^L, v\}$ is given by

$$\begin{pmatrix} \lambda_1 & 0 & 0 & -b_1\frac{\sqrt{-c}}{2} \\ 0 & \lambda_2 & 0 & -b_2\frac{\sqrt{-c}}{2} \\ 0 & 0 & \lambda_3 & -b_3\frac{\sqrt{-c}}{2} \\ b_1\frac{\sqrt{-c}}{2} & b_2\frac{\sqrt{-c}}{2} & b_3\frac{\sqrt{-c}}{2} & 0 \end{pmatrix}.$$

Using $b_1^2 + b_2^2 + b_3^2 = 1$, we get the characteristic polynomial of this matrix:

$$\begin{aligned} & x^4 + (-\lambda_1 - \lambda_2 - \lambda_3)\, x^3 \\ & \quad + \frac{1}{4}(-c + 4\lambda_1\lambda_2 + 4\lambda_1\lambda_3 + 4\lambda_2\lambda_3)\, x^2 \\ & \quad + \frac{1}{4}\left(b_1^2 c\lambda_2 + b_1^2 c\lambda_3 + b_2^2 c\lambda_1 + b_3^2 c\lambda_1 + b_3^2 c\lambda_2 + b_2^2 c\lambda_3 - 4\lambda_1\lambda_2\lambda_3\right) x \\ & \quad - \frac{c}{4}\left(b_1^2\lambda_2\lambda_3 + b_3^2\lambda_1\lambda_2 + b_2^2\lambda_1\lambda_3\right). \end{aligned}$$

Both the previous polynomial and $(x-\lambda)^3(x-\mu)$ must coincide, as they come from the same endomorphism of $L \oplus \mathbb{R}v$. Thus, by comparing the linear and independent terms of these polynomials, we obtain the following linear system in the variables b_1^2, b_2^2, b_3^2:

$$\begin{aligned} \frac{c}{4}\left((\lambda_2+\lambda_3)b_1^2 + (\lambda_1+\lambda_3)b_2^2 + (\lambda_1+\lambda_2)b_3^2\right) - \lambda_1\lambda_2\lambda_3 &= -\lambda^2(\lambda + 3\mu), \\ -\frac{c}{4}\left(b_1^2\lambda_2\lambda_3 + b_3^2\lambda_1\lambda_2 + b_2^2\lambda_1\lambda_3\right) &= \lambda^3\mu, \\ b_1^2 + b_2^2 + b_3^2 &= 1. \end{aligned}$$

The determinant of the matrix of this linear system is $c^2(\lambda_1-\lambda_2)(\lambda_3-\lambda_1)(\lambda_2-\lambda_3)/16 \neq 0$, so the system has a unique solution. Using the relations among λ, μ, λ_1, λ_2 and λ_3 that the equality of the characteristic polynomials imposes for the quadratic and cubic terms, namely,

$$\begin{aligned} \lambda_1 + \lambda_2 + \lambda_3 &= 3\lambda + \mu, \\ -\frac{c}{4} + \lambda_1\lambda_2 + \lambda_1\lambda_3 + \lambda_2\lambda_3 &= 3\lambda(\lambda + \mu), \end{aligned}$$

one can check, after some elementary but long calculations, that the solution to the linear system above is given by:

$$b_i^2 = -\frac{4(\lambda-\lambda_i)^3(\lambda_i-\mu)}{c(\lambda_{i+1}-\lambda_i)(\lambda_i-\lambda_{i+2})}, \qquad i = 1,\,2,\,3, \quad \text{(indices modulo 3)}.$$

Since μ and λ are different from any λ_i, $i \in \{1,2,3\}$, we conclude that $b_i \neq 0$ for all $i \in \{1,2,3\}$, whence $h(p) = 3$. This finishes the proof of Theorem 3.4 (iii).

Now assume $\varphi = \pi/2$, that is, $w = 0$ (recall that we are still assuming that $\tilde{S}$ has two distinct eigenvalues λ, $\mu \neq 0$ at q). In this case, (6) yields

$$r_2 = \frac{\sqrt{-c-4\lambda^2}}{2}, \qquad w = 0, \qquad r_3 = \frac{\lambda}{\sqrt{-c-4\lambda^2}}.$$

Then, the vectors of $T_\lambda(q) \ominus \mathbb{R}e_2$ and $T_\mu(q)$ are orthogonal to v and $J\xi^L$, project via π_{*q} onto the principal curvature spaces of λ and μ respectively, and these projections are orthogonal to $J\xi$. So, in this case, we have $h(p) = 2$. Defining l_1 and l_2 as above, the shape operator of M at p restricted to span$\{\pi_* l_1, \pi_* l_2\}$, with respect to the basis $\{\pi_* l_1, \pi_* l_2\}$, turns out to be

$$\begin{pmatrix} \lambda + r_2 r_3 & r_2^2 \\ 1 + r_3^2 & \lambda + r_2 r_3 \end{pmatrix} = \begin{pmatrix} \frac{3\lambda}{2} & -\frac{c}{4} - \lambda^2 \\ \frac{c+3\lambda^2}{c+4\lambda^2} & \frac{3\lambda}{2} \end{pmatrix}.$$

Thus, the eigenvalues of the shape operator of M at p restricted to span$\{\pi_* l_1, \pi_* l_2\}$ are

$$\frac{1}{2}\left(3\lambda \pm \sqrt{-c-3\lambda^2}\right).$$

These eigenvalues are different and also different from λ.

For $\lambda = \sqrt{-c}/(2\sqrt{3})$ we have $\mu = \left(3\lambda + \sqrt{-c-3\lambda^2}\right)/2 = \sqrt{-3c}/2$, hence $g(p) \in \{2, 3\}$, and the principal curvatures are 0 (with multiplicity one), $\sqrt{-3c}/2$, and possibly $\sqrt{-c}/(2\sqrt{3})$. The possibility $g(p) = 2$ arises if $T_\lambda(q) = \mathbb{R}e_1$, and in this case λ is not a principal curvature of M at p. We have $g(p) = 3$ otherwise. The Hopf vector has nontrivial projections onto the principal curvature spaces corresponding to 0 and μ. This corresponds to Theorem 3.4 (iia).

For $\lambda \neq \sqrt{-c}/(2\sqrt{3})$ we get $g(p) \in \{3, 4\}$. We have $g(p) = 3$ if $T_\lambda(q) = \mathbb{R}e_1$, that is, if λ is not a principal curvature of M at p, and $g(p) = 4$ otherwise. The principal curvatures $(3\lambda \pm \sqrt{-c-3\lambda^2})/2$ have both multiplicity one, and the Hopf vector has nontrivial projection onto their corresponding principal curvature spaces. This corresponds to case (i) if $g(p) = 4$ and to case (iib) if $g(p) = 3$. This finishes the proof if $\tilde{S}$ has two distinct principal curvatures λ and μ.

Finally, assume that $\tilde{M}$ has just one principal curvature $\lambda \geq 0$ at q. In this case, calculations are very similar to what we have just obtained if $w = 0$. Thus, we get

$$\lambda \in \left(-\frac{\sqrt{-c}}{2}, \frac{\sqrt{-c}}{2}\right), \qquad r_2 = \frac{\sqrt{-c-4\lambda^2}}{2}, \qquad r_3 = \frac{\lambda}{\sqrt{-c-4\lambda^2}}.$$

Arguing as in the case $\varphi = \pi/2$ above, we obtain $h(p) = 2$ and $g(p) = 3$ (for dimension reasons $T_\lambda(q) = \mathbb{R}e_1$ cannot happen now). The principal curvatures of M at p are $\left(3\lambda \pm \sqrt{-c-3\lambda^2}\right)/2$ and λ. The first two have multiplicity one and the Hopf vector has nontrivial projection onto their corresponding principal curvature spaces. Now we can have $\lambda = 0$, and then, the other principal curvatures would be $\pm\sqrt{-c}/2$. If $\lambda \neq \sqrt{-c}/(2\sqrt{3})$, this corresponds to case (iia) again. If $\lambda = \sqrt{-c}/(2\sqrt{3})$, then we also get case (iib), although now $\sqrt{-3c}/2$ has multiplicity one and $\lambda = \sqrt{-c}/(2\sqrt{3})$ is definitely a principal curvature of M at p. □

Remark 3.5 Theorem 3.4 implies that, for points of Type III, the principal curvatures of isoparametric hypersurfaces in $\mathbb{C}H^n$ and their multiplicities must coincide (at that precise point) with those of the homogeneous examples in cases (iv) and (v) in Theorem 1.1, except for some particular cases which we would like to point out here (we assume the notation given in the proof of Theorem 3.4):

A. Theorem 3.4(iia) for $g(p) = h(p) = 2$: this happens if $T_\lambda(p) = \mathbb{R}e_1$.
B. Theorem 3.4(iib) if λ is not a principal curvature of M at p, that is, if $T_\lambda(q) = \mathbb{R}e_1$.
C. Theorem 3.4(iii) for $g(p) = h(p) = 3$ (this happens whenever $T_\lambda(q) \ominus \mathbb{R}e_2 = T_\mu(q) \ominus \mathbb{R}w = 0$) or if $g(p) = 4$ and λ is not a principal curvature of M at p (equivalently, if $T_\lambda(q) = \mathbb{R}e_1$).

The three cases in Remark 3.5 are ruled out by a different method in [7]. Here we content ourselves with Theorem 3.4 which, together with Proposition 3.2, implies Theorem 1.3. The multiplicities of these principal curvatures are the same as in the homogeneous examples except for the three possibilities above. This stronger result is a consequence of Theorem 1.1 but cannot be proved with the method used in this paper. Proposition 1.4 is also obtained from Theorem 3.4.

Acknowledgements The authors have been supported by projects MTM2016-75897-P (AEI/FEDER, UE), ED431F 2017/03, GRC2013-045 and MTM2013-41335-P with FEDER funds (Spain). The second author has received funding from a Juan de la Cierva fellowship (Spain), from the ICMAT Severo Ochoa project SEV-2015-0554 (MINECO, Spain), and from the European Union's Horizon 2020 research and innovation programme under the Marie Skłodowska-Curie grant agreement No. 745722. The third author has been supported by an FPU fellowship and by Fundación Barrié de la Maza (Spain).

References

1. J. Berndt, M. Brück, Cohomogeneity one actions on hyperbolic spaces. J. Reine Angew. Math. **541**, 209–235 (2001)
2. J. Berndt, J.C. Díaz-Ramos, Homogeneous hypersurfaces in complex hyperbolic spaces. Geom. Dedicata **138**, 129–150 (2009)
3. J. Berndt, H. Tamaru, Cohomogeneity one actions on noncompact symmetric spaces of rank one. Trans. Amer. Math. Soc. **359**, 3425–3438 (2007)
4. É. Cartan, Familles de surfaces isoparamétriques dans les espaces à courbure constante. Ann. Mat. Pura Appl. IV. Ser. **17**, 177–191 (1938)
5. Q.S. Chi, Isoparametric hypersurfaces with four principal curvatures III. J. Differ. Geom. **94**, 469–504 (2013)
6. J.C. Díaz-Ramos, M. Domínguez-Vázquez, Inhomogeneous isoparametric hypersurfaces in complex hyperbolic spaces. Math. Z. **271**, 1037–1042 (2012)
7. J.C. Díaz-Ramos, M. Domínguez-Vázquez, V. Sanmartín-López, Isoparametric hypersurfaces in complex hyperbolic spaces. Adv. Math. **314**, 756–805 (2017)

8. M. Domínguez-Vázquez, Isoparametric foliations on complex projective spaces. Trans. Amer. Math. Soc. **368**, 1211–1249 (2016)
9. D. Ferus, H. Karcher, H.F. Münzner, Cliffordalgebren und neue isoparametrische Hyperflächen. Math. Z. **177**, 479–502 (1981)
10. J. Hahn, Isoparametric hypersurfaces in the pseudo-Riemannian space forms. Math. Z. **187**, 195–208 (1984)
11. W.-Y. Hsiang, H.B. Lawson, Minimal submanifolds of low cohomogeneity. J. Differ. Geom. **5**, 1–38 (1971)
12. A. Kollross, A classification of hyperpolar and cohomogeneity one actions. Trans. Amer. Math. Soc. **354**(2), 571–612 (2002)
13. M. Lohnherr, H. Reckziegel, On ruled real hypersurfaces in complex space forms. Geom. Dedicata **74**, 267–286 (1999)
14. R. Miyaoka, Isoparametric hypersurfaces with $(g, m) = (6, 2)$. Ann. Math. (2) **177**, 53–110 (2013)
15. B. O'Neill, The fundamental equations of a submersion. Michigan Math. J. **13**, 459–469 (1966)
16. B. O'Neill, *Semi-Riemannian geometry with applications to relativity*, Academic Press, (1983)
17. B. Segre, Famiglie di ipersuperficie isoparametriche negli spazi euclidei ad un qualunque numero di dimensioni. Atti Accad. Naz. Lincei Rend. Cl. Sci. Fis. Mat. Natur. (6) **27**, 203–207 (1938)
18. R. Takagi, On homogeneous real hypersurfaces in a complex projective space. Osaka J. Math. **10**, 495–506 (1973)

Future Completion of a Spacetime and Standard Causal Constructions

Stacey (Steven) G. Harris

Abstract The future completion of a spacetime, equipped with the future chronological topology, is ripe for consideration of constructions and properties typically expressed for a spacetime. Joining elements with a causal curve is problematic, but some progress is reported. We look at trying to do global hyperbolicity in the future completion $\hat{M}$ of M; this works well if M has a compact Cauchy surface, but otherwise, not so well.

Keywords Causal boundary · Topology of causal boundary · Cauchy surface Standard static spacetime

1 Introduction

The intent of this note is to examine the extent to which various causal constructions and properties for spacetimes can be applied to the future completion of a strongly causal spacetime M, that is to say, to $\hat{M} = M \cup \hat{\partial}(M)$, M together with its future-causal boundary, considered as a topological space using the future chronological topology.

Although there is good reason to want to consider M together with its full causal boundary—future- and past-causal boundaries combined in some way, such as with the Szabados relation or with the technique (also employing the Szabados relation) found in [3]—looking at M together with just the future-causal boundary has several advantages: It is simpler, it has known categorical naturalness and universality, and the future chronological topology provides the remarkable property of future-quasi-compactness (to be detailed below). Sufficient success in this examination may lead to expansion of this examination to M with the fuller causal boundary.

Stacey (Steven) G. Harris (✉)
Department of Mathematics and Statistics, Saint Louis University,
St. Louis, MO 63103, USA
e-mail: harrissg@slu.edu ; stacey.harris@slu.edu

M.A. Cañadas-Pinedo et al. (eds.), *Lorentzian Geometry and Related Topics*,
Springer Proceedings in Mathematics & Statistics 211,
DOI 10.1007/978-3-319-66290-9_7

The causal boundary construction was introduced in 1972 with [4]. The categorically natural and universal nature of this construction—in terms of chronological sets (i.e., sets with something that passes for a chronology relation)—was established in 1998 in [5], though this was confined to the future-causal boundary alone (or, equivalently, the past-causal boundary alone); the amalgamation of the future- and past-causal boundaries has thus far eluded a categorical treatment. The topology of the 1972 paper has severe problems, and a far simpler topology—applicable just to the addition of the future-causal boundary—was suggested in 2000 in [6]; this is called the future chronological topology. It is the future chronological topology that will be considered here.

A related but different topology—applicable to the use of the entire causal boundary, future and past amalgamated—has been more recently advocated in [2, 3]; that is not the focus of this note, but it is alluded to in the last topic examined.

2 Definitions and Properties

The approach followed here is presented in detail in [5, 6]. The basic notion is that of a *chronological set*: a set X with a partial order $\ll$, called the *chronology relation*, satisfying a few axioms: for all x, $x \not\ll x$; for some countable subset $D \subset X$, for all $x \ll y$, there is some $d \in D$ with $x \ll d \ll y$ (we say D is *chronologically dense* in X); and for all x, there is some y with $y \ll x$ (alternatively, we could use there is some y with $y \ll x$ or $x \ll y$, if we wished to consider future and past boundaries together; but that is not what will be done in this note). Then we define *past* and *future* operators as usual, respectively $I^-(x) = \{y \mid y \ll x\}$ and $I^+(x) = \{y \mid x \ll y\}$. A *past set* is a set P which obeys $I^-[P] = P$, where $I^-[S] = \bigcup_{s \in S} I^-(s)$; then anything of the form $I^-[S]$ is a past set. A past set P is *indecomposable* iff it cannot be expressed as the union of two past sets, $P = P_1 \cup P_2$, unless $P_1 \subset P_2$ or $P_2 \subset P_1$; this is equivalent to saying that for all $x_1, x_2 \in P$, there is some $y \in P$ with both $x_1, x_2 \in I^-(y)$. An indecomposable past set is also called an IP. A *future chain* is a sequence $\{x_n\}$ satisfying, for all n, $x_n \ll x_{n+1}$; and a point x is a *future limit* of a future chain $\sigma = \{x_n\}$ if $I^-(x) = I^-[\sigma]$. Then every IP P can be expressed as the past of a future chain σ, $P = I^-[\sigma]$; and either all such future chains have a future limit, in which case P is called a PIP, or no such future chains have a future limit, in which case P is called a TIP or a *boundary* IP. Nice properties a chronological set might have: X is *past-regular* if for all x, $I^-(x)$ is an IP (hence, a PIP); X is *future-complete* if every future chain has a future limit (i.e., there are no TIPs).

It is evident that every (strongly causal) spacetime is a past-regular chronological set. What is important is that the construction of the future-causal boundary and the addition of that to a spacetime can be precisely mirrored for any chronological set, producing another chronological set. Specifically:

For any chronological set X, we define the *future-causal boundary* of X, denoted $\hat{\partial}(X)$, to be the collection of boundary IPs (or TIPs) of X; and we define the *future completion* of X, denoted $\hat{X}$, to be the set $\hat{X} = X \cup \hat{\partial}(X)$, equipped with the following partial order, $\ll_{\hat{X}}$: for $x, y \in X$ and $P, Q \in \hat{\partial}(X)$,

$$\begin{aligned} x \ll_{\hat{X}} y &\iff x \ll y \\ x \ll_{\hat{X}} Q &\iff x \in Q \\ P \ll_{\hat{X}} y &\iff \text{for some } z \ll y,\ P \subset I^-(z) \\ P \ll_{\hat{X}} Q &\iff \text{for some } z \in Q,\ P \subset I^-(z) \end{aligned}$$

Then $(\hat{X}, \ll_{\hat{X}})$ is a future-complete chronological set (Theorem 5 in [5]); and if X is past-regular, so is $\hat{X}$. With appropriate precautions, the construction of $\hat{X}$ from X is functorial (in a category of chronological sets whose morphisms preserve the chronology relation and future limits of future chains) into a subcategory of future-complete chronological sets, and it is left-adjoint to the forgetful functor (Theorem 14 in [5]); in other words, future completion is the unique natural way to imbed a chronological set into a future-complete one.

The *future chronological topology* for any chronological set was introduced in [6]; this is defined not conventionally, in terms of open sets, but in terms of a limit operator $\hat{L}$ on sequences (it is the existence of a countable chronologically dense set in X that makes it possible to use just sequences, not nets). This is most easily stated for past-regular chronological sets: For any sequence $\sigma = \{x_n\}$, $x \in \hat{L}(\sigma)$ iff

(1) for all $y \ll x$, $y \ll x_n$ for all n sufficiently large, and
(2) for any IP $P \supsetneq I^-(x)$, for some $z \in P$, $z \not\ll x_n$ for all n sufficiently large.

As is shown in [1] (Proposition 5.1), this is equivalent to the following: For any sequence of sets $\{A_n\}$ define the liminf and limsup of the sequence as, respectively, $\text{LI}(\{A_n\}) = \bigcup_{n=1}^{\infty} \bigcap_{k=n}^{\infty} A_k$ (the set of points in all but finitely many A_n) and $\text{LS}(\{A_n\}) = \bigcap_{n=1}^{\infty} \bigcup_{k=n}^{\infty} A_k$ (the set of points in infinitely many A_n); then $x \in \hat{L}(\sigma)$ iff

(1) $I^-(x) \subset \text{LI}(\{I^-(x_n)\})$ and
(2) $I^-(x)$ is maximal (in the subset relation) among all IPs in $\text{LS}(\{I^-(x_n)\})$.

Then a set $A \subset X$ is closed in the future chronological topology for X iff for all sequences σ lying in A, $\hat{L}(\sigma) \subset A$.

The future chronological topology has a number of desirable properties:

(1) Points are closed, and future limits of future chains are precisely the topological limits (Propositions 2.1 and 2.2 of [6]). The topology need not be Hausdorff, even for the future completion of a strongly causal spacetime, but this reveals interesting geometric (or physical) properties of the spacetime (Sect. 2.1 of [1]).
(2) For a strongly causal spacetime, the future chronological topology is precisely the manifold topology (Theorem 2.3 of [6]).

(3) The obvious injection $X \to \hat{X}$ is a homeomorphism onto its image, which is dense in $\hat{X}$ (Corollary 2.5 of [6]).
(4) For M a strongly causal spacetime, $\hat{\partial}(M)$ is closed in $\hat{M}$ (Proposition 2.6 of [6]).
(5) The future completion $\widehat{\mathbb{L}^n}$ of Minkowski space $\mathbb{L}^n$ is homeomorphic to the standard conformal embedding into the Einstein static space, $\mathbb{E}^n = \mathbb{L}^1 \times \mathbb{S}^{n-1}$—something notably untrue about the original topology proposed in [4] (Proposition 5.1 of [6]).
(6) Restricted to past-distinguishing chronological sets with purely spacelike boundaries (i.e., $\hat{\partial}(X)$ closed in $\hat{X}$ and no TIPs $P \prec_{\hat{X}} Q$—meaning $P \subsetneq Q$, as explicated below), future completion is categorical, natural, and universal in a category of topological chronological sets (Theorem 2.9 of [6]).
(7) For a largish number of spacetimes M, including multiply warped products with $\mathbb{L}^1$, standard static spacetimes, and stationary spacetimes, the future chronological topology for $\hat{M}$ is practicably calculable ([1–3, 6–8]; though [2, 3] are actually concerned with a different topology on the full causal boundary, it builds off the $\hat{L}$ operator).

But perhaps the most remarkable property of the future chronological topology is the following: A chronological set X with a topology is *future-quasi-compact* if all sequences $\{x_n\}$ with a common element in their past (some $x \ll x_n$ for all n) have a convergent subsequence. Then we have the following result.

Theorem 2.1 *Let X be a past-regular, future-complete chronological set. Then X is future-quasi-compact in the future chronological topology.*

Indeed, for any sequence σ in X there is a subsequence σ^∞ such that for any point α in the common past of σ, $\hat{L}(\sigma^\infty)$ contains an element β with $I^-(\alpha) \subset I^-(\beta)$.

Proof Theorem 5.11 of [1]. □

The following material on a causality relation in $\hat{X}$ is new.

We can also impute a *causality relation* $\prec_{\hat{X}}$ to $\hat{X}$: Assuming X to be past-regular, let us represent any $x \in X$ by the PIP $I^-(x)$. Then for any IPs P and Q we define

$$P \prec_{\hat{X}} Q \iff P \subsetneq Q$$

where, for instance, for $x \in X$, $x \prec_{\hat{X}} Q$ is interpreted as meaning $I^-(x) \prec_{\hat{X}} Q$. (Note that in case X has a causality relation $\prec$ on its own—notably, X being a spacetime—we might well have $x \prec_{\hat{X}} y$ even though $x \not\prec y$; however, in case X is a globally hyperbolic spacetime, this won't happen.) This comports well with $\ll_{\hat{X}}$, in case X has the property (called being *past-determined* in [5]) that $I^-(x) \subset I^-(y)$ and $y \ll z$ implies $x \ll z$:

Proposition 2.2 *Let X be a past-regular, past-determined chronological set. Then for $A, B, C \in \hat{X}$,*

$$A \prec_{\hat{X}} B \prec_{\hat{X}} C \implies A \prec_{\hat{X}} C$$
$$A \ll_{\hat{X}} B \prec_{\hat{X}} C \implies A \ll_{\hat{X}} C$$
$$A \prec_{\hat{X}} B \ll_{\hat{X}} C \implies A \ll_{\hat{X}} C$$

Proof The first implication is obvious. For the other two, we have eight cases each to consider. Let x, y, z be in X and P, Q, R be boundary IPs of X.

If $x \ll y$ and $I^-(y) \subsetneq I^-(z)$, or if $I^-(x) \subsetneq I^-(y)$ and $y \ll z$, then $x \ll z$.

If $x \ll y$ and $I^-(y) \subsetneq R$, or if $I^-(x) \subsetneq I^-(y)$ and $y \in R$, then $x \in R$ (for the latter hypothesis, pick $w \in R$ with $y \ll w$ and use past-determination).

If $x \in Q$ and $Q \subsetneq I^-(z)$, or if $I^-(x) \subsetneq Q$ and $Q \subset I^-(w)$ for some $w \ll z$, then $x \ll z$.

If $x \in Q$ and $Q \subsetneq R$ or if $I^-(x) \subsetneq Q$ and $Q \subset I^-(w)$ for some $w \in R$, then $x \in R$.

If $P \subset I^-(w)$ for some $w \ll y$ and $I^-(y) \subsetneq I^-(z)$, or if $P \subsetneq I^-(y)$ and $y \ll z$, then $P \subset I^-(v)$ for some $v \ll z$ (for the former hypothesis, pick v with $w \ll v \ll y$).

If $P \subset I^-(w)$ for some $w \ll y$ and $I^-(y) \subsetneq R$, or if $P \subsetneq I^-(y)$ and $y \in R$, then $P \subset I^-(v)$ for some $v \in R$.

If $P \subset I^-(w)$ for some $w \in Q$ and $Q \subsetneq I^-(z)$, or if $P \subsetneq Q$ and $Q \subset I^-(w)$ for some $w \ll z$, then $P \subset I^-(w)$ for some $w \ll z$.

If $P \subset I^-(w)$ for some $w \in Q$ and $Q \subsetneq R$, or if $P \subsetneq Q$ and $Q \subset I^-(w)$ for some $w \in R$, then $P \subset I^-(w)$ for some $w \in R$. □

(Of course, all definitions and properties are time-symmetric, with $\check{L}$ acting on $\check{X} = X \cup \check{\partial}(X)$ for the past chronological topology.)

3 Results

We will now consider how some causal constructions in spacetimes might be extended to the future completions of spacetimes.

First, consider any past-regular, past-determined chronological set X. Suppose $c : [a, b] \to \hat{X}$ is a curve, continuous in the future chronological topology. Let us call c *future-timelike* or *future-causal*, respectively, if for all $s < t$ in $[a, b]$, $c(s) \ll_{\hat{X}} c(t)$ or $c(s) \prec_{\hat{X}} c(t)$. (And these definitions make sense even for functions defined on arbitrary subsets of an interval.) Then we have a pair of obvious questions: For $\alpha, \beta \in \hat{X}$,

(1) does $\alpha \ll_{\hat{X}} \beta$ imply the existence of a future-directed timelike curve in $\hat{X}$ from α to β?
(2) does $\alpha \prec_{\hat{X}} \beta$ imply the existence of a future-directed causal curve in $\hat{X}$ from α to β?

Clearly, these questions hardly even make sense unless X admits curves in the first place. Our focus, then, should be for X being a strongly causal spacetime. Even in that case, question (1) can have a negative answer: Consider X being $\mathbb{L}^2$ with the following set removed: $\{(x, t) \mid 0 \leq t \leq 1 \text{ and } x \leq t\}$; let $\alpha = I^-((0, 0))$ (a TIP) and $\beta = (1, 2)$. Then $\alpha \ll_{\hat{X}} \beta$, but there is no future-directed timelike curve in $\hat{X}$ from α to β; there is a future-directed causal curve—the null line segment from $I^-((0, 0))$ to $I^-((1, 1))$, followed by the vertical line to $(1, 2)$—but the concatenation of a causal curve with a timelike curve in $\hat{X}$ does not necessarily have a timelike curve between its endpoints, as this example shows.

But question (2) has a more interesting, if as yet incomplete, answer:

Theorem 3.1 *Let M be a strongly causal spacetime satisfying $I^-(x) \subsetneq I^-(y) \implies x \prec y$ (so we need not distinguish between $\prec_M$ and $\prec_{\hat{M}}$). For any $\alpha \prec \beta$ in $\hat{M}$, there is a set of points in $\hat{M}$, with order-type that of the rationals ranging from 0 to 1, future-causal, going from α to β.*

Proof Let $I^-(\alpha)$ and $I^-(\beta)$ be generated by timelike curves, respectively c_α and c_β, in M (an IP P is generated by a causal curve c if $I^-[c] = P$; in case $P \in \hat{\partial}(M)$, $I^-(P)$ effectively coincides with P; more precisely, $I^-_{\hat{M}}(P) \cap M = P$, but the same timelike curves in M generate both P and $I^-_{\hat{M}}(P)$); we can parametrize both of these on $[0, \infty)$. As $I^-[c_\alpha] \subset I^-[c_\beta]$, there is a future-timelike curve $\bar{c}$ from $c_\alpha(0)$ to some $c_\beta(\bar{t})$. Let C be the concatenation $C = c_\beta|_{[\bar{t},\infty)} \cdot \bar{c}$, parametrized on $[0, 1)$, so $C : [0, 1) \to M$ is a future-timelike curve from $c_\alpha(0)$ to β, with an obvious continuous extension of $C(1) = \beta$. As $I^-[c_\alpha] \neq I^-[c_\beta]$, there is some t_1^2 with $c_\beta(t_1^2) \notin I^-[c_\alpha]$.

Let $Q = \{q_n\}$ be any countable dense subset of $[0, 1]$ with $q_1 = 0$, $q_2 = 1$. We will define a map $\sigma^Q : Q \to \hat{M}$ which is future-causal. We do this through an inductive process: we will inductively define points $\{\gamma_n^Q\}_{n\geq 1}$ in $\hat{M}$ with $\gamma_i^Q \prec \gamma_j^Q$ for $q_i < q_j$; then we let $\sigma^Q(q_n) = \gamma_n^Q$ for each n.

We let $\gamma_1^Q = \alpha$ and $\gamma_2^Q = \beta$. Suppose $\{\gamma_1^Q, \ldots, \gamma_n^Q\}$ have been defined, satisfying $\gamma_i^Q \prec \gamma_j^Q$ for $q_i < q_j$ and also $C(q_k) \ll \gamma_k^Q$ for all $k \leq n$; in particular, each γ_i^Q is the future endpoint of a timelike curve c_i from $C(q_i)$ to γ_i^Q, parametrized on $[0, \infty)$, and for $q_i < q_j$ there is some t_i^j with $c_j(t_i^j) \notin I^-(\gamma_i^Q)$. To define γ_{n+1}^Q, we locate i and j among $\{1, \ldots, n\}$ such that $q_i < q_{n+1} < q_j$ and no q_k comes between q_i and q_{n+1} or between q_{n+1} and q_j, for $k \leq n$. As $I^-[c_i] \subset I^-[c_j]$, we have, for each m, there is a future-timelike curve τ_m from $c_i(m)$ to some $c_j(p_m)$, and we can take $p_m \geq m$; let us parametrize τ_m on $[q_i, q_j]$. Let us choose the parametrization inductively:

We let $\tau_0 = C|_{[q_i,q_j]}$. Suppose τ_m has been parametrized. We know all of $c_j|_{[p_m,\infty)}$ lies in $I^+(\tau_m(q_{n+1}))$, so τ_{m+1} enters $I^+(\tau_m(q_{n+1}))$; then choose the parametrization of τ_{m+1} so that $\tau_{m+1}(q_{n+1}) \in I^+(\tau_m(q_{n+1}))$. We also know $c_j(t_i^j) \notin I^-[c_i]$; for m sufficiently large—say, $m \geq m_0$—τ_{m+1} enters $I^+(c_j(t_i^j))$, so we can also arrange that $\tau_{m+1}(q_{n+1}) \notin I^-[c_i]$ (helping us to arrange for $c_{n+1}(t_i^{n+1})$). But we also want to choose $\tau_{m+1}(q_{n+1})$ so that, for some t_{n+1}^j, $\tau_{m+1}(q_{n+1}) \not\gg c_j(t_{n+1}^j)$. Combining the requirements: We need to locate a point on τ_{m+1} in $I^+(\tau_m(q_{n+1}))$ and in $I^+(c_j(t_i^j))$ (when $m \geq m_0$) but not in $I^+(c_j(t_{n+1}^j))$ for some t_{n+1}^j. So we just need

(a) that where τ_{m+1} enters $I^+(\tau_m(q_{n+1}))$, it isn't already in $I^+(c_j(t))$ for all t, and
(b) likewise for where τ_{m+1} enters $I^+(c_j(t_i^j))$.

If criterion (a) were violated, that would mean all of c_j would be in the past of some point on τ_{m+1}—including $c_j(p_{m+1})$, the future endpoint of τ_{m+1}, clearly an impossibility. In particular, any $t' > p_{m+1}$ is a candidate for t_{n+1}^j by this criterion.

In consideration of criterion (b): We know that τ_{m+1} starts on c_i and thus not in $I^+(c_j(t_i^j))$ (as $c_j(t_i^j) \notin I^-(\gamma_i^Q)$); so τ_{m+1} actually enters $I^+(c_j(t_i^j))$ at some point x_{m+1}. And x_{m+1} cannot be in $I^+(c_j(t'))$ for any $t' > t_i^j$, because $I^+(c_j(t'))$ would then be a neighborhood of x_{m+1}, meaning τ_{m+1} is actually in $I^+(c_j(t'))$, hence, in $I^+(c_j(t_i^j))$, for points of τ_{m+1} preceding x_{m+1}. Thus, any $t' > t_i^j$ will suffice for t_{n+1}^j by criterion (b).

So we take t_{n+1}^j to be any $t' > \max\{p_{m+1}, t_i^j\}$, and we choose the parametrization on τ_{m+1} so that $\tau_{m+1}(q_{n+1})$ is a point in $I^+(\tau_m(q_{n+1})) \cap I^+(c_j(t_i^j))$ that is not in $I^+(c_j(t_{n+1}^j))$.

Then $\{\tau_m(q_{n+1})\}_{m\geq 0}$ is a future chain, so it has a future limit; we take that to be γ_{n+1}^Q, with c_{n+1} generating γ_{n+1}^Q by connecting the points of the future chain; and we select t_i^{n+1} so that $c_{n+1}(t_i^{n+1})$ is not in $I^-[c_i]$ (e.g., take $c_{n+1}(t_i^{n+1}) = \tau_{m_0}(q_{n+1})$). Then $C(q_{n+1}) \ll \gamma_{n+1}^Q$. We clearly have $c_i \subset I^-(\gamma_{n+1}^Q)$, and with $c_{n+1}(t_i^{n+1})$ not in $I^-[c_i] = I^-(\gamma_i^Q)$ we have $I^-(\gamma_i^Q) \subsetneq I^-(\gamma_{n+1}^Q)$, so $\gamma_i^Q \prec \gamma_{n+1}^Q$. We also have each $\tau_{m+1}(q_{n+1}) \subset I^-[c_j]$, and, with $c_j(t_{n+1}^j) \notin I^-[c_{n+1}]$, $I^-(\gamma_{n+1}^Q) \subsetneq I^-(\gamma_j^Q)$, so $\gamma_{n+1}^Q \prec \gamma_j^Q$. This completes the inductive step, thus defining all $\{\gamma_n^Q\}$.

We now have a map $\sigma^Q : Q \to \hat{M}$, obeying $q < q' \implies \sigma^Q(q) \prec \sigma^Q(q')$, with $\sigma^Q(0) = \alpha$ and $\sigma^Q(1) = \beta$. □

What is unclear is whether the various rational-ordered future-causal arcs from α to β can be fitted together to make a curve. What makes this problematic is that the construction offered above does not clearly guarantee that the image of the function σ^Q has the topological quality of the rationals; it could conceivably have gaps in it.

Is it feasible to consider whether $\hat{M}$ is, in some sense, globally hyperbolic? Strong causality is somewhat difficult to consider, as the essence of strong causality is the existence of a fundamental neighborhood system $\{U_n\}$ about each point, having the correct property (which we could express, for instance, as no $x, y \in U_n$ and $z \notin U_n$ with $x \ll z \ll y$); the problem here is getting hold of a fundamental neighborhood system for an arbitrary point α in $\hat{\partial}(M)$. If there is no $\beta \succ \alpha$, then we just take $U_n = \hat{I}^+(x_n)$, where $\{x_n\}$ is a future chain with α as future limit. (Note that in any past-regular chronological set X, for all $x \in X$, $I^+(x)$ is open in the future chronological topology on X: If we have $\{x_n\}$ any sequence in $X - I^+(x)$ and $x_\infty \in \hat{L}(\{x_n\})$, then we must also have $x_\infty \in X - I^+(x)$, since otherwise, $x \in I^-(x_\infty)$, so $x \ll x_n$, i.e., $x_n \in I^+(x)$, for n sufficiently large.)

If there is some $\beta \gg \alpha$, then a likely-appearing plan is to find a past-chain $\{y_n\}$ of points in M approaching α and with $\{x_n\}$ as before, and take $U_n = \hat{I}^+(x_n) \cap \hat{I}^-(y_n)$ (with $\hat{I}^\pm$ denoting the respective operators in $\hat{M}$). That, however, won't quite work if $\hat{I}^-(y_n)$ fails to be open, which can happen: Consider $M = \mathbb{L}^2 - \{(0,t) \mid t \geq 0\}$, and let $P = I^-(0,0)$, a TIP in M, and let $x_n = (1/n, 0)$ and $y = (-1, 2)$; then $P \in \hat{L}(\{x_n\})$, but $P \in \hat{I}^-(y)$, while for all n, $x_n \notin \hat{I}^-(y)$, so $\hat{M} - \hat{I}^-(y)$ is not closed in $\hat{M}$. Still, some modification of U_n as above may work.

But if we have a null boundary around α—$\beta \succ \alpha$, $\hat{I}^+(\alpha) = \emptyset$—then it is far less easy to pick out a fundamental neighborhood system; this is what brought the original [4] topology to grief (see end discussion in Sect. 5.1 of [6]).

Setting aside the issue of how to cast strong causality in $\hat{M}$, we have a strong case for saying that $\hat{M}$ is always, in at least a partial sense, globally hyperbolic: we can emulate the compactness of $I^+(x) \cap I^-(y)$ for $x \ll y$. Indeed, we have something even stronger: relative sequential compactness of $\hat{I}^+(x)$ for any x:

Proposition 3.2 *Let X be any past-regular chronological set (for instance, a strongly causal spacetime). Then for any $x \in X$, any sequence in $I^+(x)$ has a subsequence which, in $\hat{X}$, is convergent.*

Proof Any sequence in $I^+(x)$ has x in its common past, so Theorem 2.1 yields a subsequence which converges in $\hat{X}$. □

It is well worth noting, however, that this does not imply $\hat{M}$ has a Cauchy surface: Consider $M = \mathbb{L}^2 - [0, \infty) \times \{0\}$, and consider the curves $c_1 : (0, \infty] \to \hat{M}$, $c_1(s) = (1, s)$ (with $(1, \infty)$ interpreted as the TIP i^+) and $c_2 : (-\infty, 0] \to \hat{M}$, $c_2(s) = (1, s)$ (with $(1, 0)$ interpreted as the TIP $I^-((1, 0)))$. Then c_1 and c_2 are both inextendible timelike curves in $\hat{M}$; but there is no acausal curve that intersects them both. It is clear that this is not due to any failure of strong causality in $\hat{M}$.

And the situation is even worse if we look, as we arguably should, for a Cauchy surface to intersect any inextendible causal curve in $\hat{M}$, even one purely in $\hat{\partial}(M)$: Consider $M = \mathbb{L}^2$ and c^+, c^- the two null lines making up $\mathcal{I}^+$. Any spacelike curve in $\hat{M}$ has a choice between intersecting both c^+ and c^- and thereby failing to intersect some endless timelike curves in M, or intersecting all timelike curves in M by extending to spacelike infinity, i.e., missing both c^+ and c^-; no acausal curve in $\hat{M}$ will catch both types of intextendible causal curves (other than the rather exceptional curve which is all of $\hat{\partial}(M)$, which seems unlikely as a reasonable candidate for Cauchy surface).

But all is not lost for finding Cauchy surfaces in $\hat{M}$.

Theorem 3.3 *Let M be a globally hyperbolic spacetime with a compact Cauchy surface K. Then K is also a Cauchy surface for $\hat{M}$, i.e., every inextendible causal curve in $\hat{M}$ intersects K exactly once.*

Proof We first need a lemma:

Lemma 3.4 *Let M be a globally hyperbolic spacetime with a compact Cauchy surface K and c a past-inextendible causal curve in $\hat{M}$; then, traveling to the past, c enters M.*

Proof of Lemma 3.4 We can find a sequence of points $\{\alpha_k\}$ on c with the properties that for each k, $\alpha_{k+1} \prec \alpha_k$, and that for all α on c, for some k, $\alpha_k \prec \alpha$; let's call any such sequence a *past-causal coextensive chain* for c. Suppose c remains in $\hat{\partial}(M)$; then for all k, α_k is a TIP in M, so there is a timelike curve σ_k such that $\alpha_k = I^-[\sigma_k]$; we can consider each σ_k extended to the past so that it is past-intextendible in M.

Then as each σ_k is also future-inextendible in M (having its future endpoint in $\hat{\partial}(M)$), it intersects the Cauchy surface K at some point y_k. Since K is compact, there is a subsequence $\{k_i\}$ with $\{y_{k_i}\}$ having a limit $y \in K$. Pick a point $z \ll y$ in M. Then as $I^+(z)$ is a neighborhood of y in M, for i sufficiently large, $y_{k_i} \in I^+(z)$, so $z \in I^-[\sigma_{k_i}]$, i.e., $z \ll \alpha_{k_i}$. It follows that for all k, $z \ll \alpha_k$ (indeed, for all $\alpha \in c$, $z \ll \alpha$). Then by future-quasi-compactness of $\hat{M}$, we obtain a subsequence $\{\alpha_{m_n}\}$ and a point $\alpha_\infty \in \hat{L}_{\hat{M}}(\{\alpha_{m_n}\})$. We know that $\alpha_\infty \in \hat{\partial}(M)$ (since otherwise, it has a neighborhood in M that c does not enter).

We now show that α_∞ is a past extension of c, in violation of the hypothesis that c is past-inextendible:

We know

(1) for all $\theta \in I^-_{\hat{M}}(\alpha_\infty)$, $\theta \ll \alpha_{m_n}$ for n sufficiently large
(2) for any IP $P \supset I^-_{\hat{M}}(\alpha_\infty)$, if for all $\theta \in P$, $\theta \ll \alpha_{m_n}$ for infinitely many n, then $P = I^-_{\hat{M}}(\alpha_\infty)$

We want to know if the same holds true for any other past-causal coextensive chain $\{\beta_i\}$ for c. Consider (1) for $\{\beta_i\}$: For any $\theta \in I^-_{\hat{M}}(\alpha_\infty)$, for all i, there is some n with $\alpha_{m_n} \prec \beta_i$; we know (from (1) for $\{\alpha_{m_n}\}$) that $\theta \ll \alpha_{m_n}$, from which it follows that $\theta \ll \beta_i$. Now consider (2) for $\{\beta_i\}$: For any IP $P \supset I^-_{\hat{M}}(\alpha_\infty)$, suppose for all $\theta \in P$, $\theta \ll \beta_i$ for infinitely many i; then for all n, there is some i with $\beta_i \prec \alpha_{m_n}$, from which it follows $\theta \ll \alpha_{m_n}$. We conclude, via (2) for $\{\alpha_{m_n}\}$, that $P = I^-_{\hat{M}}(\alpha_\infty)$.

Since we have $\alpha_\infty \in \hat{L}_{\hat{M}}(\sigma)$ for any past-causal coextensive sequence σ for c, α_∞ is a past extension of c, contrary to hypothesis. Ergo, c must enter M. □

And a second lemma:

Lemma 3.5 *Let M be a globally hyperbolic spacetime and c a causal curve in M. Then c cannot have a past endpoint on $\hat{\partial}(M)$.*

Proof of Lemma 3.5 Let $c : (t^-, t^+) \to M$ be future-directed, and suppose we have $P = \lim_{t \to t^-} c(t)$ with $P \in \hat{\partial}(M)$. Then we can represent P as $P = I^-[\sigma]$ for a future-directed timelike curve $\sigma : (s^-, s^+) \to M$. Note for any sequence $\{t_n\} \to t^-$, we have $P \in \hat{L}(\{c(t_n)\})$, so in particular, for all $x \in P$, $x \in I^-(c(t_n))$ for n sufficiently large; it follows that for all $x \in P$, $x \in I^-(c(t))$ for all t. Thus, for all $t > t^-$ and $s < s^+$, $J_{s,t} = J^-(c(t)) \cap J^+(\sigma(s)) \neq \emptyset$, as it includes $\sigma|_{(s,s^+)}$. By global hyperbolicity, each $J_{s,t}$ is compact, so there is a point $p \in M$ that lies in $\bigcap_{t>t_-} \bigcap_{s>s^+} J_{s,t}$.

We use the future chronological topology to show p is a past endpoint of c: For any sequence $\{t_n\} \to t^-$, for any $x \ll p$, $x \ll c(t_n)$ for n sufficiently large (since $p \in J_{s,t_n}$, any s); in other words, $I^-(p) \subset \mathrm{LI}(\{c(t_n)\})$. Note that $P \subset I^-(p)$, since for all $s < s^+, \sigma(s) \prec p$, as $p \in J_{s,t}$ (for any t). And finally note that $I^-(p) \subset \mathrm{LS}(\{c(t_n)\})$, as $p \in J_{s,t_n}$ (any s). Then it follows, since $P \in \hat{L}(\{c(t_n)\})$, that $I^-(p) = P$. Thus, $I^-(p) \in \hat{L}(\{c(t_n)\})$ But the only way for $I^-(p)$ to be in $\hat{L}(\{c(t_n)\})$ for all sequences $\{t_n\} \to t^-$ is for p to be a past endpoint of c in the manifold topology. But, as $\hat{\partial}(M)$

is closed in $\hat{M}$, that is incompatible with c also approaching a point in $\hat{\partial}(M)$ for its past endpoint. □

Now let $c : (t^-, t^+) \to \hat{M}$ be an inextendible causal curve. By Lemma 3.4, for some $t_0 \in (t^-, t^+)$, c, going to the past, exits $\hat{\partial}(M)$ at $c(t_0)$. By Lemma 3.5, once that happens, c cannot re-enter $\hat{\partial}(M)$, i.e., $c|_{(t^-,t_0)}$ lies in M. Also by Lemma 3.5, $c(t)$ cannot be in M for $t > t_0$; therefore, $c|_{[t_0,t^+)}$ lies in $\hat{\partial}(M)$ and $c|_{(t^-,t_0)}$ lies in M. Thus, c being inextendible in $\hat{M}$ implies $c|_{(t^-,t_0)}$ is inextendible in M. Then with K a Cauchy surface in M, there is some $t_1 \in (t^-, t_0)$ with $c(t_1) \in K$. □

Due to the inherent interest in putting a topology on the full causal boundary—future- and past-causal boundaries combined in an appropriate manner—and the notable research contained in this direction by what might be called the Andalusian school of researchers, seen in such publications as [2, 3], it is worth noting an important context in which the "Andalusian" topology and the future chronological topology coincide: standard static spacetimes. So more definitions are called for now.

The most cogent way to combine IPs and IFs appears to be by using the Szabados relation (referencing [11, 12]), introduced in 1988: For P an IP and F an IF, define $P \sim_{\mathrm{Sz}} F$ iff P is contained in the common past of F and is maximal (using subset relation) among such IPs, and F is contained in the common future of P and is maximal among such IFs; we will say that P and F are *Szabados mates* of one another, if this relation obtains between them, and that (P, F) is a *Szabados pair*. It is easy to check that for any point x in a strongly causal spacetime, $I^-(x)$ and $I^+(x)$ are Szabados mates, and neither has any other Szabados mates. Although one might consider $\hat{\partial}(M) \cup \check{\partial}(M)/\sim_{\mathrm{Sz}}$ as the causal boundary, Marolf and Ross in 2003 ([10]) adopted a perhaps more sophisticated notion: $\bar{\partial}(M) = \{(P, F) \in \hat{\partial}(M) \times \check{\partial}(M) \mid P \sim_{\mathrm{Sz}} F\} \cup \{(P, \emptyset) \in \hat{\partial}(M) \times \{\emptyset\} \mid P \text{ has no Szabados mate}\} \cup \{(\emptyset, F) \in \{\emptyset\} \times \check{\partial}(M) \mid F \text{ has no Szabados mate}\}$; and we can consider the causal completion of M, $\bar{M} = M \cup \bar{\partial}(M)$, in a unified manner by treating each point of x as the Szabados pair $(I^-(x), I^+(x))$, i.e., expanding that collection of Szabados pairs in the definition of $\bar{\partial}(M)$ to $\{(P, F) \mid P \sim_{\mathrm{Sz}} F\}$. Then the Andalusian topology (by which I mean that found in [2, 3]) defines a limit operator $\bar{L}$ on sequences in $\bar{M}$ by $\bar{L}(\{(P_n, F_n)\})$ contains

(1) (P, F) iff $P \in \hat{L}(\{P_n\})$ and $F \in \check{L}(\{F_n\})$
(2) $(P, \emptyset)$ iff $P \in \hat{L}(\{P_n\})$ and $\check{L}(\{F_n\}) = \emptyset$
(3) $(\emptyset, F)$ iff $\hat{L}(\{P_n\}) = \emptyset$ and $F \in \check{L}(\{F_n\})$

We need notation for working in a standard static spacetime: Let $c : [\alpha, \omega) \to N$ be a unit-speed curve in a Riemannian manifold (N, h). Then the Busemann function associated to c is $b_c : N \to \mathbb{R}$ defined by $b_c(x) = \lim_{s\to\infty}(s - d(x, c(s)))$, where d is the distance function on N (this function is allowed to have the constant value ∞). Let $M = N \times \mathbb{L}^1$ be the product spactime. For any $f : N \to \mathbb{R}$ (possibly ∞-valued), let $\mathrm{P}(f)$ be the past of the graph of f, i.e., $\mathrm{P}(f) = \{(x, t) \in M \mid t < f(x)\}$ (so $\mathrm{P}(f) = M$ in case $f = \infty$); similarly, $\mathrm{F}(f)$ is the future of the graph of f. For $c : [\alpha, \omega) \to N$ a unit-speed curve, let $\hat{c} : [\alpha, \omega) \to M$ be the future-null curve

$\hat{c}(s) = (c(s), s)$; then $\mathrm{P}(b_c)$ is the IP $I^-[\hat{c}]$. Dually, for $\check{c}(s) = (c(s), -s)$, $\mathrm{F}(-b_c)$ is the IF $I^+[\check{c}]$.

Note that the future chronological topology on $\hat{M}$ for $M = N \times \mathbb{L}^1$ has $P \in \hat{L}(\{P_n\})$, with IPs P and P_n expressed via Busemann functions as $P = \mathrm{P}(b)$ and $P_n = \mathrm{P}(b_n)$, iff $b \leq \liminf_n b_n$ and b is maximal among Busemann functions b' satisfying $b' \leq \limsup_n b_n$ (and dually for IFs); see Sect. 2 of [1].

Lemma 3.6 *Let (N, h) be a Riemannian manifold and $M = N \times \mathbb{L}^1$ the product spacetime. Let $c : [\alpha, \omega) \to N$ be a unit-speed curve, and let $P = \mathrm{P}(b_c)$; then the following are equivalent:*

(1) P has a Szabados mate
(2) $\omega < \infty$
(3) b_c is bounded above.

Proof From Proposition 4 in [7], we know $\omega < \infty$ iff b_c is bounded above.

Consider $\omega < \infty$. Let $F = \mathrm{F}(2\omega - b_c)$; then $P \sim_{\mathrm{Sz}} F$. On the other hand, consider $\omega = \infty$; then (x, t) is in the common future of P iff for all y, $(x, t) \gg (y, b_c(y))$, i.e., for all y, $t - b_c(y) > d(x, y)$, i.e., for all y, $t > d(x, y) + b_c(y)$. But since b_c is unbounded above, this cannot be true for any finite t; thus, P has no Szabados mate. □

Theorem 3.7 *Consider a sequence of IPs $\{P_n\}_n$ in $M = N \times \mathbb{L}^1$ with $P_n \sim_{\mathrm{Sz}} F_n$ (where we allow $F_n = \emptyset$). Suppose $P \in \hat{L}(\{P_n\}_n)$; then either P has no Szabados mate and $\check{L}(\{F_n\}_n) = \emptyset$, or P has a Szabados mate F and $F \in \check{L}(\{F_n\}_n)$.*

In other words, the future chronological topology on $\hat{M}$ coincides with the Andalusian topology on $\bar{M}$, restricted to $\hat{M}$.

Proof We can express the IPs concerned as $P = \mathrm{P}(b)$ and $P_n = \mathrm{P}(b_n)$ using Busemann functions. Then $F_n = \mathrm{F}(f_n)$ for $f_n = 2\omega_n - b_n$ with $\omega_n = \sup_N(b_n)$ (which works even if P_n is unbounded, for then $\omega_n = \infty$, and $\mathrm{F}(\infty) = \emptyset$). Similarly, let $f = 2\omega - b$ with $\omega = \sup_N(b)$, and $F = \mathrm{F}(f)$ is the Sabados mate of P (if it has one). We have $b \leq \liminf_{n\to\infty} b_n$ and b is maximal among Busemann functions b' satisfying $b' \leq \limsup_n b_n$. This gives us

$$\begin{aligned} f &= 2\omega - b \\ &\geq 2\omega - \liminf_n b_n \\ &= \limsup_n (2\omega - b_n) \\ &= \limsup_n (2(\omega - \omega_n) + f_n) \end{aligned}$$

We cannot have $\omega < \liminf_n \omega_n$, as, for consider: We can without harm assume each P_n is the past of a point (x_n, ω_n), and that $\{\omega_n\}$ has a limit ω_∞. Assume $\omega_\infty < \infty$. As $\{P_n\}$ have some point (x_0, t_0) in their common past, all x_n are within $\omega_\infty - t_0$ of x_0; hence, we can assume $\{x_n\}$ is a Cauchy sequence. Let $c : [\alpha, \omega_\infty) \to N$ be a unit-speed curve joining $\{x_n\}$ by (almost) minimizing arcs; then b_c violates the maximality of b. With $\omega_\infty = \infty$, create a violation similarly with $\{x_n\}$ divergent. It follows that $f \geq \limsup_n f_n$. (This manifestly works for ω infinite or finite.) We next need to show f has the requisite minimality among negatives of Busemann functions.

For ω finite, for any f', the negative of a Busemann function, obeying $f \geq f' \geq \liminf_n f_n$, we obtain

$$\begin{aligned} b &= 2\omega - f \\ &\leq 2\omega - f' \\ &\leq \limsup_n (2\omega - f_n) \\ &= \limsup_n (2\omega - 2\omega_n + b_n) \\ &\geq 2(\omega - \liminf_n \omega_n) + \limsup_n b_n \end{aligned}$$

We cannot have $\omega > \liminf_n \omega_n$, for that would admit $(x, t) \in P$ that violates $(x, t) \in P_n$ for infinitely many n; in other words, $\omega = \liminf_n \omega_n$. Therefore, we have $b \leq 2\omega - f' \leq \limsup_n b_n$, and the maximality of b gives us $b = 2\omega - f'$, or $f' = f$. This is the required minimality for f so that $F \in \check{L}(\{F_n\}_n)$.

For $\omega = \infty$, we need to show $\liminf_n f_n = \infty$. Since b is unbounded above, for all $K > 0$, there is some x_K with $b(x_K) > K$. This forces $b_n(x_K) > K$ for all n sufficiently large (from P lying in $\mathrm{LI}(\{P_n\}_n)$); say, $b_n(x_K) > K$ for $n > n_K$. It follows that $f_n > K$ for all $n > n_K$, since $\inf_N(f_n) = \sup_N(b_n)$. And from that we learn $\liminf_n f_n = \infty$, as desired. □

And the proof of Theorem 3.7 yields this technical result:

Proposition 3.8 *In a product spacetime $M = N \times \mathbb{L}^1$, suppose we have IPs $P = P(b)$ and $\{P_n = P(b_n)\}_n$ with $P \in \hat{L}(\{P_n\}_n)$. Let $\omega = \sup_N(b)$ and $\omega_n = \sup_N(b_n)$. Then $\omega = \liminf_{n\to\infty} \omega_n$.* □

Acknowledgements These researches were conducted or commenced at VIII International Meeting on Lorentzian Geometry. The author wishes particularly to thank Miguel Sánchez, José Flores, and Jónatan Herrera for helpful discussions.

References

1. J.L. Flores, S.G. Harris, Topology of the causal boundary for standard static spacetimes. Class. Quantum Grav. **24**, 1211–1260 (2007)
2. J.L. Flores, J. Herrera, M. Sánchez, *Gromov, Cauchy, and Causal Boundaries for Riemannian, Finslerian and Lorentzian Manifolds*, vol. 226, no. 1064. (Memoirs American Mathematical Society, 2013)

3. J.L. Flores, J. Herrera, M. Sánchez, On the final definition of the causal boundary and its relation with the conformal boundary. Adv. Theor. Math. Phys. **15**, 991–1057 (2011)
4. R.P. Geroch, E.H. Kronheimer, R. Penrose, Ideal points in space-time. Proc. Roy. Soc. Lond. A **327**, 545–567 (1972)
5. S.G. Harris, Universality of the future chronological boundary. J. Math. Phys. **39**, 5427–5445 (1998)
6. S.G. Harris, Topology of the future chronological boundary: universality for spacelike boundaries. Class. Quantum Grav. **17**, 551–603 (2000)
7. S.G. Harris, Causal boundary for standard static spacetimes. Nonlinear Anal. **47**, 2971–2981 (2001)
8. S.G. Harris, Discrete group actions on spacetimes. Class. Quantum Grav. **21**, 1209–1236 (2004)
9. S.W. Hawking, G.F.R. Ellis, *Large Scale Structure of Space-Time*. (Cambridge University Press, Cambridge, 1972)
10. D. Marolf, S.R. Rolf, A new recipe for causal completion. Class. Quantum Grav. **20**, 4085–4117 (2003)
11. L.B. Szabados, Causal boundary for strongly causal spacetimes. Class. Quantum Grav. **5**, 121–134 (1988)
12. L.B. Szabados, Causal boundary for strongly causal spacetimes II. Class. Quantum Grav. **6**, 77–91 (1989)

Some Criteria for Wind Riemannian Completeness and Existence of Cauchy Hypersurfaces

Miguel Ángel Javaloyes and Miguel Sánchez

Abstract Recently, a link between Lorentzian and Finslerian Geometries has been carried out, leading to the notion of *wind Riemannian structure* (WRS), a generalization of Finslerian Randers metrics. Here, we further develop this notion and its applications to spacetimes, by introducing some characterizations and criteria for the completeness of WRS's. As an application, we consider a general class of spacetimes admitting a time function t generated by the flow of a complete Killing vector field (generalized standard stationary spacetimes or, more precisely, SSTK ones) and derive simple criteria ensuring that its slices $t =$ constant are Cauchy. Moreover, a brief summary on the Finsler/Lorentz link for readers with some acquaintance in Lorentzian Geometry, plus some simple examples in Mathematical Relativity, are provided.

Keywords Globally hyperbolic spacetime · Killing vector field · Stationary and SSTK spacetimes · Finsler metrics · Randers and kropina metrics · Zermelo navigation · Wind finslerian structure

2010 Mathematics Subject Classification Primary 53C60 · 53C22

1 Introduction

In the last years, a fruitful link between Lorentzian and Finslerian geometries has been refined more and more; indeed, ramifications to different areas such as control theory, have also appeared. In this framework, our purpose here is twofold. First, a

M.Á. Javaloyes (✉)
Departamento de Matemáticas, Universidad de Murcia, Campus de Espinardo, 30100 Murcia, Espinardo, Spain
e-mail: majava@um.es

M. Sánchez
Departamento de Geometría y Topología, Facultad de Ciencias, Universidad de Granada, Campus Fuentenueva S/n, 18071 Granada, Spain
e-mail: sanchezm@ugr.es

M.A. Cañadas-Pinedo et al. (eds.), *Lorentzian Geometry and Related Topics*, Springer Proceedings in Mathematics & Statistics 211,
DOI 10.1007/978-3-319-66290-9_8

brief summary on the subject for an audience of Lorentzian geometers is provided. Then, a new application for Lorentzian Geometry will be obtained. Namely, we will consider a big class of spacetimes admitting a time function t (generated by the flow of a complete Killing vector field $K = \partial_t$), and we will characterize when its slices $t =$ constant are Cauchy hypersurfaces, providing also simple criteria that ensure this property.

The refinements of the Lorentz/Finsler link can be understood in three steps:

(1) First, consider a product $\mathbb{R} \times M$ endowed with a standard Lorentzian product metric $g = -dt^2 + g_0$ (see below for more precise notation and definitions). Clearly, all the properties of the spacetime will be encoded in the Riemannian metric g_0. Even more, if one considers a standard static metric $g = -\Lambda dt^2 + g_0$ ($\Lambda > 0$), then all the conformal properties of the spacetime can be studied in the conformal representative g/Λ and, thus, they are encoded in the Riemannian metric $g_R = g_0/\Lambda$ (see for example [4, Theorem 3.67] or [15] and references therein).
(2) Second, the previous case can be generalized by allowing t-independent cross-terms, i.e., the stationary metric $g = -\Lambda dt^2 + 2\omega dt + g_0$ ($\Lambda > 0, \omega$ 1-form on M). In this case, the conformal properties are provided by a Randers metric (formula (8) below), which is a special class of Finsler metric characterized by a Riemannian metric g_R and a vector field W with $|W|_R < 1$ (according to the interpretation of Zermelo problem, [2, Prop. 1.1]). This correspondence is carried out in full detail in [9] and many related properties can be seen in [7, 8, 11, 12, 16].
(3) Finally, suppress the restriction $\Lambda > 0$ in the previous case (SSTK splitting). Now, the conformal structure is still characterized by a pair (g_R, W), but the restriction $|W|_R < 1$ does not apply. From the Finslerian viewpoint, this yields a *wind Riemannian structure* (WRS), which is a generalization of Randers metrics; even more, this also suggests the subsequent generalization of all Finsler metrics: *wind Finslerian structures*. Such new structures were introduced and extensively studied in [10] and further developments and applications for Finslerian geometry have been carried out in [17, 20].

The study of SSTK splittings in [10] includes quite a few topics. Among them, relativists may be interested in a link between the well-known (and nonrelativistic) problem of *Zermelo navigation* and the relativistic *Fermat principle*; indeed, this link allows one to solve both problems beyond their classical scopes; moreover, it shows connections with the so-called *analogue gravity* [3]. On the other hand, global properties of causality of SSTK spacetimes are neatly characterized by their Finslerian counterparts. So, the exact step in the ladder of causality of SSTK spacetimes is described sharply in terms of the associated WRS. In particular, the fact that the slices $t =$ constant are Cauchy hypersurfaces becomes equivalent to the (geodesic) completeness of the WRS.

Even though such results are very accurate, a difficulty appears from a practical viewpoint. WRS's are not standard known elements, as they have been introduced only recently. Therefore, to determine whether they satisfy or not some geometrical properties may be laborious. Due to this reason, our purpose here is to introduce some simple notions and results which allow one to check easily properties of WRS's. As a first approach, we will focus on results about completeness because, on the one hand, completeness has a direct translation to spacetimes in terms of Cauchy hypersurfaces and, on the other, it is a basic natural property with applications to Finslerian Geometry (see [17]). However, the introduced tools are expected to be applicable for other properties too, and some hints are made in the examples at the end.

Our task is organized as follows. In Sect. 2, we give an overview on the Lorentz/Finsler correspondence for readers with some knowledge on Lorentzian Geometry (compare with the overview in [17], written for a more Finslerian audience). More precisely, in Sect. 2.1, we describe the class of spacetimes to be studied (generalized standard stationary or, more precisely, SSTK spacetimes). The generality of this class, which includes many typical relativistic spacetimes, is stressed, and the way to obtain a (nonunique) *SSTK splitting* is detailed. Then, in Sect. 2.2, the relation between the conformal classes of spacetimes and the properties of associated Finslerian structures is introduced gradually, with increasing generality in the Finslerian tools: Riemannian/Randers/Randers-Kropina/WRS. In Sect. 2.3, as a toy application of the correspondence, we consider a causally surprising example of static spacetime constructed recently by Harris [15], and we show its Finslerian counterpart, explaining the corresponding curious properties which appear in the distance of the associated Finsler manifold.

Section 3 makes both, to introduce a geometric element for the practical study of WRS's and to prove our main result on completeness. Sect. 3.1 explains the precise technical notions on WRS balls, geodesics and completeness, extracted from [10]. Then, in Sect. 3.2 we introduce a new key ingredient, the *extended conic Finsler metric* $\bar{F}$ associated with any WRS. We emphasize that, as analyzed in [10], any WRS Σ determines both, a *conic Finsler* metric F and a *Lorentz–Finsler* one F_l. The former differs from a standard Finsler metric only in the fact that its domain is just an open conic region of the tangent bundle (this is a possibility with independent interest, see [18]). However, our aim here is to show that, for any such Σ, the conic metric F admits a natural extension $\bar{F}$ to the boundary of the conic region; moreover, $\bar{F}$ has an associated exponential, distance-type function (called here $\bar{F}$*-separation*), Cauchy sequences, etc. Then, in Sect. 3.3, we give our main result, Theorem 3.23, which contains a double goal: to prove that the completeness of the WRS Σ is fully equivalent to the completeness of $\bar{F}$, and to show that $\bar{F}$ satisfies a set of properties in the spirit of Hopf–Rinow Theorem. These properties will allow us to determine if $\bar{F}$ and, then, Σ, are complete.

In Sect. 4, we derive some applications using the previous Theorem 3.23. In Sect. 4.1, some simple criteria for checking whether a WRS is complete or not are provided. These criteria are stated in both, natural WRS elements and the original SSTK metric. So, one obtains also criteria which ensure whether the slices of an SSTK spacetime are Cauchy, in an easily manageable way. Finally, Sect. 4.2 ends with some further concrete examples and prospects in Mathematical Relativity.

2 A Lorentzian Overview on Wind Riemannian Structures

2.1 Revisiting SSTK Spacetimes

We will follow standard conventions and background results as in [4, 23]. In particular, a spacetime (L, g) is a time-oriented connected Lorentzian manifold $(-, +, \ldots, +)$ of dimension $n + 1, n \geq 1$. Lightlike vectors $v \in TL$ will satisfy both, $g(v, v) = 0$ and $v \neq 0$ (while null vectors would be allowed to be equal to 0) so, causal vectors, being either timelike or lightlike, also exclude 0. Except when otherwise specified, (L, g) will also be stably causal, so that it admits a *temporal function* $t : L \to \mathbb{R}$ according to [5, 27] (that is, t is smooth and onto, with timelike past-directed gradient ∇t and, in particular, a time function). Let us start with a simple result when a vector field $K \in \mathfrak{X}(L)$ complete and transversal to the slices of t can be chosen.

Proposition 2.1 *Let t be a temporal function for (L, g) and $K \in \mathfrak{X}(L)$ complete with flow $\varphi : \mathbb{R} \times L \to L$ such that $\mathrm{d}t(K) \equiv 1$. Putting $M := t^{-1}(0)$, the map*

$$\Phi : \mathbb{R} \times M \to L, \qquad (\bar{t}, x) \mapsto \varphi_{\bar{t}}(x)$$

is a diffemorphism such that: (a) $t \circ \Phi : \mathbb{R} \times M \to \mathbb{R}$ agrees with the natural projection and (b) $t \circ \Phi$ is also a temporal function for the pullback metric $\Phi^*(g)$ and time orientation induced on $\mathbb{R} \times M$ via Φ (which will be denoted simply as t). Then, $\Phi_*(\partial_t) = K$ and

$$\Phi^*(g) = -\Lambda^L \mathrm{d}t^2 + \omega^L \otimes \mathrm{d}t + \mathrm{d}t \otimes \omega^L + g_0^L, \tag{1}$$

where Λ^L, ω^L and g_0^L are, respectively, the smooth real function $g(K, K) \circ \Phi$, a one form whose kernel includes ∂_t and a positive semi-definite metric tensor whose radical is spanned by ∂_t, all of them defined on $\mathbb{R} \times M$.

Proof Φ is onto because $z = \Phi(t(z), \varphi_{-t(z)}(z))$ for all $z \in L$, and one-to-one because the equality $\mathrm{d}t(K) \equiv 1$ forbids the unique integral curve of K through z to close. That equality also implies $\Phi(\{t_0\} \times M) = t^{-1}(t_0)$ plus the transversality of K and the slices $t^{-1}(t_0)$, so that Φ becomes a (local) diffeomorphism, and all the other assertions follow easily. □

As already done for the natural projection $\mathbb{R} \times M \to \mathbb{R}$, the diffeomorphism Φ will be omitted with no further mention in the remainder.

Remark 2.2 (1) As any timelike vector field T satisfies that $\mathrm{d}t(T)$ cannot vanish, the normalized vector $K = T/\mathrm{d}t(T)$ satisfies $\mathrm{d}t(K) \equiv 1$; in particular, one can choose $K = \nabla t / g(\nabla t, \nabla t)$. However, Proposition 2.1 shows that the completeness of K is much more difficult to obtain, even taking into account that, in general, K is not assumed to be timelike.

Indeed, no complete K can exist in a stably causal spacetime (L, g) such that L is not a smooth product manifold $\mathbb{R} \times M$ as, for example, the spacetime obtained by removing two points from Lorentz–Minkowski spacetime $\mathbb{L}^2$. However, the completeness of K may not be achieved even when L is a product. Indeed, this is the case of $\mathbb{L}^2 \setminus \{0\}$: this spacetime is diffeomorphic to $\mathbb{R} \times S^1$ and its natural coordinate $t = x^0$ is a temporal function; nevertheless, it contains the non-homeomorphic slices $t = 0$ and $t = 1$.

(2) The choice $K = \partial_t + 2\partial_x$ in the globally hyperbolic strip

$$L = \{(t, x) \in \mathbb{L}^2 : x = 2t + \lambda, \forall\lambda \in (-1, 1), \forall t \in \mathbb{R}\}$$

shows that a complete choice of K may be possible even when no complete timelike choice exists. However, in a globally hyperbolic spacetime, one can always choose a temporal function t whose levels are Cauchy hypersurfaces [5]. In this case, the choice $K = \nabla t / g(\nabla t, \nabla t)$ suggested above is necessarily complete (notice that temporal functions are assumed to be onto, and the integral curves of K must cross all the slices of t, due to its Cauchy character). As a last observation, notice that the onto character of a temporal function can be deduced when a complete K satisfying $\mathrm{d}t(K) \equiv 1$ exists.

(3) A straightforward computation shows that a metric on $\mathbb{R} \times M$ written as in (1) is Lorentzian if and only if for each tangent vector u to $\mathbb{R} \times M$ such that $g_0^L(u, u) \neq 0$,

$$\Lambda + \frac{\omega^L(u)^2}{g_0^L(u, u)} > 0, \tag{2}$$

(see for example [10], around formula (28)).

In what follows, we will be interested in the case that K in Proposition 2.1 is a Killing vector field. In this case, all the metric elements in the proof of Proposition 2.1 (plus the bound (2)) are independent of the flow of K and, thus, of the coordinate t, yielding directly:

Corollary 2.3 *For any spacetime (L, g) endowed with a temporal function t and a complete Killing vector field K such that $\mathrm{d}t(K) \equiv 1$, the splitting $L = \mathbb{R} \times M$ in Proposition 2.1 can be sharpened metrically into*

$$g = -(\Lambda \circ \pi)\mathrm{d}t^2 + \pi^*\omega \otimes \mathrm{d}t + \mathrm{d}t \otimes \pi^*\omega + \pi^* g_0, \tag{3}$$

where Λ, ω, and g_0 are, respectively, a smooth real function (the "lapse"), a one form (the "shift"), and a Riemannian metric on M, $\pi : \mathbb{R} \times M \to M$ is the natural projection, and π^* the pullback operator. Moreover, the relation

$$\Lambda + |\omega|_0^2 > 0 \tag{4}$$

holds, being $|\omega|_0$ the pointwise g_0-norm of ω.

Following [10], let us introduce the notion of SSTK spacetime.

Definition 2.4 A spacetime (L, g) is *standard with a space-transverse Killing vector field (SSTK)* if it admits a (necessarily nonvanishing) complete Killing vector field K and a spacelike hypersurface S (differentiably) transverse to K which is crossed exactly once by every integral curve of K.

Corollary 2.5 *A spacetime (L, g) is SSTK if and only if it admits a temporal function t and a complete Killing vector field K such that* $\mathrm{d}t(K) \equiv 1$. *In this case, the global splitting provided by Corollary 2.3 will be called an* SSTK splitting.

Proof By [10, Proposition 3.3] (L, g) is an SSTK spacetime if and only if it admits an SSTK splitting and, therefore, it admits t and $K = \partial_t$ as in the statement. The converse follows from Corollary 2.3. □

Recall that a vector field X (and, then, the full spacetime) is called *stationary* (resp., *static*; *stationary-complete*; *static-complete*) when it is Killing and timelike (resp., additionally: the orthogonal distribution $X^\perp$ is involutive; X is complete; both conditions occur).[1] When $\Lambda > 0$ in (3) the spacetime is called *standard stationary* and if, additionally, $\omega = 0$, standard static. Any stationary or static spacetime can be written locally as a standard one. A stationary-complete spacetime is (globally) standard stationary if and only if it satisfies the mild causality condition of being distinguishing [19] (or, as pointed out in [15, Proposition 2.13], and taking into account [15, Proposition 1.2], future distinguishing). In this case, the spacetime is not only stably causal but also causally continuous; however, the conditions to ensure that a static-complete spacetime is standard static are more involved, see [14, 28].

2.2 *Conformal Geometry and the Appearance of Finslerian Structures*

Now, let us consider the conformal structure for an SSTK splitting (3). This is equivalent to compute the (future-directed) lightlike directions and, because of t-independence, we can consider just the points on the slice $M = t^{-1}(0)$. Thus, the relevant vectors at each $p \in M$ can be written with natural identifications as $u_p = \partial_t|_p + v_p$ where $v_p \in T_pM$ and one must assume:

$$0 = g(u_p, u_p) = -\Lambda(p) + 2\omega(v_p) + g_0(v_p, v_p) = 0 \qquad (5)$$

The next result yields a geometric picture of the setting (see Fig. 1).

[1] Sometimes, our stationary spacetimes are called *strictly stationary* in the literature about Mathematical Relativity, and the name *stationary* is used for a Killing vector field K that is timelike at some point (see for example [21, Definition 12.2]). Indeed, the name *SSTK spacetime* is introduced just to avoid confusions with the previous ones: no restriction on the causal character of K is assumed (whenever its flow is temporal) but the global structure must split.

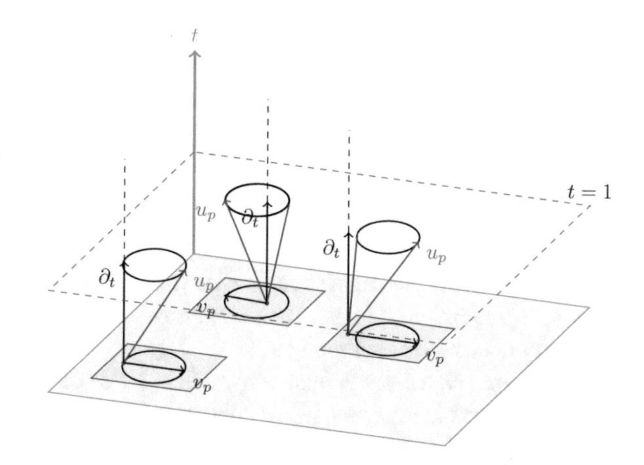

Fig. 1 We show the lightlike vectors $u_p = \partial_t|_p + v_p$ of (T_pL, g_p) with v_p tangent to $M = t^{-1}(0)$. There are three different possibilities according to the causal character of ∂_t. All the other lightlike vectors are proportional to one of these. The subset of vectors v_p forms an ellipsoid in T_pM

Lemma 2.6 *The set Σ_p which contains all $v_p \in T_pM$ satisfying* (5) *is a g_0-sphere of center W_p, where $g_0(W_p, \cdot) = -\omega_p$, and (positive) radius $r_p := \sqrt{\Lambda(p) + |\omega_p|_0^2}$ $= \sqrt{\Lambda(p) + |W_p|_0^2}$ (recall* (4)*).*

Proof Putting $w_p = v_p - W_p$, one has:

$$g_0(w_p, w_p) = g_0(W_p, W_p) - 2g_0(W_p, v_p) + g_0(v_p, v_p) = |\omega_p|_0^2 + 2\omega(v_p) + g_0(v_p, v_p),$$

so, (5) holds if and only if $g_0(w_p, w_p) = \Lambda(p) + |\omega_p|_0^2$. □

Remark 2.7 As we will be interested only in the conformal structure of the spacetime, we can choose the conformal metric $\tilde{g} = \Omega g$ was $\Omega = 1/\left(\Lambda + |\omega|_0^2\right)$, and its corresponding lapse $\tilde{\Lambda}$, shift $\tilde{\omega}$ and Riemannian metric $\tilde{g}_0$ will satisfy

$$\tilde{r}_p := \sqrt{\tilde{\Lambda}(p) + |\tilde{\omega}_p|_{\tilde{g}_0}^2} \equiv 1. \tag{6}$$

From the definition of W_p, its independence of conformal changes becomes apparent. Consistently, we can attach the conformally invariant Riemannian metric

$$g_R = g_0/\left(\Lambda + |\omega|_0^2\right) \tag{7}$$

to the SSTK splitting. Indeed, then $\tilde{\Lambda} = \Omega\Lambda$ and $|\tilde{\omega}|_{\tilde{g}_0}^2 = \Omega|\omega|_0^2$, thus, Σ_p will always be given by a g_R-sphere of radius 1.

Let us call the pair (g_R, W) composed by a Riemannian metric g_R and a vector field $W \in \mathfrak{X}(M)$ on M, *Zermelo data*. We can summarize and systematize the previous results as follows:

Proposition 2.8 *(1) For each SSTK splitting* (3), *there exists a smooth hypersurface $\Sigma \subset TM$ and Zermelo data (g_R, W), both univocally determined and invariant under pointwise conformal transformations $g \mapsto \Omega g$, $\Omega > 0$, such that:*

(i) Σ is transverse to each tangent space T_pM, $p \in M$, and all the lightlike directions on M are spanned by the vectors $\partial_t + v$ such that $v \in \Sigma$.
(ii) At each point $p \in M$, $\Sigma_p := \Sigma \cap T_pM$ is the g_R-sphere of center W_p and radius 1.

(2) Conversely, for each Zermelo data (g_R, W) on M, there exists an SSTK splitting (unique up to pointwise conformal transformations) whose associated Zermelo data by the previous point (1) are (g_R, W).

(3) Moreover, a smooth hypersurface $\Sigma \subset TM$ can be written as the set of all the unit g_R-spheres with center W_p at each point $p \in M$ for some Zermelo data (g_R, W) if and only if Σ is transverse to all T_pM, $p \in M$ and each $\Sigma_p := \Sigma \cap T_pM$ is an ellipsoid in the coordinates induced by any basis of T_pM.

Proof First of all, the transversality of a hypersurface Σ constructed from Zermelo data (g_R, W) as in (3) can be proved as follows. Let F_R be the g_R-norm, namely $F_R(v) = \sqrt{g_R(v, v)}$ for $v \in TM$. Its indicatrix $\Sigma_R = F_R^{-1}(1)$ (i.e., the set of all its unit vectors) must be tranverse to all the tangent spaces. Otherwise, as 1 is a regular value of F_R, there would be some $v \in TM \cap \Sigma_R$ such that $(\mathrm{d}F_R)_v(w_v) = 0$ for some $w_v \in T_v(T_pM) \subset T(TM)$, where $v \in T_pM$ and w_v are not tangent to the unit g_R-sphere on T_pM. Thus, all the vectors tangent to T_pM, when looked as elements of (the vertical space in) $T_v(TM)$ lie in the kernel of $\mathrm{d}(F_R)_v$. In particular, this happens to $v_v := \mathrm{d}(v + \lambda v)/\mathrm{d}t|_0$, so,

$$0 = (\mathrm{d}F_R)_v(v_v) = \left.\frac{\mathrm{d}F_R(v + \lambda v)}{\mathrm{d}\lambda}\right|_{\lambda=0} = \left.\frac{\mathrm{d}(1+\lambda)}{\mathrm{d}\lambda}\right|_{\lambda=0} F_R(v) = 1,$$

a contradiction. Then, notice that the pointwise translation $TM \to TM, u_p \mapsto u_p + W_p$ provided by the vector field W does preserve the smooth fiber bundle struture of TM (namely, its structure as an affine bundle, even though not as a linear bundle). Therefore, $\Sigma = \Sigma_R + W$ must remain transverse, as required.

Now, part (1) follows just by applying pointwise the previous lemma and remark, and (2) by constructing the SSTK splitting with $g_0 = g_R$, $\omega = -g(W, \cdot)$ and $\Lambda = 1 - g_R(W, W)$. For (3), the necessary condition is now straightforward, and the sufficient one follows by taking W as the centroid of each ellipsoid and the hypersurface $\Sigma - W$ as the unit sphere bundle for[2] g_R. □

Definition 2.9 A *wind Riemannian structure (WRS)* on a manifold is any hypersurface Σ embedded in TM which satisfies the equivalent conditions in Proposition 2.8 (3), namely for some (univocally determined) Zermelo data (g_R, W), one has $\Sigma = S_R + W$, where S_R is the indicatrix (unit sphere bundle) of g_R.

---

[2]The smoothness of the Zermelo data (g_R, W) follows from the transversality of Σ, [10, Proposition 2.15]; see also Sect. 2.2 of this reference (especially around Example 2.16) for further discussions on transversality.

Such a definition admits natural extensions: $\Sigma \subset M$ is a *wind Finslerian structure* when $\Sigma = S + W$, where S is the indicatrix for a Finsler metric F (now, (F, W) is determined univocally if one imposes additionally that W provides the centroid of S at each point); then Σ_p is a *wind Minkowski structure* at each $p \in M$. The properties of wind Finslerian structures and norms as well as the relation with classical *Zermelo navigation problem* are studied extensively in [10] (see also [17]).

Let us recall the following particular cases for an SSTK splitting.

2.2.1 Static Case

$\omega \equiv 0$. Now, $W \equiv 0$, $g_R = g_0/\Lambda$ (necessarily $\Lambda > 0$) and the spacetime is conformal to the product $(\mathbb{R} \times M, -dt^2 + g_R)$. It is well known that the global hyperbolicity of the spacetime is equivalent to the completeness of g_R as well as to the fact that M is a Cauchy hypersurface, [4, Theorems 3.67, 3.69]. Moreover, the spacetime is always causally continuous, and it will be causally simple if and only if g_R is convex (see the next case).

2.2.2 Stationary Case

$\Lambda > 0$ (K timelike). Now, Σ_p is a g_0-sphere of center W_p and radius $r_p > \|W_p\|_0$ (recall Lemma 2.6), that is, the g_R-norm of W_p is smaller than one. Therefore, the zero tangent vector is always included in the interior of each sphere Σ_p and the hypersurface Σ can be regarded as the indicatrix of a Finsler metric F of Randers type; concretely,

$$F(v) = \frac{\omega(v)}{\Lambda} + \sqrt{\frac{g_0(v, v)}{\Lambda} + \left(\frac{\omega(v)}{\Lambda}\right)^2}. \tag{8}$$

Then, the (future-directed) SSTK lightlike directions on M are neatly described by the vectors $\partial_t + v$ such that $F(v) = 1$.

Recall that, in general, a Finsler metric is *not reversible*, that is, F behaves as a pointwise norm which is only positive homogeneous ($F(\lambda v) = |\lambda| F(v)$ is ensured only for $\lambda \geq 0$). As a consequence, F induces a (possibly nonsymmetric) *generalized distance* d_F and one must distinguish between *forward* open balls $B_F^+(p, r) = \{q \in M : d_F(p, q) < r\}$ and *backward* ones $B_F^-(p, r) = \{q \in M : d_F(q, p) < r\}$. Moreover, even though geodesics make the usual natural sense, the reverse parametrization of a geodesic may not be a geodesic.

Standard stationary spacetimes were systematically studied in [9] using (8) (choosing a conformal representative so that $\Lambda \equiv 1$), including its causal hierarchy. Indeed, the existence of the temporal function t implies that standard stationary spacetimes are stably causal. The other higher steps in the standard ladder of causality are neatly characterized by the Randers metric F as follows (see [9]):

(1) $M = t^{-1}(0)$ is a Cauchy hypersurface if and only if F is complete.
(2) $\mathbb{R} \times M$ is globally hyperbolic if and only if the intersections of the closed balls $\bar{B}^+_F(p, r) \cap \bar{B}^-_F(p', r')$ are compact for all $p, p' \in M, r, r' > 0$.
(3) $\mathbb{R} \times M$ is causally simple if and only if F is convex, in the sense that for each $(p, q) \in M \times M$, there exists a (non-necessarily unique) minimizing F-geodesic from p to q.
(4) $\mathbb{R} \times M$ is always causally continuous, [19].

It is also worth pointing out that the causal boundary of such a stationary spacetime can be described in terms of the natural (Cauchy, Gromov) boundaries of the Finslerian manifold (M, F) (see [12] for a thorough study). On the other hand, a description of the conformal maps of the stationary spacetime in Finslerian terms can be found in [16] and some links between the flag curvature of the Randers metric and the conformal invariants of the spacetime are developed in [11].

2.2.3 Nonnegative Lapse

$\Lambda \geq 0$ (K causal). When $\Lambda(p) = 0$ then the g_R-norm of W_p is equal to 1 and Σ_p contains the zero vector. Then, F becomes a *Kropina* metric; this is a singular type of Finslerian metric $F(v) = -g(v, v)/(2\omega(v))$ which applies only to $v \in T_pM$ such that $\omega(v) < 0$. Indeed, one can rewrite (8) as:

$$F(v) = \frac{g_0(v, v)}{-\omega(v) + \sqrt{\Lambda g_0(v, v) + \omega(v)^2}}, \tag{9}$$

which makes sense even when Λ vanishes and will be called a *Randers-Kropina metric*. Now, the SSTK lightlike directions on M are again described by the vectors of the form $\partial_t + v$ with $F(v) = 1$ with the caution that, whenever $\Lambda = 0$, the direction ∂_t must be included (as $0 \in \Sigma_p$) and F is applied only on vectors v with $\omega(v) < 0$.

The Randers-Kropina metric defines an *F-separation d_F* formally analogous to the generalized distance of the stationary case. Its properties are carefully studied in [10, Sect. 4]. Some important differences between the F- separation d_F and the generalized distance in the standard stationary case are: (i) $d_F(p, q)$ is infinite if there is no admissible curve α from p to q (where *admissible* means here satisfying $\omega(\alpha'(s)) < 0$ whenever $\Lambda(\alpha(s)) = 0$), and (ii) the continuity of d_F is ensured only outside the diagonal; indeed, d_F is discontinuous on (p, p) whenever $d_F(p, p) > 0$ (and in this case $\Lambda(p) = 0$ necessarily), [10, Theorem 4.5, Proposition 4.6]. However, Randers-Kropina metrics admit geodesics analogous to the Finslerian ones.

The ladder of causal properties of the spacetime can be characterized in terms of the properties of F-separation, and it becomes formally analogous to the conclusions (a)–(d) of the stationary case, [10, Theorem 4.9].

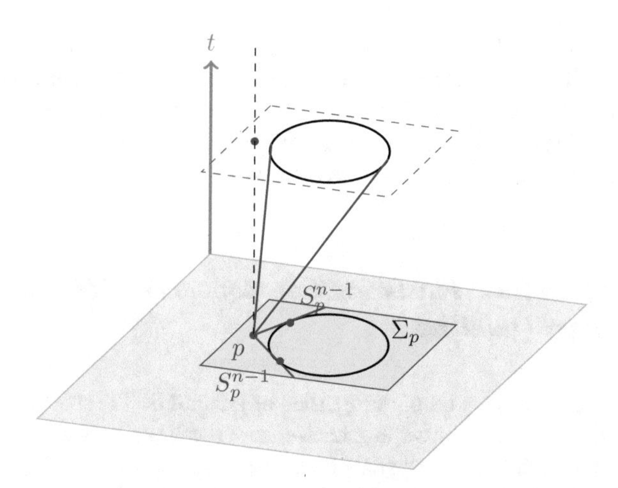

Fig. 2 The diagram shows the g_R-sphere Σ_p determined by the projections of lightlike vectors with the time coordinate equal to 1. When $\Lambda(p) < 0$, the half lines starting at 0 and intersecting Σ_p form a conic region. The intersection of the boundary of this conic region with Σ_p is an $(n-1)$-dimensional g_R-sphere S_p^{n-1} (in the figure, only two points) which divides Σ_p in two connected components

2.2.4 General SSTK Case

(K may be spacelike). Now, when $\Lambda(p) < 0$ one has $g_R(W_p, W_p) > 1$, that is, the zero vector is not included in the solid ellipsoid enclosed by Σ_p. The half lines starting at 0 and tangent to the g_R-sphere Σ_p, provide a cone. Such a cone is tangent to Σ_p in an $(n-1)-$sphere S_p^{n-1}, and $\Sigma_p \setminus S_p^{n-1}$ has two connected pieces. One of them is convex (when looked inside the cone from infinity), and can be described as the open set containing the vectors v inside the cone such that $F(v)$ (computed by using the expression (9)) is equal to 1, see Fig. 2. The other connected part is computed analogously by putting $F_l(v) = 1$ where, now,

$$F_l(v) = -\frac{g_0(v, v)}{\omega(v) + \sqrt{\Lambda g_0(v, v) + \omega(v)^2}} \; (= -F(-v)). \tag{10}$$

The part $F(v) \equiv 1$ in the regions $\Lambda < 0$ and $\Lambda \geq 0$ matches naturally. So, F behaves as a *conic* Finsler metric on all M; indeed, the conic region where F is defined is properly the interior of a cone when $\Lambda(p) < 0$, an open half plane when $\Lambda(p) = 0$ and all T_pM otherwise (see [18] for a systematic study of conic Finsler metrics). However, the concaveness of $F_l(v) \equiv 1$ makes F_l to behave as a Lorentz–Finsler metric, in the sense that it yields a reverse triangle inequality similar to the Lorentzian one, that is, $F_l(v + w) \geq F_l(v) + F_l(w)$ for all $v, w \in T_pM$ in the conic domain of definition.

So, the SSTK lightlike directions on the hypersurface M are described by the vectors of the form $\partial_t + v$ taking into account that, when $\Lambda(p) < 0$, one has to choose vectors with either $F(v) = 1$ or $F_l(v) = 1$, including those in S_p^{n-1} (which corresponds with the limit case $F(v) = F_l(v) = 1$). Again, the causal properties of the SSTK spacetime can be described using the Finslerian elements F, F_l or, directly, by means of the hypersurface Σ. However, this general case is much subtler than the previous ones, and it will be sketched in Sect. 3.

2.3 An Application: Finslerian Consequences of Harris' Stationary Quotients

Even though we will focus on the properties of general SSTK spacetimes linked to wind Riemannian structures, we emphasize now the links between the conformal geometry of standard stationary spacetimes and the geometry of Randers spaces, with applications also to arbitrary Finsler manifolds. Apart from the applications explained in the point (2) of Sect. 2.2, several links introduced in [9] include the behavior of completeness under projective changes (see below) and properties on the differentiability of the distance function to a closed subset (see also [25]) as well as on the Hausdorff measure of the set of cut points. Now, we can add a new application by translating a recent result by Harris on group actions [15] on static/stationary spacetimes to the Finslerian setting, namely:

> There exists a Randers manifold (M, R) such that not all its closed symmetrized balls are compact but its universal covering $(\tilde{M}, \tilde{R})$ satisfies that all its closed symmetrized balls are compact.

In order to understand the subtleties of this result, recall the following.

Remark 2.10 (1) Of course, such a property cannot hold in the Riemannian case (M, g_R), because the closed g_R-balls are compact if and only g_R is complete, and this property holds if and only if the universal covering $(\tilde{M}, \tilde{g}_R)$ is complete.

(2) The key in the Randers case is that the compactness of the closed symmetrized R-balls (which is equivalent to the compactness of the intersections between closed forward and backward R-balls) does not imply R-geodesic completeness (nor the completeness of R in any of the equivalent senses of the Finslerian Hopf–Rinow result). Indeed, as shown in [9] for Randers metrics (and then extended to the general Finslerian case in [22]), the compactness of the closed symmetrized R-balls is equivalent to the existence of a complete Randers metric R^f which is related to R by means of a *trivial projective transformation* (i.e., $R^f = R + df$ for some function f on M such that $R + df > 0$ on $TM \setminus \mathbf{0}$). As will be apparent below, the existence of such a function $\tilde{f}$ in the universal covering $\tilde{M}$ does not imply the existence of an analogous function f in the manifold M, as $\tilde{f}$ is not necessarily projectable.

Next, let us derive the Randers result from Harris' ones.

Example 2.11 Let start with [15, Example 3.4(b)], which exhibits a globally hyperbolic standard static spacetime L' admitting a group of isometries G such that the quotient $L = L'/G$ is static-complete (i.e., it admits a complete static vector field) and causally continuous, but not globally hyperbolic. A static-complete causally continuous spacetime is not necessarily a standard static spacetime; nevertheless, as any distinguishing stationary-complete space is standard stationary [19], so is L too. Then, we can write the standard stationary splitting $L = \mathbb{R} \times M$ with associated Randers metric F, as explained in Sect. 2.2. As L is not globally hyperbolic, (M, F) cannot satisfy the property of compactness of closed symmetrized balls. However, its universal Lorentzian covering $\tilde{L} = \mathbb{R} \times \tilde{M}$ inherits a Randers metric $\tilde{F}$ which must satisfy such a property of compactness (indeed, $\tilde{L}$ must also be the universal covering of L' and, so, globally hyperbolic). We emphasize that, being $\tilde{L}$ static-complete and simply connected, it can be written as a standard static spacetime, [26, Theorem 2.1(1)]; however, the splitting $\tilde{L} = \mathbb{R} \times \tilde{M}$ we are using is only standard stationary but not standard static (otherwise, $\tilde{F}$ would be Riemannian).

3 A New Characterization of WRS Completeness

3.1 *WRS Geodesics and Completeness Versus SSTK Cauchy Hypersurfaces*

In what follows, let Σ be a WRS on M with associated Zermelo data (g_R, W) and SSTK splitting $(\mathbb{R} \times M, g)$, being $K = \partial_t$ Killing, $\omega = -g(K, \cdot)$ and the conformal normalization $\Lambda + |\omega|_0^2 \equiv 1$ chosen, so that $g_0 = g_R$, according to (6) and (7).

3.1.1 First Definitions

The region where K is, resp., timelike ($\Lambda > 0$), lightlike ($\Lambda = 0$), or spacelike ($\Lambda < 0$) will be called of *mild wind* (as $\|W\|_R < 1$), *critical wind*, and *strong wind*; the latter will be denoted M_l as it contains the Lorentz–Finsler metric F_l (besides the conic Finsler one F, defined on all M). Moreover, we define the (possibly signature-changing) metric

$$h = \Lambda g_0 + \omega \otimes \omega, \tag{11}$$

which is Riemannian (resp. degenerate, of signature $(+, -, \ldots, -)$) in the mild (resp. critical, strong) wind region. Remarkably, the conformal metric h/Λ in the region $\Lambda \neq 0$ is the induced metric on the orbit space of $\mathbb{R} \times M$ obtained by taking the quotient by the flow of K (see [9, Proposition 3.18]); clearly, h/Λ is not conformally invariant (however, h/Λ^2 is). In any case, $-h$ (as well as h/Λ and $-h/\Lambda^2$) becomes a Lorentzian metric on M_l, and it is naturally time-oriented because the future timelike vectors of $\mathbb{R} \times M$ project onto a single timecone of $-h$.

Let us define the following subsets of TM. First, put

$$A \cup A_E := \{v \in TM : v \text{ is the projection of a lightlike vector of the SSTK splitting}\}.$$

Inside this set, we will consider two non-disjoint subsets A and A_E. The set A will contain the interior of all the conic domains where the conic Finsler metric F in (9) is naturally defined; then, $A_l := TM_l \cap A$ will also be the natural open domain for the Lorentz–Finsler metric F_l in (10). The set A_E will contain all the continuously extended domain of F_l in A_l, plus the zeroes of the critical region; then, $A \cup A_E$ can also be regarded as the extended domain for F plus the zeroes of the critical region. Explicitly, $A_p = A \cap T_pM$ and $(A_E)_p = A_E \cap T_pM$ are, at each $p \in M$:

$$v_p \in A_p \Longleftrightarrow \begin{cases} v_p \neq 0 & \text{when } \Lambda(p) > 0 \\ -\omega_p(v_p)(= g_R(W_p, v_p)) > 0 & \text{when } \Lambda(p) = 0 \\ -\omega_p(v_p) > 0 \ \text{ and } \ h(v_p, v_p) > 0 & \text{when } \Lambda(p) < 0 \end{cases}$$
$$v_p \in (A_E)_p \Longleftrightarrow \begin{cases} 0_p & \text{when } \Lambda(p) = 0 \\ -\omega_p(v_p) > 0 \ \text{ and } \ h(v_p, v_p) \geq 0 & \text{when } \Lambda(p) < 0 \end{cases}$$

Remark 3.1 F and F_l can be extended continuously from their open domains A, A_l to $A \cup (A_E \cap TM_l)$ and $A_E \cap TM_l$, respectively. Indeed, the metric h (see (11)) vanishes in $(A_E \cap TM_l) \setminus A_l$, making equal the expressions (9) and (10) for F and F_l there. However, the introduced notation will take also into account the following subtlety which occurs for the zeroes of the critical region.

We will work typically with lightlike curves $\tilde{\alpha}(t) = (t, \alpha(t))$ in the spacetime parameterized with the t-coordinate and then, necessarily, either $F(\alpha'(t)) \equiv 1$ or $F_l(\alpha'(t)) \equiv 1$ whenever α' does not vanish. However, when $\tilde{\alpha}$ is parallel to K at some point $\tilde{\alpha}(t_0) = (t_0, p)$, $p \in M$, then p belongs to the critical region and $\tilde{\alpha}'$ projects onto the zero vector $\alpha'(t_0) = 0_p$, which belongs to the indicatrix Σ_p. Clearly, F_p and $(F_l)_p$ cannot be extended continuously to 0_p; however, whenever the F-length (resp. F_l-length) of lightlike curves as $\tilde{\alpha}$ above is considered, we will put $F(0_p) = 1$ (resp. $F_l(0_p) = 1$) as this is the continuous extension of the function $t \to F(\alpha'(t))$ (resp. $t \to F_l(\alpha'(t))$).

Summing up, we adopt the following convention: the *extended domain* of F and F_l is all $A \cup A_E$, with $F(0_p) = F_l(0_p) = 1$ whenever $\Lambda(p) = 0$ (even if F and F_l are not continuous there) and with $F_l(v_p) = \infty$ on the mild and critical regions when $v_p \neq 0$.

Remark 3.2 Both pseudo-Finsler metrics F and F_l are Riemannianly lower bounded (according to [18, Definition 3.10]), that is, there exists a Riemannian metric h_R in M such that $\sqrt{h_R(v, v)} \leq F(v), F_l(v)$ for every $v \in A_E \cup A$. This is observed in [10, below Definition 2.24] and it is easy to check directly, because it is sufficient to prove that the property holds locally (see [18, Remark 3.11 (1)]).

3.1.2 Wind Curves and Balls

A (piecewise smooth) curve $\tilde{\alpha}(t) = (t, \alpha(t))$ is causal (necessarily future-directed) in the SSTK splitting if and only if the velocity $\alpha'(t)$ is *Zermelo-bounded*, in the sense that $\alpha'(t)$ always belongs to the closed g_R-ball of $T_{\alpha(t)}M$ with center $W_{\alpha(t)}$ and radius 1. Even though the curve $\tilde{\alpha}$ is always regular, the velocity of α may vanish either in the region of mild wind or in the region of critical wind. Nevertheless, in Riemannian Geometry, it is natural to work with regular curves in order to reparametrize all the curves with the arc-parameter. In our approach, one can avoid the use of non-regular curves α in the region of mild wind but, for critical wind, the appearance of curves with vanishing velocity becomes unavoidable; indeed, as emphasized in the definition of $A \cup A_E$, the zero vector 0_p is the projection of a lightlike vector if (and only if) p lies in the critical region. Accordingly, we will say that a (piecewise smooth) curve α is Σ*-admissible* if its velocity lies in $A \cup A_E$ and a *wind curve* if, additionally, it is Zermelo-bounded. Even though we will work with piecewise smooth curves as usual, the order of differentiability can be lowered in a natural way. This happens when considering causal continuous curves [4, p. 54]; in this case, the natural assumption would be to consider locally H^1-curves (for causal curves and, then, for wind ones), because, under this regularity, being future-directed continuous causal becomes equivalent to the existence of an almost everywhere future-directed causal velocity [6, Theorem 5.7]).

The previous definitions yield directly the following characterization of wind curves.

Proposition 3.3 *A (piecewise smooth) curve $\alpha : I \subset \mathbb{R} \to M$, with I an interval, is a wind curve if and only if its* graph *$\tilde{\alpha}$ in the associated SSTK splitting defined as*

$$\tilde{\alpha}(t) = (t, \alpha(t)), \qquad \forall t \in I$$

is a causal curve. In this case,

$$\ell_F(\alpha|_{[a,b]}) \leq b - a \leq \ell_{F_l}(\alpha|_{[a,b]}). \tag{12}$$

for each $a, b \in I$, with $a < b$.

Now, for each $p, q \in M$, let

$$C^{\Sigma}_{p,q} = \{\text{wind curves starting at } p \text{ and ending at } q\}.$$

The *forward* and *backward wind balls* of center $p_0 \in M$ and radius $r > 0$ associated with the WRS Σ are, resp:

$$B^+_\Sigma(p_0, r) = \{x \in M :\ \exists\, \gamma \in C^{\Sigma}_{p_0,x},\ \text{s.t. } r = b_\gamma - a_\gamma \text{ and } \ell_F(\gamma) < r < \ell_{F_l}(\gamma)\},$$
$$B^-_\Sigma(p_0, r) = \{x \in M :\ \exists\, \gamma \in C^{\Sigma}_{x,p_0},\ \text{s.t. } r = b_\gamma - a_\gamma \text{ and } \ell_F(\gamma) < r < \ell_{F_l}(\gamma)\},$$

where a_γ and b_γ are the endpoints of the interval of definition of each γ. These balls are open [10, Remark 5.2] and their closures are called *(forward, backward) closed wind balls*, denoted $\bar{B}^+_\Sigma(p_0, r)$, $\bar{B}^-_\Sigma(p_0, r)$. Between these two types of balls, the *forward* and *backward c-balls* are defined, resp., by:

$$\hat{B}^+_\Sigma(p_0, r) = \{x \in M : \exists\, \gamma \in C^\Sigma_{p_0,x}, \text{ s.t. } r = b_\gamma - a_\gamma \text{ (so, } \ell_F(\gamma) \le r \le \ell_{F_l}(\gamma))\},$$
$$\hat{B}^-_\Sigma(p_0, r) = \{x \in M : \exists\, \gamma \in C^\Sigma_{x,p_0}, \text{ s.t. } r = b_\gamma - a_\gamma \text{ (so, } \ell_F(\gamma) \le r \le \ell_{F_l}(\gamma))\}$$

for $r > 0$; for $r = 0$, by convention $\hat{B}^\pm_\Sigma(p_0, 0) = p_0$ (so that, consistently with our conventions, if $0_{p_0} \in \Sigma_{p_0}$, then $p_0 \in \hat{B}^\pm_\Sigma(p_0, r)$ for all $r \ge 0$). When Σ is the indicatrix of a Randers metric, then $B^\pm_\Sigma(p_0, r)$ coincides with the usual (forward or backward) open balls. Such balls have a neat interpretation [10, Proposition 5.1]:

Proposition 3.4 *For the associated SSTK splitting:*

$$I^+(0, x_0) = \cup_{s>0}\{0 + s\} \times B^+_\Sigma(x_0, s),$$
$$I^-(0, x_0) = \cup_{s>0}\{0 - s\} \times B^-_\Sigma(x_0, s),$$
$$J^+(0, x_0) = \cup_{s\ge 0}\{0 + s\} \times \hat{B}^+_\Sigma(x_0, s),$$
$$J^-(0, x_0) = \cup_{s\ge 0}\{0 - s\} \times \hat{B}^-_\Sigma(x_0, s).$$

It is worth emphasizing that the c-balls make a proper natural sense even for a Riemannian metric (see [10, Example 2.28]). Indeed, the property of closedness for all the forward and backward c-balls, called *w-convexity*, extend naturally the notion of convexity for Riemannian and Finslerian manifolds, and becomes equivalent to the closedness of all $J^\pm(t_0, x_0)$ above and, thus, to the causal simplicity of the SSTK spacetime, [10, Theorems 4.9, 5.9].

3.1.3 Geodesics

Starting at the notions of balls on the WRS, geodesics can be defined as follows. A wind curve $\gamma : I = [a, b] \to M$, $a < b$, is called a *unit extremizing geodesic* if

$$\gamma(b) \in \hat{B}^+_\Sigma(\gamma(a), b - a) \setminus B^+_\Sigma(\gamma(a), b - a). \tag{13}$$

Then, a curve is an *extremizing geodesic* if it is an affine reparametrization of a unit extremizing geodesic, and it is a *geodesic* if it is locally an *extremizing geodesic*.

However, geodesics for a WRS can be looked simply as the projections on M of the future-directed lightlike pregeodesics of the associated SSTK spacetime $(\mathbb{R} \times M, g)$ parametrized proportionally to the t coordinate [10, Theorem 5.5]. As K is Killing, any spacetime lightlike geodesic ρ has the relevant invariant $C_\rho = g(\rho'(t), K)$. Reparametrizing with the t coordinate, we have a lightlike pregeodesic $\tilde{\gamma}(t) = (t, \gamma(t))$ and its projection is a unit Σ-geodesic which belongs to one of the following cases:

(1) $C_\rho < 0$: γ is a geodesic of the conic Finsler metric F on M, and $\gamma'(t)$ lies always in A.
(2) $C_\rho > 0$: γ is a geodesic of the Lorentz–Finsler metric F_l on M_l, and $\gamma'(t)$ lies always in $A_l := TM_l \cap A$.
(3) $C_\rho = 0$. We have two subcases:

 (3a) γ is an *exceptional geodesic*, constantly equal to some point p_0 with $\Lambda(p_0) = 0$ and $d\Lambda_{p_0}$ vanishing on the kernel of ω_p, and
 (3b) γ is a *boundary geodesic*, included in the closure of M_l and it satisfies: (i) whenever γ remains in M_l, it is a lightlike pregeodesic of the Lorentzian metric h in (11), reparametrized so that $F(\gamma') \equiv F_l(\gamma')$ is a constant $c = 1$, and (ii) γ can reach the boundary ∂M_l (which is included in the critical region $\Lambda = 0$) only at isolated points $s_j \in I$, $j = 1, 2, ...,$ where $\gamma'(s_j) = 0$ (in this case, $d\Lambda$ does not vanish on all the Kernel of $\omega_{\gamma(s_j)}$).

Even if normal neighborhoods do not make sense in general for WRS's, all the geodesics departing from a given point x_0 (or more generally from points close to x_0) have length uniformly bounded from below by a positive constant.

Proposition 3.5 *Let (M, Σ) be a wind Riemannian structure and $x_0 \in M$, then there exists $\varepsilon > 0$ and a neighborhood U_0 of x_0, such that the unit Σ-geodesics departing from $x \in U_0$ are defined on $[0, \varepsilon)$ and they are extremizing.*

Proof The same proof of [10, Proposition 6.5] works in this case just by replacing x_0 with $x \in U_0$ where, as in that proposition, U_0 is the neighborhood obtained in [10, Lemma 6.4]. □

The WRS is *(geodesically) complete* when its inextendible geodesics are defined on all $\mathbb{R}$. The next result proves, in particular, that its extendibility as a geodesic becomes equivalent to continuous extendibility.

Proposition 3.6 *Let $\gamma : [a, b) \to M$, $b < \infty$, be a Σ-geodesic. If there exists a sequence $\{t_n\} \nearrow b$ such that $\{\gamma(t_n)\}_n$ converges to some $p \in M$, then γ is extendible beyond b as a Σ-geodesic.*

Proof Apply Proposition 3.5 to $p = x_0$ to obtain the corresponding $\varepsilon > 0$. Then, for some n_0 the length of γ between t_{n_0} and b is smaller than ε, and Proposition 3.5 can be claimed again to extend γ beyond b. □

Notice that, from the spacetime viewpoint, *the completeness of the WRS geodesics means that the they cross all the slices $t = constant$ of the SSTK spacetime.* This observation and Proposition 3.4 underlie the following result [10, Theorem 5.9(iv)]:

Theorem 3.7 *The following assertions are equivalent:*

(i) *A slice S_t (and, then every slice) is a spacelike Cauchy hypersurface.*
(ii) *All the c-balls $\hat{B}^+_\Sigma(x,r)$ and $\hat{B}^-_\Sigma(x,r)$, $r>0$, $x\in M$, are compact.*
(iii) *All the (open) balls $B^+_\Sigma(x,r)$ and $B^-_\Sigma(x,r)$, $r>0$, $x\in M$, are precompact.*
(iv) *Σ is geodesically complete.*

It is also worth pointing out that the other causal properties of the spacetime (as being globally hyperbolic even if the slices S_t are not Cauchy) can also be characterized in terms of tidy properties of the corresponding WRS, extending the results (1)–(4) for the stationary case (Sect. 2.2.2).

3.2 The Extended Conic Finsler Metric $\bar{F}$

In [18], the properties of distances associated with pseudo-Finsler metrics were studied in full generality. Indeed, the metrics in that reference were defined in an open conic subset without assuming that they can be extended to the boundary. The case of critical wind considered in Sect. 2.2.3 is included in this general case. Indeed, only the conic Finsler metric F appears there and its extension to the boundary of A in the set $TM\setminus\mathbf{0}$ would be naturally equal to infinity (this is the underlying reason why this boundary did not contain allowed directions and, so, it was not included in A_E). Bearing this infinite limit in mind, such an F has been extensively studied in [10, Sect. 4].

However, as discussed in Remark 3.1, the metrics F and F_l are continuously extendible to the boundary of A in $TM_l\setminus\mathbf{0}$. Next, we will study some elements associated with this extension of F, which will allow us to extend Theorem 3.7, adding a further characterization to those appearing there.

3.2.1 $\bar{F}$-Separation

Next, we will consider the extension of F explained in detail in Remark 3.1, in order to introduce a related *separation*, whose role has some similarities with a distance. Even though the extension of F has already been taking into account in order to compute the length of wind curves, we will introduce explicitly the notation $\bar{F}$ in order to distinguish our study from previous ones, where the F*-separation* only takes into account the open domains (for example, in the Randers-Kropina case studied in [10, Section 4] or in [18]).

Definition 3.8 The *extended conic Finsler metric* of a WRS (M,Σ) (or its associated SSTK splitting) is the map $\bar{F}: A\cup A_E\to[0,+\infty)$ given by (9) and the associated $\bar{F}$*-separation* is the map $d_{\bar{F}}: M\times M\to[0,+\infty]$ given by[3]:

[3]For simplicity, we assume here that the connecting curves constitute the set of wind curves $C^\Sigma_{p,q}$. As $\bar{F}$ is invariant under (positive) reparametrizations, one could also consider curves with

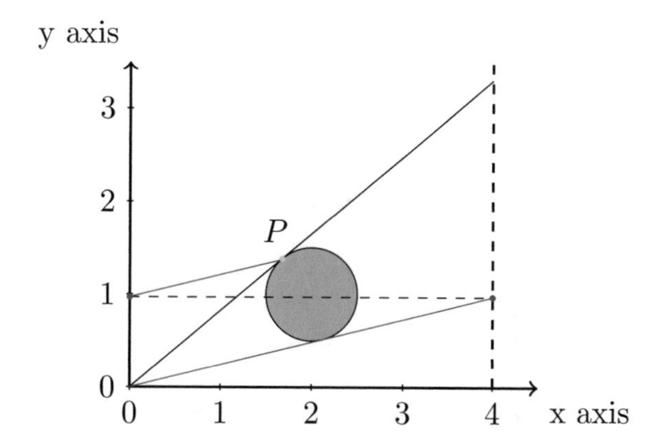

Fig. 3 Wind Riemannian structure on a torus $T^2(= \mathbb{R}^2/\sim$, with $(x, y) \sim (x', y')$ if and only if $(x - x')/4, (y - y')/4 \in \mathbb{Z})$ with non-continuous $d_{\bar{F}}$ on points at finite separation

$$d_{\bar{F}}(p, q) = \inf_{\gamma \in C^{\Sigma}_{p,q}} \ell_{\bar{F}}(\gamma)$$

for every $p, q \in M$. In particular, $d_{\bar{F}}(p, q) = +\infty$ if and only if $C^{\Sigma}_{p,q}$ is empty.

Notice that, in this definition, $\ell_{\bar{F}}(\gamma)$ is equal to the length $\ell_F(\gamma)$ in the definition of the wind balls, as we already used there the extension of F; however, the $\bar{F}$-separation is different to the F-separation d_F in [10, Sect. 4] and [18] as, now, curves with velocity in A_E are allowed, that is, $d_{\bar{F}} \leq d_F$. Consequently, let us denote

$$B^+_{\bar{F}_p}(r) = \{v \in (A_E)_p \subset T_pM : \bar{F}(v) < r\}, \qquad B^+_{\bar{F}}(p, r) = \{q \in M : d_{\bar{F}}(p, q) < r\}, \tag{14}$$

where the latter is the *(forward) $d_{\bar{F}}$-ball of radius $r > 0$ and center $p \in M$*; consistently, $\bar{B}^+_{\bar{F}}(p, r)$ is its closure and dual *backward* notions appear replacing $d_{\bar{F}}(p, q)$ with $d_{\bar{F}}(q, p)$ in (14).

For a wind Minkowski structure of strong wind on $\mathbb{R}^n$, $d_{\bar{F}}$ becomes discontinuous because it jumps from a finite and locally bounded value of $d_{\bar{F}}(0, q)$ when q belongs to $A \cup A_E$ to an infinite value when q is outside. However, $d_{\bar{F}}$ may be discontinuous even when it remains finite, resembling the behavior of the time-separation (Lorentzian distance) on a spacetime.

Example 3.9 Let us consider the WRS induced on the torus $T^2 = \mathbb{R}^2/4\mathbb{Z}$ from the wind Minkowskian structure on $\mathbb{R}^2$ whose indicatrix is the sphere of radius 1/2 centered at (2, 1), as depicted in Fig. 3. Then $d_{\bar{F}}((0, 0), P) = 1$ (P as in the Fig. 3), but the points in the red line close to P are at a distance much greater than 1. This concludes that the distance associated with $\bar{F}$ is not necessarily continuous even when it is finite and the WRS is geodesically complete.

(Footnote 3 continued)
velocities in $A \cup A_E$ even vanishing in the region of mild wind (the relevant restriction for such a curve γ would be the existence of a parametrization $\hat{\gamma}$ satisfying that $t \to (t, \hat{\gamma}(t))$ is causal), but no more generality would be obtained.

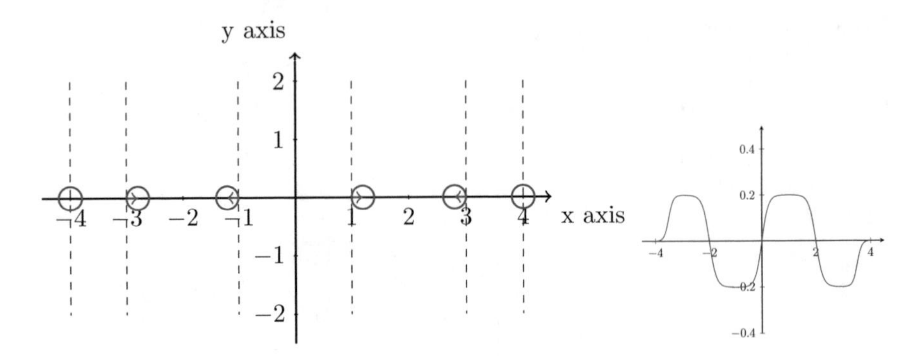

Fig. 4 Even for a Randers-Kropina metric, the $\bar{F}$-separation may present subtle properties, forbidden for a generalized distance (as $\{p_n\} \to p$ but $d_F(p_n, p) = \infty$, see Example 3.11), including "*black hole*"-type behaviors (Example 3.24)

Let us discuss some properties related to *generalized distances* (compare with [12, Section 3.1]). The first one is very simple.

Proposition 3.10 *If* $\{p_n\}$ *is a sequence in* M *such that* $d_{\bar{F}}(p, p_n) \to 0$ *or* $d_{\bar{F}}(p_n, p) \to 0$ *for some* $p \in M$*, then* $p_n \to p$.

Proof By Remark 3.2, there exists a Riemannian metric h_R such that $\sqrt{h_R(v, v)} \leq \bar{F}(v)$ for every $v \in A \cup A_E$. As a consequence, the h_R-distance from p_n to p is smaller than the $\bar{F}$-separation and it must go to 0, yielding the result. □

However, as an important difference with the case of generalized distances, the converse may not hold. The following example (a small variation of the one introduced in the Fig. 4 of [17]) will be useful for this and other purposes.

Example 3.11 Consider the WRS (g_R, W) in Fig. 4, where g_R is just the usual Euclidean metric multiplied by a factor 1/5, $W_{(x,y)} = f(x)\partial_x$ and the function f behaves as depicted in the graph, so that the lines $x = \pm 1, x = \pm 3$ have critical wind. Clearly $\{p_n = (1 + 1/n, 0)\} \to p = (1, 0)$, but $d_{\bar{F}}(p_n, p) = \infty$ for all $n \in \mathbb{N}$ (recall, however, $\{d_{\bar{F}}(p, p_n)\} \to 0$).

With this caution, however, notions related to Cauchy sequences and completeness can be maintained.

Definition 3.12 Let $\bar{F}$ be the extended conic Finsler metric associated with a WRS (M, Σ) and $d_{\bar{F}}$ its associated separation:

(1) A subset $\mathcal{A} \subset M$ is *forward* (resp. *backward*) *bounded* if there exists $p \in M$ and $r > 0$ such that $\mathcal{A} \subset B^+_{\bar{F}}(p, r)$ (resp. $\mathcal{A} \subset B^-_{\bar{F}}(p, r)$).
(2) A sequence $\{x_i\}$ is *forward* (resp. *backward*) *Cauchy* if for any $\varepsilon > 0$ there exists $N > 0$ such that $d_{\bar{F}}(x_i, x_j) < \varepsilon$ for all i, j with $N < i < j$ (resp. $N < j < i$).

(3) The space $(M, d_{\bar{F}})$ is *forward* (resp. *backward*) *Cauchy complete* if every forward (resp. backward) Cauchy sequence converges to a point q. Moreover, $(M, d_{\bar{F}})$ is *Cauchy complete* if it is both, forward and backward Cauchy complete.

It is worth noting that, even in the more restrictive framework of generalized distances, the convergence of all the forward Cauchy sequences does not imply the convergence of all the backward ones, so, to be only forward or backward Cauchy complete makes sense (see [12, Section 3] for a comprehensive study).

In spite of Example 3.11, the following property holds for converging Cauchy sequences.

Proposition 3.13 *If an $\bar{F}$-forward Cauchy sequence $\{p_n\}_n$ converges to a point $p \in M$, then $\{d_{\bar{F}}(p_n, p)\} \to 0$.*

Moreover, if a forward Cauchy sequence $\{p_n\}_n$ admits a partial subsequence $\{p_{n_k}\}_k \to p \in M$, then all the sequence converges to p.

Proof For the first assertion, the Cauchy sequence $\{p_n\}_n$ will admit a subsequence $\{p_{n_m}\}_m$ such that $d_{\bar{F}}(p_{n_m}, p_{n_{m+1}}) < 2^{-m}$. Then, by Proposition 3.4, there exists a sequence t_m such that $(t_{m+1}, p_{n_{m+1}}) \in J^+(t_m, p_{n_m})$ in the associated splitting $(\mathbb{R} \times M, g)$ and $t_{m+1} - t_m < 2^{-m}$ for every $m \in \mathbb{N}$. Now define $\bar{t} = \lim_{m\to\infty} t_m$, which is finite (since $\bar{t} - t_1 = \lim_{k\to\infty} \sum_{m=1}^{k} (t_{m+1} - t_m) < \sum_{m=1}^{\infty} 2^{-m} = 2$), consider a convex neighborhood V of $(\bar{t}, p)$ and let N be big enough such that $p_{n_m} \in V$ for $m \geq N$. As $(t_m, p_{n_m}) \in J^+(t_N, p_{n_N})$ for $m \geq N$ and the causal relation is closed in any convex neighborhood (see [24, Lemma 14.2]), it follows that $(\bar{t}, p) \in J^+(t_N, p_{n_N})$. Moreover, for $m \geq N$, $d_{\bar{F}}(p_{n_m}, p) \leq \bar{t} - t_m < \sum_{k=m}^{\infty} 2^{-m} \to 0$. Using the triangle inequality and the definitions of $\bar{F}$-forward Cauchy sequences, we conclude that $\{d_{\bar{F}}(p_n, p)\} \to 0$.

For the last assertion, recall that, then $\{d_{\bar{F}}(p_{n_k}, p)\}_k \to 0$ and, using again the triangle identity, $\{d_{\bar{F}}(p_n, p)\} \to 0$. So, convergence follows from Proposition 3.10. □

3.2.2 $\bar{F}$-Exponential and Geodesic Balls

Next, our aim will be to define an exponential map $\exp_p^{\bar{F}}$ for $\bar{F}$ at each point $p \in M$. This will extend the usual exponential for F in A (namely, constructed using the formal Christoffel symbols and the associated geodesics) and will be well defined and continuous even in the boundary of $(A_E)_p$ in the strong wind region (in this region, $\exp_p^{\bar{F}}$ will yield lightlike pregeodesics of $-h$). In order to ensure that both, F-geodesics and lightlike h-geodesics match continuously we will work with pregeodesics of the SSTK spacetime and will project them on M. However, recall that F cannot be extended continuously to the zero section in the critical region, so, this section will be excluded first and the possibilities of extension will be studied specifically.

Definition 3.14 Let Σ be a WRS on M. The $\bar{F}$*–exponential* is the map

$$\exp^{\bar{F}} : \mathcal{U} \subset A \cup (A_E \setminus \mathbf{0}) \longrightarrow M \qquad v_p \mapsto \alpha(1)$$

where, for each v_p, α is the geodesic constructed by taking the unique lightlike pregeodesic $\tilde{\alpha}$ in the associated SSTK splitting $(\mathbb{R} \times M, g)$ written as

$$\tilde{\alpha}(s) = (\bar{F}(v_p)s, \alpha(s)), \qquad s \in [0, 1],$$

(i.e., $\tilde{\alpha}$ is reparametrized proportionally to the projection $t : \mathbb{R} \times M \to \mathbb{R}$) and initial velocity

$$\tilde{\alpha}'(0) = (\bar{F}(v_p), v_p) \in T_{(0,p)}(\mathbb{R} \times M),$$

while $\mathcal{U}$ is the open subset of $A \cup (A_E \setminus \mathbf{0})$ containing all the tangent vectors v_p such that their associated pregeodesics $\tilde{\alpha}$ are defined on all $[0, 1]$.

Recall that the exact behavior of all the Σ-geodesics as either F, F_l, or boundary geodesics explained in Sect. 3.1.3 is determined in [10, Proposition 6.3] (see also Theorems 5.5 and 2.53 in that reference). Indeed, the curves α constructed in the previous definition correspond with the F-geodesics and the boundary geodesics for Σ. Bearing this in mind, the following properties are in order.

Proposition 3.15 *For each $p \in M$, consider the star-shaped domain $\mathcal{U}_p := T_pM \cap \mathcal{U}$ and the restriction $\exp_p^{\bar{F}} := \exp^{\bar{F}}|_{\mathcal{U}_p}$.*

(1) If the wind on p is mild (resp. critical; strong), then $\mathcal{U}_p \subset A_p$ being $A_p = T_pM \setminus \{0_p\}$ (resp., $\mathcal{U}_p \subset A_p$, being A_p an open half-space of T_pM; $\mathcal{U}_p \subset (A_E)_p$, being $(A_E)_p$ a solid cone without vertex).

(2) Assume that the curve α constructed for $v_p \in \mathcal{U}_p$ in Definition 3.14 remains in:

(a) the region of mild wind (and, so, Σ is the indicatrix of a Randers metric around the image of α): then, α is an F-geodesic (so that $\exp_p^{\bar{F}}$ will agree with the natural F-exponential).

(b) an open region of non-strong wind (and, so, Σ is the indicatrix of a Randers-Kropina metric whose Christoffel symbols and geodesics can be computed as in the Finslerian case on all A) then, α is a Randers-Kropina geodesic (according to [10, Sect. 4]).

(c) the region M_l of strong wind, (and, so, Σ yields, the indicatrix of both, a conic Finsler metric F and a Lorentz–Finsler one F_l, plus the Lorentz metric $-h$): then α satisfies one of the following two exclusive possibilites:

(ci) α' lies in the open cones A_l and α is an F-geodesic, or

(cii) α' lies in the boundary $A_E \setminus A_l$ of the cones (thus, being a boundary geodesic*) and it becomes a pregeodesic of the Lorentzian metric $-h$ on M_l.*

(3) The $\bar{F}$-exponential $\exp^{\bar{F}}$ is smooth on the open region $\mathcal{U} \cap A$ and continuous on all $\mathcal{U}$.

(4) The $\bar{F}$-exponential $\exp^{\bar{F}}$ can be extended continuously to the zero section away from the critical region. Moreover, the restriction of $\exp^{\bar{F}}$ to Σ (or, in general, to the hypersurface $r\Sigma$ for any $r \geq 0$) can be continuously extended to the zero section in the critical region.

Proof (1) Straightforward from the definitions of the domains.

(2) See Theorem 5.5 and 6.3 in [10] (for part (cii), recall also Lemma 3.21); this matches with the summary in Sect. 3.1.3 above.

(3) Taking into account Definition 3.14, for each $v \in \mathcal{U}$ ($v \in T_pM$), put $\tau = \bar{F}(v)$, consider the lightlike g-pregeodesic $\tilde{\alpha}(s) = (\tau s, \alpha(s))$ and reparametrize it as a geodesic $\hat{\alpha}(h) = (t(h), x(h))$ with the same initial velocity (τ, v). Then $(\tau s, \alpha(s)) = (t(h(s)), x(h(s))$ for some function $h(s)$ with $h(0) = 0$, $h'(0) = 1$. Therefore,

$$t'(h(s))\, h'(s) = \tau,$$

that is, the function h satisfies the ODE

$$h'(s) = 1/f(h(s), \tau, v), \quad h(0) = 0, \quad \text{where } f(\bar{s}, \tau, v) := \frac{1}{\tau} \frac{d}{ds}\bigg|_{s=\bar{s}} t(\exp^g_{(0,p)}(s(\tau, v))) \tag{15}$$

whenever $\exp^g_{(0,p)}$ is defined in $s(\tau, v)$. Notice that f can be regarded as a smooth function on some maximal open subset of $\mathbb{R} \times \mathbb{R}_+ \times TM$ and, as t is a temporal function, f is strictly positive when applied on (s, τ, v) whenever (τ, v), regarded as a vector in $T_{(0,p)}(\mathbb{R} \times M)$, is future-directed and causal. Summing up, for any $v = \alpha'(0) \in \mathcal{U}$,

$$\exp^{\bar{F}}_p(v) = \alpha(1) = x(\varphi(1, \bar{F}(v), v)) = \pi(\exp^g_{(0,p)}(\varphi(1, \bar{F}(v), v) \cdot (\bar{F}(v), v)), \tag{16}$$

where $s \to \varphi(s, \tau, v)$ is the solution of (15) which, obviously, depends smoothly on (τ, v). So, the expression (16) shows that $\exp^{\bar{F}}$ is smooth everywhere except at most in the boundary of A_E, because $\bar{F}$ is only continuous there.

(4) For the first assertion, recall from (15) that the map $(s, v) \to f(s, F(v), v)$ is positive homogeneous of degree 0 in v, that is, $f(s, F(\lambda v), \lambda v) = f(s, F(v), v))$ for all $\lambda > 0$. Then, $\varphi(1, \bar{F}(v), v)$ in (16) remains locally bounded outside the Kropina region (as we can take a compact neighborhood W of p which does not intersect the critical region, and consider that v varies in $\bar{F}^{-1}(1) \cap TW$, which is compact). So, when v goes to 0, the variable of the g-exponential in (16) goes to 0 and $\exp^{\bar{F}}_p(v)$ goes to p, as required.

For the second one, notice that $\bar{F}$ cannot be continuously extended to 0_p whenever p lies in the critical region. However, as $\bar{F}$ is equal to 1 on $\Sigma \setminus \mathbf{0}$, it can be extended continuously to 1 on Σ, and the continuity of the exponential in the SSTK spacetime ensures the result. Obviously, this can be extended to the case $r\Sigma$, now extending $\bar{F}$ continuously as $\bar{F}(0) = r$. □

Remark 3.16 (1) It is easy to check that $\exp^{\bar{F}}_p$ may be non-differentiable at the boundary $A_E \setminus A$, as the initial data for γ in Definition 3.14 depends on $\bar{F}(v_p)$

and $\bar{F}$ may be non-smooth on the boundary. Indeed, the root in the expression (9) becomes 0 there; in the region of strong wind, these zeroes are the lightlike vectors for the Lorentzian metric $-h$, which implies non-smoothability there.

However, direct computations in a concrete example may be illustrative. Consider as Zermelo data in $\mathbb{R}^2$ the usual scalar product and the wind vector $W = \partial_x + \partial_y$. The function $\phi(s) = \bar{F}(s\partial_x|_{(0,0)} + \partial_y|_{(0,0)})$ for $s \geq 0$ is smooth for $s > 0$ but only continuous for $s = 0$. Indeed, for each $s \geq 0$, there exists a unique point (x_s, y_s) in the convex part the indicatrix $\Sigma_{(0,0)} = \{(x, y) \in T_{(0,0)}\mathbb{R}^2 : (x-1)^2 + (y-1)^2 = 1\}$ such that $\phi(s) \cdot (x_s, y_s) = (s, 1)$. Then, $\phi(s) = 1/y_s$, $x_s = sy_s$, substituting in the indicatrix x_s, necessarily $y_s = \left(s + 1 \pm \sqrt{2s}\right)/(s^2 + 1)$, the choice of its convex part selects the positive sign for y_s and, then, $\phi(s) = (s^2 + 1)/\left(s + 1 + \sqrt{2s}\right)$.

(2) About the question of continuity in part (4), notice that, when the integral curve of K through a point p of the critical region is a geodesic, then the continuous extension of $\exp^F|_\Sigma$ to 0_p yields simply the point p. However, this does not occur when such an integral curve is not a geodesic. So, in general, it is impossible to extend continuously the full exponential $\exp_p^{\bar{F}}$ to all the zero section.

Summing up, Proposition 3.15 (4) allows one to extend continuously the domain $\mathcal{U}$ of $\exp^{\bar{F}}$ in order to include the zero section away from the critical region. What is more, the critical region will not be an obstacle to define geodesic balls of radius r, since one can extend the restriction of $\exp^{\bar{F}}$ to any $r\Sigma$.

Definition 3.17 Let Σ be a WRS on M, $p \in M$. For any $r \geq 0$ such that $\bar{B}^+_{\bar{F}_p}(r)$ (recall the notation (14)) is included in the starshaped domain $\mathcal{U}_p \cup \{0_p\}$, the *(forward) geodesic $\bar{F}$-sphere of center p and radius r* is the set

$$\mathcal{S}^+_{\bar{F}}(p, r) = \exp_p^{\bar{F}}(r\Sigma_p),$$

where $\exp_p^{\bar{F}}$ is assumed to be extended to 0_p, if necessary. Then, for any $r_0 > 0$ such that $\mathcal{S}^+_{\bar{F}}(p, r)$ is defined for all $0 \leq r < r_0$, the *(forward) geodesic $\bar{F}$-ball* and *closed (forward) geodesic $\bar{F}$-ball* of center p and radius r are, resp., the sets

$$\mathcal{B}^+_{\bar{F}}(p, r) = \cup_{0 \leq r < r_0} \mathcal{S}^+_{\bar{F}}(p, r), \qquad \bar{\mathcal{B}}^+_{\bar{F}}(p, r) = \text{closure}(\mathcal{B}^+_{\bar{F}}(p, r_0)).$$

Remark 3.18 Necessarily, $\mathcal{S}^+_{\bar{F}}(p, r)$ is compact and, whenever $\bar{B}^+_{\bar{F}_p}(r_0) \subset \mathcal{U}_p$ then

$$\bar{\mathcal{B}}^+_{\bar{F}}(p, r_0) = \mathcal{B}^+_{\bar{F}}(p, r_0) \cup \mathcal{S}^+_{\bar{F}}(p, r_0)$$

and it is compact too (indeed, the right-hand side is the projection of the compact subset in the SSTK splitting obtained by exponentianing the null vectors w tangent to $(0, p)$ such that $0 \leq dt(w) \leq r_0$ with a t-reparametrized pregeodesic analogously as in the proof of part (3) in Proposition 3.15). However, $\mathcal{B}^+_{\bar{F}}(p, r)$ is not necessarily open; in fact, this happens even for wind Minkowskian structures, as $A_p \cup (A_E)_p$ is not open away from the mild wind region.

Definition 3.19 A curve $\gamma : [a, b] \to M$, $p = \gamma(a)$, $v_p = \gamma'(a)$ is an *$\bar{F}$-geodesic* of a WRS Σ if either $v_p \in A_p \cup ((A_E)_p \setminus 0_p)$ and $\gamma(s) = \exp_p^{\bar{F}}((s-a)v_p)$ for all $s \in [a, b]$ or $\gamma'(a) = 0_p$, $\Lambda(p) = 0$ and there exists some $c > 0$ such that the curve $[a, b] \ni s \mapsto (c(s-a), \gamma(s)) \in \mathbb{R} \times M$ is a lightlike pregeodesic of the SSTK spacetime with initial condition $c\partial_t|_{(0,p)}$ at the instant a.

The previous definition extends naturally to non-compact intervals. Namely, for the case $[a, b)$ no modification is necessary, while for the case $(a, b]$ one assumes that, for all $t \in (a, b)$, the restriction $\gamma|_{[t,b]}$ is a geodesic. Indeed, *the $\bar{F}$-geodesics are the F-geodesics, boundary geodesics, and exceptional geodesics* explained in Sect. 3.1.3 (*i.e., all the Σ-geodesics except the F_l ones*). The lack of symmetry of Σ makes meaningful the following distinction, as in the case of classical Finsler metrics.

Definition 3.20 $\bar{F}$ is *forward* (resp. *backward*) *complete* if the domain of all its inextendible $\bar{F}$-geodesics is upper (resp. lower) unbounded.

Clearly, given a WRS Σ, $\bar{F}$ is forward complete if and only if the extended conic Finsler metric for the reverse WRS $\tilde{\Sigma} = -\Sigma$ is backward complete.

3.3 Main Result on the Completeness of Σ

In order to obtain our main result, let us start strengthening the relations between $\bar{F}$-balls and geodesics.

Lemma 3.21 *Let (M, Σ) be a WRS and $x_0 \in M$. For each neighborhood W_0 of x_0, there exists another neighborhood $U_0 \subset W_0$ and some $\varepsilon > 0$ such that, for every $x \in U_0$ and $0 < r < \varepsilon$:*

(i) Both, the c-balls $\hat{B}_\Sigma^+(x, r)$ and the closed $\bar{F}$-balls $\bar{B}_{\bar{F}}^+(x, r)$ are compact and included in W_0.
(ii) $\bar{B}_{\bar{F}_x}^+(r) \subset T_xM$ is included in $\mathcal{U}_x \cup \{0_x\}$ and, then, the geodesic and metric $\bar{F}$-balls coincide, that is,: $\mathcal{B}_{\bar{F}}^+(x, r) = B_{\bar{F}}^+(x, r)$ and $\bar{\mathcal{B}}_{\bar{F}}^+(x, r) = \bar{B}_{\bar{F}}^+(x, r)$.

Proof Roughly, the result follows from [10, Lemma 6.4] and Proposition 3.5. Indeed, [10, Lemma 6.4] provides directly both, the neighborhood U_0 and $\varepsilon > 0$ such that the c-balls $\hat{B}_\Sigma^+(x, r)$ are compact for every $x \in U_0$ and $0 < r < \varepsilon$. Even more, this also proves that $\cup_{r' \in [0,r]} \hat{B}_\Sigma^+(x, r')$ (which is equal to the projection on M of the compact set $J^+(0, x) \cap t^{-1}([0, r])$, recall Proposition 3.4 and the choice of U_0) is compact too. As this set includes $B_{\bar{F}}^+(x, r)$, this proves the compactness of its closure, concluding (i).

For (ii) we claim first that if U_0 is obtained as above then $[0, \varepsilon] \times U_0$ can be assumed to lie in a globally hyperbolic (and thus, causally simple) neighborhood U of the SSTK splitting satisfying $\pi(U) \subset W_0$. Indeed, $(0, x_0)$ admits an arbitrarily small globally hyperbolic neighborhood U (see [23, Theorem 2.14]), and we have just to assume that $\pi(U) \subset W_0$. Then, there exists a small neighborhood $W'_0 \subset \pi(U)$ of x_0 and some $\varepsilon' > 0$ such that $[0, \varepsilon'] \times W'_0 \subset U$. So, the claimed property follows just by repeating the step (i) imposing $U_0 \subset W'_0 (\subset W_0)$ and choosing $\varepsilon < \varepsilon'$.

Proposition 3.5 implies that $\bar{B}^+_{\bar{F}_x}(r) \setminus \{0_x\} \subset \mathcal{U}_x$ for every $x \in U_0$. Even more, for the required equalities, the inclusion $\subseteq$ follows trivially, as all the points in the geodesic ball are reached by a geodesic of $\bar{F}$- length smaller than the radius (or equal to it in the closed case, see Remark 3.18), which can be regarded as a wind curve (indeed, so is the $\bar{F}$-geodesic whenever $r \leq 1$; otherwise, it can be reparametrized affinely as a wind curve). For the inclusion $\supseteq$ in the first equality, recall that for each $y \in B^+_{\bar{F}}(x, r)$ the connecting wind curve $\gamma : [0, r'] \to M, r' < r$ yields a causal curve $[0, r'] \ni t \mapsto (t, \gamma(t))$. Then $z' = (r', \gamma(r')) \in J^+(0, x)$ and being $[0, \varepsilon] \times U_0$ included in the causally simple subset U, there exists a first point $z \in [0, \varepsilon] \times W_0$ on the integral curve of $\partial_t|_{z'}$ which belongs to $J^+(0, x)$. Then, the unique lightlike geodesic from $(0, x)$ to z projects into the required $\bar{F}$-geodesic. Moreover, then the inclusion $\supseteq$ in the second equality also follows just applying the compactness (and then closedness) of $\bar{\mathcal{B}}^+_{\bar{F}}(p, r_0)$, see Remark 3.18. □

Proposition 3.22 *Let (M, Σ) be a WRS and $\bar{F}$ the associated extended conic Finsler metric. Given $p \in M$ assume that $\exp^{\bar{F}}_p$ (resp. exponential at p for the reverse WRS $\tilde{\Sigma} = -\Sigma$) is defined in the whole domain $(A_E)_p \subset T_pM$ (resp. $-(A_E)_p \subset T_pM$) of F_p. Then, for any $q \in M$, $q \neq p$, such that $d_{\bar{F}}(p, q)$ (resp. $d_{\bar{F}}(q, p)$) is finite, there exists a minimizing $\bar{F}$-geodesic from p to q (resp. from q to p).*

Proof Consider a sequence of wind curves α_n from p to q such that $\lim_n \ell_{\bar{F}}(\alpha_n) = d_{\bar{F}}(p, q)$. Let $\tilde{\alpha}_n(t) = (t, \alpha_n(t))$ be the graph of α_n in the associated SSTK spacetime $(\mathbb{R} \times M, g)$. As the curves $\tilde{\alpha}_n$ are future-directed causal, then there exists a limit curve $\tilde{\alpha}(s) = (s, \alpha(s))$ starting at $(0, p)$ defined in a subinterval I of $[0, d_{\bar{F}}(p, q)]$, $0 \in I$, such that $\lim_n \alpha_n(s) = \alpha(s)$ for all $s \in I$ (see [4, Sect. 3.3] or [10, Lemma 5.7]). This implies that $\tilde{\alpha}$ is a lightlike pregeodesic and

$$d_{\bar{F}}(p, \alpha(s_0)) = \ell_{\bar{F}}(\alpha|_{[0,s_0]}), \tag{17}$$

for all $s_0 \in I \setminus \{0\}$. Indeed, if either $\tilde{\alpha}$ is not a lightlike pregeodesic or (17) does not hold (only the inequality $<$ should be taken into account then), we claim that there exists $\delta > 0$ such that $(s_0 - \delta, \alpha(s_0)) \in J^+(0, p)$. This follows from Proposition 3.4 if (17) does not hold. When $\tilde{\alpha}$ is not a pregeodesic, observe that $(s_0, \alpha(s_0)) \in I^+(0, p)$ (see [24, Proposition 10.46]) and, as the chronological relation is open, there exists $\delta > 0$ such that $(s_0 - \delta, \alpha(s_0)) \in I^+(0, p) \subseteq J^+(0, p)$. So, for the claimed δ, there exists a future-directed causal curve $\tilde{\rho}$ from $(0, p)$ to $(s_0 - \delta, \alpha(s_0))$. Concatenating this curve with $[s_0 - \delta, \ell_{\bar{F}}(\alpha_n) - \delta] \ni s \to (s - \delta, \alpha_n(s)) \in \mathbb{R} \times M$, we obtain future-directed causal curves $\tilde{\gamma}_n$ from $(0, p)$ to $(\ell_{\bar{F}}(\alpha_n) - \delta, q)$. Taking n big enough, one gets that the projection of $\tilde{\gamma}_n$ is a curve from p to q which has length equal to $\ell_{\bar{F}}(\alpha_n) - \delta < d_{\bar{F}}(p, q)$ in contradiction with the definition of $d_{\bar{F}}(p, q)$. Therefore, $\tilde{\alpha}$ is a lightlike pregeodesic and (17) holds, as required.

Now, the discussion in Sect. 3.1.3 implies that, being $\tilde{\alpha}$ a lightlike pregeodesic, necessarily α is either a unit geodesic for F or F_l, or a boundary geodesic (exceptional geodesics are excluded as $p \neq q$). Moreover, if α were an F_l-geodesic which is not a boundary one, then it would be included in M_l, α' could not vanish and $0 < F(\alpha') < F_l(\alpha') \equiv 1$ (recall that $F \leq F_l$ holds always and, if the equality occurred in our case, then $h(\alpha', \alpha') = 0$ at some point, that is, α would be the boundary geodesic corresponding to the initial velocity at that point). Then,

$$\ell_{\bar{F}}(\alpha|_{[0,s_0]}) = \int_0^{s_0} F(\alpha'(s))ds < d_{\bar{F}}(p, \alpha(s_0)).$$

As this is a contradiction with (17), α becomes either a boundary or F-geodesic and, thus, an $\bar{F}$-geodesic. So, by the hypothesis on $\exp_p^{\bar{F}}$, α is defined in $[0, d_{\bar{F}}(p, q)]$, which concludes the proof. □

Finally, recalling Definitions 3.12 and 3.20, we can state our main result.

Theorem 3.23 *Let $(M, \bar{F})$ be the conic Finsler manifold associated with a WRS (M, Σ). Then (M, Σ) is geodesically complete if and only if $(M, \bar{F})$ is geodesically complete. In addition, the following conditions are equivalent:*

(a) The space $(M, d_{\bar{F}})$ is forward (resp. backward) Cauchy complete.
(b) $(M, \bar{F})$ is forward (resp. backward) geodesically complete.
(c) Every closed and forward (resp. backward) bounded subset of $(M, d_{\bar{F}})$ is compact.

Moreover, any of the above conditions implies that $\bar{F}_l$ is forward (resp. backward) geodesically complete.

Finally, if $p_0 \in M$ has the following property: $d_{\bar{F}}(p_0, q)$ (resp. $d_{\bar{F}}(q, p_0)$) is finite for every $q \in M$, then the above conditions are equivalent to

(d) At $p_0 \in M$, $\exp_{p_0}^{\bar{F}}$ (resp. backward $\exp_{p_0}^{\bar{F}}$) is defined on all $(A \cup A_E)_{p_0}$ (resp. $-(A \cup A_E)_{p_0}$).

Proof Let us start with the equivalences among the displayed items. The introduced framework will allow us to use standard arguments as in [1, Theorem 6.61]. for $(a) \Rightarrow (b) \Rightarrow (c)$ and $(d) \Rightarrow (c)$. We will reason always for the forward case.

$(a) \Rightarrow (b)$. Otherwise take an incomplete geodesic $\gamma : [0, b) \to M, b < \infty$. As the sequence $\{\gamma(b - 1/m)\}_{m>1/b}$ is forward Cauchy, then it will have a limit p, obtaining so a contradiction (recall Proposition 3.6).

$(b) \Rightarrow (c)$. Let $\{p_m\}_m$ be any forward bounded sequence and let $p \in M, r > 0$ such that $\{p_m\}_m \subset B^+_{\bar{F}}(p, r)$. By hypothesis, Proposition 3.22 is applicable and, thus, $\{p_m\}_m$ lies in the geodesic ball $\mathcal{B}^+_{\bar{F}}(p, r)$. Then, the existence of a converging partial subsequence of $\{p_m\}_m$ follows because the closure of this ball is compact (recall Remark 3.18).

$(d) \Rightarrow (c)$. As in the previous case, choosing now $p = p_0$ (recall that $B^+_{\bar{F}}(p,r) \subset B^+_{\bar{F}}(p_0, r + d_{\bar{F}}(p_0, p))$ and $d_{\bar{F}}(p_0, p)$ is finite).

$(c) \Rightarrow (a)$. Let $\{p_m\}_m$ be a forward Cauchy sequence. The triangle inequality for $d_{\bar{F}}$ implies that $\{p_m\}_m$ is forward bounded. So, its closure is compact and $\{p_m\}_m$ admits a converging subsequence. Therefore, the result follows by Proposition 3.13.

For the statement about the geodesic completeness of F_l, observe that $\ell_{\bar{F}}(\alpha) \leq \ell_{\bar{F}_l}(\alpha)$. This easily implies that if $\gamma : [0, b) \to M$, $b < \infty$, is an F_l-geodesic, then the sequence $\{\gamma(t_m)\}_m$ with $\{t_m = b - 1/m\}$ is forward Cauchy for $d_{\bar{F}}$. Thus, γ is extendible to b as an F_l-geodesic (Proposition 3.6), as required.

Finally, the first statement follows because the Σ-geodesics are the geodesics of both $\bar{F}$ and F_l and we have just proved that the completeness of $\bar{F}$-geodesics implies the completeness of F_l-geodesics. □

Example 3.24 In the last theorem, the finiteness of $d(p_0, q)$ (or $d(q, p_0)$) for every $q \in M$ is necessary to obtain the equivalence between (d) and the other properties. In fact, if we consider $M = \mathbb{R}^n \setminus \{(0, 0, \dots, 0, -1)\}$, g_R the Euclidean metric and $W = (0, \dots, 0, 1)$, the corresponding Kropina metric satisfies that the forward exponential map at $\mathbf{0} = (0, 0, \dots, 0)$ is defined in the maximal domain, but the associated distance is not forward complete. Notice, however, that $d_F(\mathbf{0}, q) = +\infty$ whenever $q = (x_1, \dots, x_n)$ satisfies $x_n \leq 0$.

As a more sophisticated example, recall Example 3.11 (Fig. 4). The regions $x \geq 4$ and $x \leq -4$ are Euclidean but cannot be connected by any wind curve. So, the exponential at any point in the region $x \geq 4$ will be defined on all its tangent space, even if one removes a point of the region $x \leq -4$ (making incomplete the WRS).

Recall also that the regions $-3 < x < -1$ and $1 < x < 3$ behave as a sort of "black holes," namely, once you enter there, it is not possible to go out.

4 Some Applications for SSTK Spacetimes

4.1 Cauchy Hypersurfaces in SSTK

As a direct consequence of Theorems 3.7 and 3.23, a characterization of Cauchy hypersurfaces is obtained.

Corollary 4.1 *Let* $(\mathbb{R} \times M, g)$ *be an SSTK splitting. The following assertions are equivalent:*

(1) The slices $t =$ *constant are Cauchy hypersurfaces.*
(2) The associated WRS, Σ*, is geodesically complete.*
(3) The extended conic Finsler metric $\bar{F}$ *of* Σ *is complete (in any of the equivalent senses of Theorem 3.23).*

This precise characterization of WRS completeness/Cauchy slices may be useful for concrete examples. Indeed, incompleteness would follow just by finding an incomplete Cauchy sequence or geodesic for $\bar{F}$. Next, some criteria to ensure completeness are obtained.

Proposition 4.2 *Let Σ be a WRS on M with associated conic Finsler metric*[4] *F, and let H be an auxiliary Finsler metric such that*

$$H(v) \leq F(v), \qquad \forall v \in A. \tag{18}$$

If H is forward (resp. backward) complete, then Σ is forward (resp. backward) geodesically complete.

Proof The usual generalized distance d_H associated with H satisfies that $d_H \leq d_{\bar{F}}$. Therefore, any (forward or backward) Cauchy sequence for $\bar{F}$ will also be Cauchy for H and, as H is complete, then it will converge to some point of M. □

Recall that the inequality (18) means that the indicatrix Σ_H encloses the indicatrix Σ at each $p \in M$. In order to apply the previous criterion sharply, the indicatrix Σ_H should fit as much as possible in Σ. However, Riemannian metrics are easier to handle in practice and this may be enough in some particular cases. First criteria are the following.

Proposition 4.3 *Let Σ be a WRS with Zermelo data (g_R, W), and let $W^\flat = g_R(W, \cdot)$. Then, Σ is complete if one of the following conditions holds:*

(i) the conformal metric $h^* = g_R/(1 + |W|_R)^2$ is complete, or
(ii) the metric g_R is complete and $|W|_R$ grows at most linearly with the d_R-distance, that is, there exist $\lambda_0, \lambda_1 > 0$, $x_0 \in M$:

$$|W_x|_R \leq \lambda_0 + \lambda_1\, d_R(x_0, x) \qquad \forall x \in M.$$

Proof For (i), just notice that, at each $p \in M$, the indicatrix of h^* is a g_R-sphere of radius $|W|_R + 1$. So, it contains Σ_p and Proposition 4.2 can be applied.

The conditions in (ii) imply (i). Indeed the metric h^* is conformal now to the complete one g_R with a conformal factor $\Omega = 1/(1 + |W|_R)^2$ which decreases at most quadratically with the distance. So, it is well known that h^* is then complete (namely, if $\gamma : [0, \infty) \to M$ is any diverging curve parametrized with unit g_R-velocity, then its h^*-length satisfies $\int_0^\infty h^*(\gamma'(s), \gamma'(s))^{1/2} ds \geq \int_0^\infty \frac{ds}{1+\lambda_0+\lambda_1 s} = \infty$. □

A more accurate consequence of Proposition 4.2 is the following.

Proposition 4.4 *A WRS Σ with Zermelo data (g_R, W) is complete if so is the Riemannian metric*

$$h = g_R - \frac{1}{1 + |W|_R^2} W^\flat \otimes W^\flat,$$

[4]Notice that either using the conic Finsler metric F or the extended one $\bar{F}$ in the statement of this result are equivalent. Indeed, the inequality (18) holds when F is replaced by $\bar{F}$, since $\bar{F}$ is continuous everywhere except at most in the zeroes of the critical region, where the inequality holds trivially.

where $W^\flat$ is computed with g_R, namely, $W^\flat(v) = g_R(v, W)$ for all $v \in TM$.

Proof Taking into account that the completeness of h and $h/2$ are equivalent, we have just to check that $h(W+U, W+U) \le 2$, where U is any g_R-unit vector field:

$$
\begin{aligned}
h(W+U, W+U) &= |W|_R^2 + 1 + 2g_R(W,U) - \tfrac{\left(|W|_R^2 + g_R(W,U)\right)^2}{1+|W|_R^2} \\
&= \tfrac{1}{1+|W|_R^2}\left(\left(1+|W|_R^2\right)^2 + 2(1+|W|_R^2)\, g_R(W,U) - \left(|W|_R^2 + g_R(W,U)\right)^2\right) \\
&= \tfrac{1}{1+|W|_R^2}\left(1 + 2|W|_R^2 + 2g_R(W,U) - g_R(W,U)^2\right),
\end{aligned}
$$

where the last parenthesis can be regarded as a quadratic polynomial in $g_R(W, U)$. This takes its maximum when $g_R(W, U) = 1$, yielding so the required inequality. □

Remark 4.5 One can also take other choices of Riemannian metric, which are not conformal to g_R but may be better adapted to the shape of Σ. For example, given Σ as in the proposition above, it is not hard to check that Σ is enclosed by the indicatrix of the following metric

$$
h_\lambda = \frac{1}{\lambda^2}\left(g_R - \frac{1}{\lambda^2 - 1 + |W|_R^2} W^\flat \otimes W^\flat\right), \tag{19}
$$

where $\lambda : M \to \mathbb{R}$ is any function satisfying $\lambda > 1$ (indeed, the metric h in Proposition 4.4 is just $h = \lambda^2 h_\lambda$ for $\lambda = \sqrt{2}$). Therefore, the completeness of h_λ for such a function implies that Σ is complete too. Notice that, when λ is close to 1, the metric h_λ is very close to g_R in the directions orthogonal to W (this may be an advantage) but not in the direction of W, as $h_\lambda(W, W)$ becomes very small (this may be a disadvantage); the situation is the other way around for big λ.

Recall that the easier the application of the previous criteria, the weaker the result. Indeed, in Proposition 4.3, the criterion (ii) implies (i) (as seen explicitly in the proof), the latter implies the completeness of h (as $h^* \le h$), and the completeness of h also implies the completeness of g_R (as $g_R \ge h$); however, Proposition 4.2 can be applied even when g_R is not complete. The next examples illustrate these possibilities.

Example 4.6 Sharpness of the rough bounds. Let $(M, g_R) = \mathbb{R}^2$, $W_{(x,y)} = f(x, y)\, \partial_x$, for some smooth function f on $\mathbb{R}^2$.

Bound (ii). Choose $r \in \mathbb{R}$ and put $f(x, y) = |x|^r$ whenever $|x| \ge 1$. When $r \le 1$, the growth of W is at most linear and thus, the corresponding WRS is complete. However, if $r > 1$, the WRS is incomplete. Indeed, the inextendible curve $[0, L) \ni s \mapsto (x(s), 0)$ with $x(0) = 1$ and $x'(s) = 1 + x^r(s)$ diverges (it escapes from any compact subset), it is $\bar{F}$-unit and has finite length, $L = \int_1^\infty dx/(1 + x^r) < \infty$.

Bound (i). Choose $f \equiv 0$ except in the squares $(n - 1/n^4, n + 1/n^4) \times (-1, 1)$ where $|f|$ reaches the maximum n^2. Now, (i) is fulfilled but not (ii).

Bound with h. Put $f(x, y) = xe^y$. The metric h^* is not complete (say, the curve $[0, \infty) \ni s \mapsto (1, s)$ has finite length), but h (as well as h_λ for any constant $\lambda > 1$) is complete. Indeed,

$$h = \frac{1}{1 + x^2 e^{2y}} dx^2 + dy^2,$$

thus, if $\gamma(s) = (x(s), y(s))$ is a diverging curve and y is unbounded (resp. $|y|$ is bounded by some $C > 0$), then its length is infinite because it is lower bounded by, say, $\int |y'(s)|\, ds$ (resp. $(\int (|x'(s)| / \sqrt{1 + x(s)^2 e^{2C}}) ds > e^{-C} \int dx / \sqrt{1 + x^2})$.

Bound with Finslerian H. Proposition 4.2 should be applied when the bound with a Riemannian metric h in Propositions 4.3 and 4.4 imply a loss of sharpness. Indeed, modify the example above putting $(M, g_R) = \mathbb{R}^+ := \{x \in \mathbb{R} : x > 0\}$, $W_x = f(x)\partial_x$, for some smooth function f such that $1 - x^2 \leq f(x) \leq 2$ on $(0, 1]$ and f is 0 on $[2, \infty)$. The incompleteness of g_R close to the origin makes it impossible to apply Propositions 4.3 or 4.4. However, one can obtain the forward completeness of the WRS by applying Proposition 4.2 to the nonreversible Finsler metric:

$$H_x(v) = \begin{cases} v/3 & \forall v \geq 0 \\ -v/x^2 & \forall v \leq 0 \end{cases} \qquad x \in \mathbb{R}^+,\ v \in T_x\mathbb{R}^+.$$

The forward completeness of H follows because its unique unit geodesic $[0, L) \ni s \mapsto x(s)$ with $x(0) = 1, x'(0) < 0$, satisfies $x'(s) = -x^2(s)$, thus $L = \int_0^L ds = \int_0^1 dx/x^2 = \infty$.

Remark 4.7 In terms of the SSTK splitting (3) and using the usual index notation $\omega_i = -(g_0)_{ij} W^j$, $(g_R)_{ij} = (g_0)_{ij}/(\Lambda + \omega_k \omega^k)$, the metrics h and h^* read

$$\begin{aligned} h_{ij} &= \frac{(g_0)_{ij}}{\Lambda+\omega_k \omega^k} - \frac{1}{1+\frac{\omega_k \omega^k}{\Lambda+\omega_k \omega^k}} \frac{\omega_i \omega_j}{(\Lambda+\omega_k \omega^k)^2} \equiv (g_0)_{ij} - \frac{\omega_i \omega_j}{1+\omega_k \omega^k} \\ h^*_{ij} &= \frac{(g_0)_{ij}}{\left(\sqrt{\Lambda+\omega_k \omega^k}+\sqrt{\omega_k \omega^k}\right)^2} \equiv \frac{(g_0)_{ij}}{\left(1+\sqrt{\omega_k \omega^k}\right)^2} \end{aligned}$$

where the indices are raised and lowered with g_0 and the last expression in each equality holds under the choice $\Lambda + \omega_k \omega^k \equiv 1$ in the conformal class of g.

4.2 Further Examples: Ergospheres and Killing Horizons

Consider the Lorentzian metric g on $\mathbb{R} \times \mathbb{R}^3$ in spherical coordinates (t, r, θ, φ),

$$g = -\Lambda(r)dt^2 + dr^2 + r^2 d\theta^2 + g_{t\theta}(r)(dtd\theta + d\theta dt) + r^2 \sin^2\theta d\varphi^2$$

where, for $r \in (1/2, 3/2)$, we will choose $g_{t\theta}(r) = 1$ and $\Lambda(r) = (r-1)^m$ for some $m = 1, 2, \ldots$. As we will focus on the hypersurface $r = 1$, we will assume that the metric matches with Lorentz–Minkowski $\mathbb{L}^4$ for $|r - 1| > 2/3$.

First, notice that this is an SSTK spacetime where g_0 is the usual metric of $\mathbb{R}^2$ and $\omega = g_{t\theta}(r)d\theta$. Thus, $|\omega|_0 = |g_{t\theta}(r)|/r$, and

$$W = -\frac{1}{r^2}\partial_\theta, \qquad g_R = \frac{r^2}{1+r^2(r-1)^m} g_0 \qquad \text{when } r \in (1/2, 3/2).$$

Due to the fact that g_R is complete (it agrees with g_0 outside the compact subset $|r-1| \leq 2/3$) and $|W|_R$ is bounded, any of the criteria in the previous section implies that the slices of t are Cauchy hypersurfaces.

The (hyper)surface S given as $r = 1$ can be seen as an *ergosphere*, because the sign of Λ changes there. This hypersurface is always timelike and so, neither the exterior region $r > 1$ nor the interior one $r < 1$ are globally hyperbolic. One can also check that the slices of t are not Cauchy hypersurfaces for these regions using the $\bar{F}$-separation. Indeed, consider the region $r \in (1-\epsilon, 1+\epsilon)$ for small $\epsilon > 0$. The vector field $Z = (\partial_r - \partial_\theta)/2$ lies inside Σ because Σ is just obtained by taking the usual g_0-unit bundle and displacing it with $-\partial_\theta/r^2$; thus, the $\bar{F}$-length of Z is smaller than one. As the integral curves of Z must cross S, they yield non-compact $\bar{F}$-bounded subsets for both, the inner and the outer regions. Extending our computations (see the next example), it is not difficult to check also the lack of global hyperbolicity by using the $\bar{F}$ distance.

Even though we have focused on completeness and Cauchy hypersurfaces, other properties of causality can be studied, suggesting that $\bar{F}$ can also be useful beyond our scope in this chapter. A computation shows

$$\nabla_{\partial_t}\partial_t = \Gamma^r_{tt}\partial_r = -\frac{1}{2}g^{rr}\frac{\partial g_{tt}}{\partial r}\partial_r = \frac{1}{2}\frac{\partial \Lambda}{\partial r}\partial_r = \frac{m}{2}(r-1)^{m-1}\partial_r, \qquad r \in (1/2, 3/2). \tag{20}$$

Thus, when $m > 1$, the integral curves of ∂_t are geodesics, but when $m = 1$ they are not. This property is related to the light convexity of S (see [7] for background) and, then, to the causal simplicity of the regions $r > 1$ and $r < 1$. Indeed, for $m = 1$, the inner region $r < 1$ cannot be causally simple, as there are lightlike geodesics starting at this region that touch S and come back to the inner region (those geodesics of the spacetime with initial velocity parallel to ∂_t on S).[5] Such a property also implies the lack of w-convexity of the inner region (and, in a natural sense, the lack of $\bar{F}$-convexity of S), which characterizes causal convexity in terms of Σ.

Finally, consider the following variation of the previous example:

$$g = -\Lambda(r)dt^2 + dr^2 + g_{tr}(r)(dtdr + drdt) + r^2d\theta^2 + r^2\sin^2\theta d\varphi^2$$

where, again, we choose $\Lambda(r) = (r-1)^m$ for $m = 1, 2, \ldots$, $g_{tr}(r) = 1$ when $r \in (1/2, 3/2)$ and $\mathbb{L}^4$ when $|r-1| > 2/3$. Now, one has:

---

[5]To understand this easily, (20) implies that the integral curves of ∂_t are accelerated upward, so, geodesics with initial velocity in ∂_t should come from and go inward. Analytically, if $\rho(s) = (t(s), r(s), \theta(s), \varphi(s))$ is such a geodesic, at $s = 0$ one has $r'(0) = \theta'(0) = \varphi'(0) = 0$, thus, $r''(0) + \Gamma^r_{tt}(t'(0))^2 = 0$ and, so, $r''(0) < 0$.

$$W = -\partial_r, \qquad g_R = \frac{1}{1 + (r-1)^m} g_0 \qquad \text{when } r \in (1/2, 3/2).$$

As in the previous case, the slices of t are Cauchy hypersurfaces for the full spacetime.

Now, the surface S given as $r = 1$ is a *null hypersurface*, as g becomes degenerate there; even more, it can be regarded as a Killing horizon. If one considers only the inner $r < 1$ or outer $r > 1$ regions, again, the slices $t =$ constant are not Cauchy. However, these regions are globally hyperbolic (this property goes a bit beyond our previous study, but it shows further applications of $\bar{F}$). In fact, for, say, the region $r > 1$, a closer look at the incompleteness of $\bar{F}$ shows that $\bar{F}$ is forward incomplete but backward complete. In order to check this, the relevant curves can be taken as $(s_-, s_+) \ni s \mapsto (r(s), \theta_0, \varphi_0)$ with θ_0, φ_0 constants and $r(0) > 1$. From (9), the unit curves (necessarily $\bar{F}$-geodesics)[6] satisfy $1 = r'(s)/(-1 + \epsilon\sqrt{1 + (r-1)^m})$ where $\epsilon = \text{sign}(r'(s)) \in \{\pm 1\}$ when $r \in (1/2, 3/2)$ and they are the Euclidean unit geodesics if $|r - 1| > 2/3$. Clearly, the $\bar{F}$-geodesics with $\epsilon = 1$ (resp. $\epsilon = -1$) are forward (resp. backward) complete. Moreover, the geodesic with $\epsilon = -1$ is also clearly forward incomplete. However, the geodesics with $\epsilon = 1$ are backward complete. Indeed, assuming $r(0) = 2$ (as r' cannot vanish one can focus only in one geodesic),

$$\begin{aligned} s_- &= \int_2^1 dr/(-1 + \sqrt{(r-1)^m + 1})) \\ &= -\int_1^2 (1 + \sqrt{(r-1)^m + 1})\, dr/(r-1)^m \\ &\leq -\int_1^2 dr/(r-1)^m = -\infty, \end{aligned}$$

as required. These properties of completeness are sufficient for the compactness of the intersections between the forward and backward $\bar{F}$-balls and, then, for the compactness of the corresponding intersections of the Σ-balls, the latter property being a characterization of global hyperbolicity, as proven in[7] [10, Theorem 5.9].

Being the inner and outer regions globally hyperbolic, they are causally simple too. However, it is interesting to consider again the lightlike geodesics of the spacetime tangent to[8] S. First, a straightforward computation shows:

---

[6] Recall that all the direct computations in this example (either for $\bar{F}$ or for other more classical procedures to study global hyperbolicity) become especially simple, because the M part of the SSTK spacetime is essentially one-dimensional, as the coordinates θ, φ do not play any relevant role.

[7] In any case, the readers used to the stuff in Mathematical Relativity can reason alternatively that these regions are globally hyperbolic because they admit S as a conformal boundary with no timelike points, which is a known characterization of global hyperbolicity (see [13, Corollary 4.34] for a precise formulation of this result).

[8] The fact that they remain in S implies its light convexity with respect to the inner and outer regions and, then, the causal simplicity of these regions, see [7].

$$\nabla_{\partial_t}\partial_t = \Gamma^t_{tt}\partial_t + \Gamma^r_{tt}\partial_r = \frac{g_{tr}}{2(\Lambda + g_{tr}^2)}\frac{\partial\Lambda}{\partial r}\partial_t + \frac{\Lambda}{2(\Lambda + g_{tr}^2)}\frac{\partial\Lambda}{\partial r}\partial_r. \tag{21}$$

So, one has $\nabla_{\partial_t}\partial_t = \frac{1}{2}\frac{\partial\Lambda}{\partial r}\partial_t$ on S. This means that the integral curves of ∂_t (which are the null generators of S) become geodesics when $m > 1$ and pregeodesics when $m = 1$. These geodesics plus the ones in the previous example fulfill all the possible types of lightlike geodesics orthogonal to $K = \partial_t$, described in part (3) of Sect. 3.1.3 (see Lemma 3.21, Theorem 6.3(c) in [10] for details).

Summing up, even though the previous examples are very simple and can be handled directly by means of the explicit computations of lightlike geodesics, causal futures etc., they show the applicability of both, the general methods introduced in [10] and the additional tools and criteria introduced here, which are valid for general SSTK spacetimes with no restrictions on energy conditions, asymptotic behaviors, etc.

Acknowledgements Partially supported by Spanish MINECO/FEDER project reference MTM 2015-65430-P and Fundación Séneca (Región de Murcia) project 19901/GERM/15 (MAJ) and MTM2013-47828-C2-1-P (MS).

References

1. D. Bao, S.-S. Chern, Z. Shen, *An Introduction to Riemann-Finsler Geometry, Graduate Texts in Mathematics* (Springer, New York, 2000)
2. D. Bao, C. Robles, Z. Shen, Zermelo navigation on Riemannian manifolds. J. Differ. Geom. **66**, 377–435 (2004)
3. C. Barceló, S. Liberati, M. Visser, Analogue gravity. Living Rev. Relativ. **8**, 12 (2005)
4. J.K. Beem, P.E. Ehrlich, K. Easley, Monographs and Textbooks in Pure and Applied Mathematics, in *Global Lorentzian geometry*, vol. 202, 2nd edn. (Marcel Dekker, Inc., New York, 1996)
5. A.N. Bernal, M. Sánchez, Smoothness of time functions and the metric splitting of globally hyperbolic spacetimes. Comm. Math. Phys. **257**(1), 43–50 (2005)
6. A.M. Candela, J.L. Flores, M. Sánchez, Global hyperbolicity and Palais-Smale condition for action functionals in stationary spacetimes. Adv. Math. **218**, 515–536 (2008)
7. E. Caponio, A.V. Germinario, M. Sánchez, Convex regions of stationary spacetimes and randers spaces. Applications to lensing and asymptotic flatness. J. Geom. Anal. **26**(2), 791–836 (2016)
8. E. Caponio, M.A. Javaloyes, A. Masiello, On the energy functional on Finsler manifolds and applications to stationary spacetimes. Math. Ann. **351**, 365–392 (2011)
9. E. Caponio, M.A. Javaloyes, M. Sánchez, On the interplay between Lorentzian causality and Finsler metrics of Randers type. Rev. Mat. Iberoamericana **27**, 919–952 (2011)
10. E. Caponio, M.A. Javaloyes, M. Sánchez, *Wind Finslerian Structures: From Zermelo's Navigation to the Causality of Spacetimes*, arXiv:1407.5494v4 [math.DG]
11. G. Gibbons, C. Herdeiro, C. Warnick, M. Werner, Stationary metrics and optical Zermelo-Randers-Finsler geometry. Phys. Rev. D **79**, 044022 (2009)
12. J.L. Flores, J. Herrera, M. Sánchez, Cauchy and causal boundaries for Riemannian, Finslerian and Lorentzian manifolds. Mem. Amer. Math. Soc. **226**(1064), vi+76 (2013)
13. J.L. Flores, J. Herrera, M. Sánchez, On the final definition of the causal boundary and its relation with the conformal boundary. Adv. Theor. Math. Phys. **15**(4), 991–1057 (2011)

14. M. Gutiérrez, B. Olea, Uniqueness of static decompositions. Ann. Global Anal. Geom. **39**(1), 13–26 (2011)
15. S. G. Harris, Static- and Stationary-complete spacetimes: algebraic and causal structures. Class. Quantum Gravity **32**(13), 135026 (2015)
16. M.A. Javaloyes, L. Lichtenfelz, P. Piccione, Almost isometries of non-reversible metrics with applications to stationary spacetimes. J. Geom. Phys. **89**, 38–49 (2015)
17. M.A. Javaloyes, M. Sánchez, *Wind Riemannian spaceforms and Randers metrics of constant flag curvature*, Eur. J. Math. **3** (2017) 1225–1244 (topical issue on *Finsler metrics: New Methods and Perspectives*), arXiv: 1701.01273
18. M.A. Javaloyes M. Sánchez, On the definition and examples of Finsler metrics. Ann. Sc. Norm. Super. Pisa, Cl. Sci. **XIII**(5), 813-858 (2014)
19. M.A. Javaloyes, M. Sánchez, A note on the existence of standard splittings for conformally stationary spacetimes. Class. Quantum Gravity **25**(16), 168001 (2008)
20. M.A. Javaloyes, H. Vitório, *Some properties of Zermelo navigation in pseudo-Finsler metrics under an arbitrary wind.* Houston J. Math., provisionally accepted, arXiv:1412.0465
21. M. Ludvigsen, *General Relativity*, A Geometric Approach (Cambridge University Press, Cambridge, 2004)
22. V. Matveev, Can we make a Finsler metric complete by a trivial projective change? Recent trends in Lorentzian Geometry, in *Springer Proceedings in Mathematics & Statistics*, vol. 26 (2013), pp. 231–243
23. E. Minguzzi, M. Sánchez, The causal hierarchy of spacetimes, in *Recent Developments in pseudo-Riemannian Geometry*, ESI Letters in Mathematical Physics (European Mathematical Society, Zürich, 2008), pp. 299–358
24. B. O'Neill, Semi-Riemannian geometry with applications to relativity, in *Pure and Applied Mathematics*, vol. 103 (Academic Press Inc. [Harcourt Brace Jovanovich Publishers], New York, 1983)
25. S.V. Sabau, M. Tanaka, *The cut locus and distance function from a closed subset of a Finsler manifold*, Houston J. Math. **42**(4), 1157–1197 (2016) arXiv:1207.0918
26. M. Sánchez, On causality and closed geodesics of compact Lorentzian manifolds and static spacetimes. Diff. Geom. Appl. **24**, 21–32 (2006)
27. M. Sánchez, Causal hierarchy of spacetimes, temporal functions and smoothness of Geroch's splitting. A revision. Mat. Contemp. **29**, 127–155 (2005)
28. M. Sánchez, J.M.M. Senovilla, A note on the uniqueness of global static decompositions. Class. Quantum Gravity **24**(23), 6121–6126 (2007)

Extending Translating Solitons in Semi-Riemannian Manifolds

Erdem Kocakuşaklı and Miguel Ortega

Abstract In this paper, we recall some general properties and theorems about Translating Solitons in Semi Riemannian Manifolds. Moreover, we investigate those which are invariant by the action of a Lie group of isometries of the ambient space, by paying attention to the behavior close to the singular orbit (if any) and at infinity. Then, we provide some related examples.

Keywords Translating solitons · Semi-riemannian manifolds · ODE · Boundary problem

1 Introduction

Given a smooth manifold M, assume a family of smooth immersions in a semi-Riemannian manifold $(\mathbf{M}, \mathbf{g})$, $F_t : M \to \mathbf{M}$, $t \in [0, \delta)$, $\delta > 0$, with mean curvature vector $\vec{H}_t$. The initial immersion F_0 is called a solution to the *mean curvature flow* (up to local diffeomorphism) if

$$\left(\frac{d}{dt}F_t\right)^{\perp} = \vec{H}_t, \tag{1}$$

where $\perp$ means the orthogonal projection on the normal bundle. In the Euclidean and Minkowski space, there is a famous family of such immersions, namely, translating

Dedicated to Prof. Ceferino Ruiz on his retirement.

E. Kocakuşaklı (✉)
Faculty of Science, Department of Mathematics, University of Ankara Tandogan, Ankara, Turkey
e-mail: kocakusakli@ankara.edu.tr

M. Ortega
Institute of Mathematics Department of Geometry and Topology, University of Granada, Granada, Spain
e-mail: miortega@ugr.es

M.A. Cañadas-Pinedo et al. (eds.), *Lorentzian Geometry and Related Topics*, Springer Proceedings in Mathematics & Statistics 211,
DOI 10.1007/978-3-319-66290-9_9

solitons. A submanifold is called *translating soliton* in the Euclidean Space when its mean curvature $\vec{H}$ satisfies the following equation:

$$\vec{H} = v^{\perp}, \tag{2}$$

for some constant unit vector $v \in \mathbb{R}^{n+1}$. Indeed, if a submanifold $F : M \to \mathbb{R}^{n+1}$ satisfies this condition, then it is possible to define the forever flow $\Gamma : M \times [0, +\infty) \to \mathbb{R}^{n+1}$, $\Gamma(p,t) = F_t(p) = F(p) + tv$. Clearly,

$$\left(\frac{d}{dt}F_t\right)^{\perp} = v^{\perp} = \vec{H}.$$

This justifies our definition. The same situation holds in Minkowski Space. Until now such solutions have been almost exclusively studied in the case where the ambient space is the Euclidean (or the Minkowski) space. For a good list of known examples, see [6]. Probably, the most famous examples are the Grim Reaper curve in $\mathbb{R}^2$ and the translating paraboloid and translating catenoid, [2]. Also, in [7] there are some examples with complicated topology. Recently, in [5], the authors studied those translating solitons in Minkowski three-space with rotational symmetry.

If one wants to generalize (2), the simplest way is to choose a parallel vector field. But manifolds admitting such a vector field are locally a product $M \times \mathbb{R}$. Thus, in [4], the authors introduce the notion of (graphical) translating solitons on a semi-Riemannian product $M \times \mathbb{R}$. Needles to say, their study include the Riemannian case. When the translating soliton is the graph of map u defined on (an open subset of) M, the corresponding partial differential equation that u must satisfy is obtained in [4]. Although this paper includes more results, one of the main concern is the study of translating solitons which are invariant under the action of a Lie group by isometries on M. This action is very easily extended to $M \times \mathbb{R}$. The authors focused on the case when the quotient map is an open interval, $M/\Sigma \equiv I \subset \mathbb{R}$. This is so because among the classical examples, the translating paraboloid and the translating catenoid are constructed this way in [2].

In this paper, we would like to continue the study of [4] by further developing some ideas.

Firstly, in the Preliminaries Section we recall some known results that we will use later. Among them, we include a summary of [4], where we find the first steps of translating solitons in product spaces $M \times \mathbb{R}$ with a product metric $g_M + \varepsilon dt^2$, with $\varepsilon = \pm 1$ and g_M the metric on M. Since the manifold M might not be complete, we can almost say that we are dealing with a *semi-Riemannian cohomogeneity of degree one* Σ*-manifold*, since there is a Lie group acting by isometries and the quotient is a 1-dimensional manifold (see [1].) In this setting, we can construct our translating solitons from the solutions to an ODE. This is clarified in the new Algorithm 1, which was not included in [4].

Section 3 is devoted to studying the already mentioned ODE, from two points of view. One of them is solving a boundary problem. Indeed, given $h \in C^1(a, b)$ such that $\lim_{s \to a} h(s) = +\infty$, and $\varepsilon, \tilde{\varepsilon} \in \{1, -1\}$, consider

$$w'(s) = (\tilde{\varepsilon} + \varepsilon w^2(s))(1 - w(s)h(s)), \quad w(a) = 0.$$

We show in Theorem 1 that there exists a solution under a not very restrictive condition on function h. The reason to consider this problem is the following. In the Euclidean Space $\mathbb{R}^{n+1}$, graphical translating solitons which are invariant by $SO(n)$ and touching the axis or rotation (in other words, *rotationally invariant*) arise from the solution of the following boundary problem:

$$w'(s) = \big(1 + w^2(s)\big)\Big(1 - \frac{n-1}{s} w(s)\Big), \quad w(0) = 0.$$

In fact, the solution gives rise to the famous example known as Translating Paraboloid. Clearly, function $h : (0, +\infty) \to \mathbb{R}$, $h(s) = (n-1)/s$, so that we are studying a much more general problem by choosing any C^1 function h satisfying simple conditions on the boundary.

The existence of solutions in Theorem 1 is just local, i. e., in a small interval $[a, a + \delta)$. Thus, the second point of view consists of the extension of our solutions. In this way, in Propositions 1 and 3 of Sect. 4, we show that for $\varepsilon\tilde{\varepsilon} = 1$ and $h > 0$ or $h < 0$, it is possible to extend the solution to the interval $[s_0, b)$, where $s_0 \in (a, b)$ is the chosen initial point. In Proposition 2, we show some reasonable conditions under which, the solutions defined on $[s_0, b)$ admit $\lim_{s \to b} w(s) \in \mathbb{R}$.

In Sect. 5 we pay attention to manifolds admitting a Lie group Σ acting by isometries, such that the orbits are (CMC) hypersurfaces, the quotient manifold is an interval $[a, b)$, and the mean curvature of the orbits tend to infinity when approaching the singular orbit. For example, this is the case of the Euclidean Space $\mathbb{R}^n$ under the action of $SO(n)$. Then, we apply our previous computations to obtain solutions (denoted by w) to the corresponding boundary problem. Next, we use Algorithm 1 to primitives of them, namely $f = \int w$, to obtain Σ-invariant translating solitons.

Last, but not least, we show some examples in Sect. 6. On one hand, we exhibit translating solitons in $\mathbb{H}^n \times \mathbb{R}$ whose invariant subsets are horospheres. Except for one case, all of them are not entire, in the sense that they admit finite time blow ups. Also, we make a study on the round sphere, where we obtain translating solitons defined on the whole sphere but removing one or two points.

2 Preliminaries

The following results can be found in [4]. Assume that (M, g) is a connected semi-Riemannian manifold of dimension $n \geq 2$ and index $0 \leq \alpha \leq n - 1$. Given $\varepsilon = \pm 1$, we construct the semi-Riemannian product $\widetilde{M} = M \times \mathbb{R}$ with metric $\langle , \rangle = g + \varepsilon dt^2$. The vector field $\partial_t \in \chi(\widetilde{M})$ is obviously Killing and unit, spacelike when $\varepsilon = +1$ and timelike when $\varepsilon = -1$. Now let $F : \Gamma \to \widetilde{M}$ be a submanifold with mean curvature vector $\vec{H}$. Denote by $\partial_t^\perp$ the normal component of ∂_t along F.

Definition A With the previous notation, we will call F a (vertical) translating soliton of mean curvature flow, or simply, a translating soliton, if $\vec{H} = \partial_t^\perp$.

In this paper, we will focus on graphical translating solitons. Namely, given $u \in C^2(M)$, we construct its graph map $F : M \to M \times \mathbb{R} =: \widetilde{M}$, $F(x) = (x, u(x))$. Let ν be the upward normal vector along F with $\varepsilon' = \text{sign}(\langle \nu, \nu \rangle) = \pm 1$.

Let Σ be a Lie group acting by isometries on M and $\pi : M \to I$ be a submersion, I and open interval, such that the fibers of π are orbits of the action. In addition, assume that π is a semi-Riemannian submersion with constant mean curvature fibers. For each $s \in I$, $\pi^{-1}\{s\} \cong \Sigma$ is a hypersurface with constant mean curvature. The value of the mean curvature of $\pi^{-1}\{s\}$ is denoted by $h(s)$. Then, we have a function $h : I \to \mathbb{R}$. We will say that *function h represents the mean curvature of the orbits.* Given a map $F : M \to P$, where P is another set, is Σ-invariant when $F(\sigma \cdot x) = F(x)$ for any $\sigma \in \Sigma$ and any $x \in M$.

Theorem A *Let (M, g) be a connected semi-Riemannian manifold. Let Σ be a Lie group acting by isometries on M and $\pi : (M, g_M) \to (I, \widetilde{\varepsilon} ds^2)$ be a semi-Riemannian submersion, I an open interval, such that the fibers of π are orbits of the action, with function h representing the mean curvature of the orbits. Take $u \in C^2(M, \mathbb{R})$ and consider its graph map*

$$F : M \to M \times \mathbb{R},\ F(x) = (x, u(x))$$

for any $x \in M$. Then, F is a Σ-invariant translating soliton if, and only if, there exists a solution $f \in C^2(I, \mathbb{R})$ to

$$f''(s) = (\widetilde{\varepsilon} + \varepsilon(f')^2(s))(1 - h(s)\, f'(s)) \tag{3}$$

such that $u = f \circ \pi$.

The following results study some conditions to obtain translating solitons which cannot be globally defined, in the sense that they are not entire graphs. Instead, they are defined on some smaller subsets, and converging to infinity.

Corollary A *Under the same conditions, assume that $\varepsilon\widetilde{\varepsilon} = -1$. Then, given $s_0 \in I$, $f_1 \in (-1, 1)$ and $f_o \in \mathbb{R}$, there exists a solution $f : I \to \mathbb{R}$ to* (3) *such that $f(s_o) = f_0$ and $f'(s_o) = f_1$.*

Corollary B *Let $\varepsilon = 1$ and $\widetilde{\varepsilon} = -1$. Take $c \in I$. Consider any $\lambda > 1$.*

1. *Let f be a solution* (3) *such that $f'(c) = \lambda$. If $h(s) \leq 0$ for any $s \geq c$, $s \in I$, and* $\sup(I) > c + \coth^{-1}(\lambda) = a$*, then f admits a finite time blow up before a.*
2. *Let f be a solution* (3) *such that $f'(c) = -\lambda$. If $h(s) \leq -1$ for any $s \geq c$, $s \in I$, and* $\sup(I) > c + \dfrac{\coth^{-1}(\lambda)}{\lambda - 1} = a$*, then f admits a finite time blow up before a.*

Corollary C *Let $\varepsilon = \widetilde{\varepsilon} = -1$ and $c \in I$. Consider any $\lambda > 0$ such that* $\sup(I) > c + \frac{1}{\lambda} = a$.

1. *Let f be a solution* (3) *such that $f'(c) = -\lambda$. If $h(s) > 0$ for any $s \geq c, s \in I$, then f admits a finite time blow up before a.*
2. *Let f be a solution* (3) *such that $f'(c) = \lambda$. If $h(s) > \frac{1}{f'(c)}$ for any $s \geq c, s \in I$ then f admits a finite time blow up before a.*

Corollary D 1. *If $\varepsilon = \widetilde{\varepsilon} = 1$, consider any $\lambda > 0$ such that* $\sup(I) > c + \frac{1}{\lambda}$. *Take any solution f to* (3). *If either $f'(c) = \lambda$ and $h(s) < 0$ for any $s \geq c, s \in I$, or $f'(c) = -\lambda$ and $h(s) < \frac{1}{f'(c)}$ for any $s \geq c, s \in I$, then f admits a finite time blow up before a.*
2. *If $\varepsilon = -1$ and $\widetilde{\varepsilon} = 1$, consider any $\lambda > 1$. Let f be a solution to* (3) *such that $f'(c) = -\lambda$. If $h(s) > 0$ for any $s \geq c, s \in I$, and* $\sup(I) > c + \coth^{-1}(\lambda) = a$, *then f admits a finite time blow up before a.*
3. *If $\varepsilon = -1$ and $\widetilde{\varepsilon} = 1$, consider any $\lambda > 1$. Let f be a solution to* (3) *such that $f'(c) = \lambda$. If $h(s) > 1$ for any $s \geq c, s \in I$, and* $\sup(I) > c + \frac{\coth^{-1}(\lambda)}{\lambda - 1} = a$, *then f admits a finite time blow up before a.*

Until now, we are recalling known results. But for the sake of clarity, we now introduce a method to construct a translating soliton in a manifold foliated by the orbits of the action of a Lie group acting by isometries.

Algorithm 1 Let (M, g) be semi-Riemannian manifold, Σ a Lie subgroup of $\mathrm{Iso}(M, g)$, and I open interval. Choose $\varepsilon \in \{\pm 1\}$. The metric in $M \times \mathbb{R}$ is $\langle , \rangle = g + dt^2$.

1. Assume $\phi : M \to (\Sigma/K) \times I$ is a diffeomorphism, for some subgroup K, such that its restriction $\pi : M \to I$ satisfies $|\nabla\pi|^2 \neq 0$.
2. By a change of variable, recompute π (and ϕ) to obtain $|\nabla\pi|^2 = \tilde{\varepsilon} = \pm 1$.
3. For each $s \in I$ compute the mean curvature $h(s)$ of the fiber $\pi^{-1}\{s\} \subset M$. Note $\pi^{-1}\{s\} \cong \Sigma/K$.
4. Solve the following problem for some initial values in an interval $J \subset I$,

$$f''(s) = (\widetilde{\varepsilon} + \varepsilon(f'(s))^2)(1 - f'(s)h(s)).$$

5. The translating soliton can be constructed by one of the following equivalent ways:

$$F : (\Sigma/K) \times J \to M \times \mathbb{R}, \ F(\sigma, s) = (\phi^{-1}(\sigma, s), f(s)).$$
$$\overline{F} : \phi^{-1}((\Sigma/K) \times J \to M \times \mathbb{R}, \ \overline{F}(x) = (x, f(\pi(x))).$$

We will use the following tools in order to solve our ODE with singularities. See [9] for details. We consider the following linear ODE,

$$\dot{y} = Ay, \; y \in \mathbb{R}^n \tag{4}$$

From elementary linear algebra, we can find a linear transformation T which transforms the linear equation (4) into block diagonal form

$$\begin{bmatrix} \dot{u} \\ \dot{v} \\ \dot{w} \end{bmatrix} = \begin{bmatrix} A_s & 0 & 0 \\ 0 & A_u & 0 \\ 0 & 0 & A_c \end{bmatrix} \begin{bmatrix} u \\ v \\ w \end{bmatrix} \tag{5}$$

where $T^{-1}y = (u, v, w) \in R_s \times R_u \times R_c, s + u + c = n, A_s$ is a $s \times s$ matrix having eigenvalues with negative real part, A_u is a $u \times u$ matrix having eigenvalues with positive real part, and A_c is a $c \times c$ matrix having eigenvalues with zero real part. Moreover, we know that

$$\begin{aligned} \dot{u} &= A_s u + R_s(u, v, w) \\ \dot{v} &= A_u v + R_u(u, v, w) \\ \dot{w} &= A_c w + R_c(u, v, w) \end{aligned} \tag{6}$$

where $R_s(u, v, w)$, $R_u(u, v, w)$ and $R_c(u, v, w)$are the first s, u and c components, respectively, of the vector $T^{-1}R(T^y)$.

Theorem B *Suppose* (6) *is* C^r, $r \geq 2$. *Then the fixed point* $(u, v, w) = 0$ *of* (6) *possesses a* C^r *s–dimensional local, invariant stable manifold,* W^s_{loc} (0)*, a* C^r *u–dimensional local, invariant unstable manifold,* W^u_{loc} (0) *and a* C^r *c–dimensional local, invariant center manifold* $W^c_{loc}(0)$*, all intersecting at* $(u, v, w) = 0$. *These manifolds are all tangent to the respective invariant subspaces of the linear vector field* (5) *at the origin and, hence, are locally representable as graphs. In particular, we have*

$$W^s_{loc}(0) = \left\{ \begin{array}{c} (u, v, w) \in \mathbb{R}^s \times \mathbb{R}^u \times \mathbb{R}^c \mid v = h^s_v(u), w = h^s_w(u); \\ Dh^s_v(0) = 0, Dh^s_w(0) = 0; \; |u| \textit{ sufficiently small} \end{array} \right\}$$

$$W^u_{loc}(0) = \left\{ \begin{array}{c} (u, v, w) \in \mathbb{R}^s \times \mathbb{R}^u \times \mathbb{R}^c \mid u = h^u_u(v), w = h^u_w(v); \\ Dh^u_u(0) = 0, Dh^u_w(0) = 0; \; |v| \textit{ sufficiently small} \end{array} \right\}$$

$$W^c_{loc}(0) = \left\{ \begin{array}{c} (u, v, w) \in \mathbb{R}^s \times \mathbb{R}^u \times \mathbb{R}^c \mid u = h^c_u(w), v = h^c_v(w); \\ Dh^c_u(0) = 0, Dh^c_v(0) = 0; \; |w| \textit{ sufficiently small} \end{array} \right\}$$

where $h^s_v(u)$, $h^s_w(u)$, $h^u_u(v)$, $h^u_w(v)$;, $h^c_u(w)$ *and* $h^c_v(w)$ *are* C^r *functions. Moreover, trajectories in* W^s_{loc} (0) *and* W^u_{loc} (0) *have the same asymptotic properties as trajectories in* E^s *and* E^u*, respectively.*

3 Solution to a Boundary Problem with Singularity

Theorem 1 *Given $a \in \mathbb{R}$, $b \leq +\infty$, $\varepsilon, \tilde{\varepsilon} \in \{1, -1\}$, choose $q \in C^1[a, b)$ such that $q(a) = 0$, $q(s) \neq 0$ for any $s > a$, $\tilde{\varepsilon}q'(a) \geq 0$, and define $h : (a, b) \to \mathbb{R}$ given by $h = 1/q$. Then, the boundary problem*

$$w'(s) = (\tilde{\varepsilon} + \varepsilon w^2(s))(1 - w(s)h(s)), \quad w(a) = 0 \tag{7}$$

has a solution $w : [a, a + \delta) \to \mathbb{R}$ for a suitable small $\delta > 0$.

Proof It is well known that it is possible to extend q a little in the following way. For some $\rho > 0$, there exists a (nonunique) $\tilde{q} \in C^1(a - \rho, b)$ such that $\tilde{q}(s) = q(s)$ for any $s \geq a$. Then, we simply work on the interval $I = (a - \rho, b)$. The extension is not going to be crutial, because we really just care for $s \in I$, $s \geq a$. We consider the following autonomous vector field:

$$X : I \times \mathbb{R} \to \mathbb{R}^2, \; X(s, x) = \big(q(s), (\tilde{\varepsilon} + \varepsilon x^2)(q(s) - x)\big).$$

Note that $X(a, 0) = (0, 0)$. Moreover, at $(a, 0)$ the linearlization is

$$DX(a, 0) = \begin{bmatrix} q'(a) & 0 \\ \tilde{\varepsilon}q'(a) & -\tilde{\varepsilon} \end{bmatrix}.$$

Since $\tilde{\varepsilon}q'(a) \geq 0$, there are two eigenvalues $\lambda_1 = q'(a)$ and $\lambda_2 = -\tilde{\varepsilon}$, with different sign or $q'(a) = 0$, with corresponding eigenvectors $v_1 = (\tilde{\varepsilon} + q'(a), \tilde{\varepsilon}q'(a))^t$, $v_2 = (0, 1)^t$. By Theorem B, there exists a one-dimensional manifold (of fixed point), around $(a, 0)$, whose tangent space at $(a, 0)$ is spanned by v_1, which is a graph in a small interval around s_0, namely

$$W = \{(s, x) \in I \times \mathbb{R} : x = w(s), |s - a| < \delta\}$$

for some function w defined on a small interval $(-\delta, \delta)$. This means that our dynamical system has a solution

$$\alpha : (-\delta, \delta) \to W, \alpha(t) = (s(t), x(t)), \text{ with } \alpha'(t) = X(\alpha(t))$$

such that $\alpha(0) = (a, 0)$, $\alpha'(0) = \lambda v_1$ for some $\lambda \in \mathbb{R}$, $\lambda \neq 0$, and $x(t) = w(s(t))$. We compose with the inverse of s, so that $w(s) = x(t(s))$. Moreover, since $X(\alpha(t)) = \alpha'(t)$, we have $\alpha'(t) = (s'(t), x'(t)) = \big(q(s(t)), (\tilde{\varepsilon} + \varepsilon x(t)^2)(q(s(t)) - x(t)\big)$, and so for $s > a$,

$$\begin{aligned} w'(s) = x'(t(s))t'(s) = \frac{x'(t(s))}{s'(t)} &= \frac{(\tilde{\varepsilon} + \varepsilon x(t(s))^2)(q(s) - x(t(s)))}{q(s)} \\ &= (\tilde{\varepsilon} + \varepsilon w(s)^2)(1 - h(s)w(s)). \end{aligned}$$

According to [9, p. 35], when $q'(a) \neq 0$, there is also uniqueness of solution. When $q'(a) = 0$, solutions *only differ by exponentially small functions of the distance from the fixed point.* □

4 Extension of Solutions

Along this section, we will always assume the following:

(H) *Given* $a < b \leq +\infty$, *take* $s_0 \in (a, b)$. *Consider* $h \in C^1(a, b)$ *such that* $h > 0$.

Proposition 1 *Assume (H).*

1. *For each* $w_0 \in \mathbb{R}$, *the initial value problem*

$$w'(s) = (1 + w^2(s))(1 - h(s)w(s)), \quad w(s_0) = w_0, \tag{8}$$

   has a unique C^2*-solution* w *on* $(s_0 - \rho, b)$, *for some* $\rho > 0$.
2. *If* $b = +\infty$, *then* $\lim_{s \to b} h(s)w(s) = 1$.

Proof First of all, there exist $\rho > 0$ and $w : (s_0 - \rho, s_0 + \rho) \to \mathbb{R}$ a solution, and we wish to extend it to b. In this proof, we will use the classical result of extension of solutions, which can be found for example on [3, p 15], without saying explicitly. Note that function $F : (a, b) \times \mathbb{R}$, $F(s, x) = (1 + x^2)(1 - h(s)x)$, is continuous, so it will be bounded on compact domains. Thus, we call

$$J = \{s \in [s_0, b) : \text{ there exists } w : [s_0, s) \to \mathbb{R}\}.$$

Note that $[s_0, s_0 + \delta) \subset J$. We want to show that $supJ = b$, so we take $s_1 \in J$ such that $s_0 < s_1 < b$.

If $w(s_1) \leq \frac{1}{h(s_1)}$, then $w'(s_1) \geq 0$. There exists $\delta_1 > 0$ such that $a < s_1 - \delta_1 < s_1 + \delta_1 < b$ and $(s_1, w(s_1))$ is an interior point of $[s_1 - \delta_1, s_1 + \delta_1] \times [w(s_1) - 1, 1 + 1/h(s_1)]$, where F is bounded. Thus, we can extend w a little.

If $w(s_1) > \frac{1}{h(s_1)} > 0$, then $w'(s_1) < 0$. There exists $\delta_1 > 0$ such that $a < s_1 - \delta_1 < s_1 + \delta_1 < b$ and $(s_1, w(s_1))$ is an interior point of $[s_1 - \delta_1, s_1 + \delta_1] \times [0, 1 + w(s_1)]$, where F is bounded. Thus, we can extend w a little.

Note also that if for some $s_1 \in J$, $w(s_1) \geq 0$, then $w(s) \geq 0$ for any $s > s_1, s \in J$. Indeed, if for some $s_2 > s_1$, $w(s_2) < 0$, by the continuity of w and w', there exists $s_3 \in (s_1, s_2)$ such that $w'(s_3) < 0$ and $w(s_3) < 0$. But by (8), $w'(s_3) > 0$, which is a contradiction.

Now, we call $\tilde{s} = \sup(J) \leq b$. Assume that $\tilde{s} < b$. Firstly, if w is bounded on a small interval $[\tilde{s} - \delta_1, \tilde{s}]$, by our previous computations, we can extend w a little to $[s_0, \tilde{s} + \delta)$, which is a contradition. Then, w cannot be bounded when s approaches $\tilde{s}$. Thus, there are two possibilities:

(a) There is a sequence $s_n \to \tilde{s}$ such that $w(s_n) \nearrow +\infty$. For some m natural number $w(s_n) \geq M$ for any $n \geq m$. By (8), $w'(s_m) < 0$ and $w'(s_{m+1}) < 0$. Since $w(s_m) < w(s_{m+1})$, then w attains its maximum on $[s_m, s_{m+1}]$ at a point $t \in (s_m, s_{m+1})$. But then, $w'(t) = 0$, and again by (8), $w(t) = 1/h(t) < M \leq w(s_m) < w(t)$. This is a contradiction.
(b) There is a sequence $s_n \to \tilde{s}$ such that $w(s_n) \searrow -\infty$. For some m natural number $w(s_n) \leq -1$ for any $n \geq m$. By (8), $w'(s_m) > 0$ and $w'(s_{m+1}) > 0$. Since $w(s_m) > w(s_{m+1})$, then w attains its minimum on $[s_m, s_{m+1}]$ at a point $t \in (s_m, s_{m+1})$. But then, $w'(t) = 0$, and again by (8), $w(t) = 1/h(t) > w(s_{m+1}) > w(t)$. This is a contradiction.

Therefore, the only possibility is $\sup(J) = b$.

Next, we want to study the behavior of w when $b = +\infty$.
Case A: Assume that there exist $s_1 \geq s_0$ and $M > 0$ such that for each $s \geq s_1$, $1 - h(s)w(s) \geq M$. Then, for each $s \geq s_1$, $\frac{w'(s)}{1+w^2(s)} \geq M$. If we integrate this inequality, we have $\arctan(w(s)) - \arctan(w(s_1)) \geq M(s - s_1)$. Hence, for each $s \geq s_1$, we know $w(s) \geq \tan(Ms - Ms_1 + \arctan(w(s_1)))$. Clearly, there exists s big enough and m a natural number such that $(2m + 1)\pi/2 = Ms - Ms_1 + \arctan(w(s_1))$. This is a contradiction.

Case B: Assume that there exist $s_1 > s_0$ and $M < 0$ such that for each $s \geq s_1$, $1 - h(s)w(s) \leq M < 0$. If we change $v = -w$, then $v'(s) = -w'(s)$ and $v'(s) = (1 + v^2)(-1 - h(s)v(s))$. On the other hand, $1 - h(s)w(s) = 1 + h(s)v(s) \leq M \leq 0$, and therefore $\frac{v'(s)}{1+v^2(s)} = -1 - h(s)v(s)) \geq -M > 0$, for any $s \geq s_1$. Next, we repeat the steps of case A. □

By Theorem 1 and Proposition 1, we obtain the following result.

Corollary 1 *Assume (H), and in addition* $\lim_{s\to a} h(s) = +\infty$ *and* $\lim_{s\to a} \frac{h'(s)}{h^2(s)} = h_1 > 0$. *Then, the boundary problem* (7) *has a unique globally defined solution* $w \in C^1[a, b)$.

Proposition 2 *Assuming (H), suppose in addition* $b < +\infty$ *and there exists* $\lim_{s\to b} h(s) = +\infty$.

1. *If there exists* $\lim_{s\to b} \frac{h'(s)}{h^2(s)} = h_1 \in [0, +\infty)$, *there is a solution to* (8) *for certain* w_0 *such that* $\lim_{s\to b} w(s) = 0$ *and* $\lim_{s\to b} w'(s) = \frac{1}{1+h_1}$.
2. *If for some* $M > 0$ *and* $s_1 \in [s_0, b)$, *it holds* $w(s) \geq M$ *for every* $s \geq s_1$, *then there exists* $\lim_{s\to b} w(s) = w_1 \geq M$ *and* $\lim_{s\to b} w'(s) = -\infty$.
3. *If there exist* $M < 0$, $s_1 \in [s_0, b)$ *such that for every* $s \geq s_1$, $w(s) \leq M$, *then there exist* $\lim_{s\to b} w(s) = w_1 \leq M$ *and* $\lim_{s\to b} w'(s) = +\infty$.

Proof We check item 1. We consider the map $\phi : [0, b - s_0] \to [s_0, b]$ given by $\phi(u) = b - u$, and the function $\tilde{h} : (0, b - s_0] \to \mathbb{R}$, $\tilde{h}(u) = h(\phi(u))$. Note that ϕ also can be seen $\phi : [s_0, b] \to [0, b - s_0]$. Clearly, $\tilde{h} > 0$, $\lim_{u\to 0} \tilde{h}(u) = \lim_{s\to b} h(s) =$

$+\infty$, so we can define the function $q : [0, b - s_0] \to \mathbb{R}$, $q(u) = 1/\tilde{h}(u)$ when $u > 0$ and $q(0) = 0$. Moreover,

$$\lim_{u \to 0} q'(u) = \lim_{u \to 0} \frac{h'(\phi(u))}{h^2(\phi(u))} = h_1 \geq 0.$$

Thus, $q \in C^1[0, b - s_0)$. By Theorem 1, but using $\tilde{h}$ instead of h, there is a solution $z : [0, b - s_0) \to \mathbb{R}$ to problem (8), such that $z(0) = 0$. We define now the function $w : [s_0, b] \to \mathbb{R}$ given by $w(s) = -z(\phi(u))$. In particular, $\lim_{s \to b} w(s) = 0$ and $w'(s) = z'(\phi(u)) = (1 + w^2(s))(1 - h(s)w(s))$. Moreover, $\lim_{s \to b} w'(s) = \lim_{u \to 0} z'(u)$. Note that

$$\lim_{u \to 0} z'(u) \left(1 - \frac{1}{\frac{\tilde{h}'(u)}{\tilde{h}^2(u)}}\right) = \lim_{u \to 0} z'(u) - \lim_{u \to 0} \frac{z'(u)}{\frac{\tilde{h}'(u)}{\tilde{h}^2(u)}} = \lim_{u \to 0} z'(u) + \lim_{u \to 0} \frac{z(u)}{1/\tilde{h}(u)}$$
$$= \lim_{u \to 0} \Big((1 + z^2(u))(1 - \tilde{h}(u)z(u)) + \tilde{h}(u)z(u)\Big) = 1.$$

Therefore

$$\lim_{s \to b} w'(s) = \lim_{s \to 0} z'(u) = \frac{1}{1 + h_1}.$$

Now, we check 2. We assume there exists $M > 0$, $s_1 \in (s_0, b)$ such that for every $s \in [s_0, b)$ we know $w(s) \geq M$. There exists $s_2 \in [s_1, b)$ such that $h(s) \geq \dfrac{2}{M}$ for every $s \in [s_2, b)$. Therefore, $1 - h(s)w(s) \leq -1$. By (8) we obtain $w'(s) < 0$ for every $s \in [s_2, b)$. By using $w(s) \geq M$ and $w'(s) < 0$ for every $s \in [s_2, b)$, there exists $\lim_{s \to b} w(s) = w_1 \geq M$. Now, by (8), we calculate the limit

$$\lim_{s \to b} w'(s) = \lim_{s \to b} \left[(1 + w^2(s))(1 - h(s)w(s))\right] = -\infty.$$

Finally, item 3. We assume there exist $M < 0$, $s_1 \in (s_0, b)$ such that for every $s \in [s_0, b)$ we know $w(s) \leq M$. We know that $h(s) > 0$ and $w(s) < 0$. Therefore, by (8), we obtain $w'(s) > 0$ for every $s \in [s_1, b)$, namely $w(s)$ is increasing. By using $w(s) \leq M$ and $w'(s) > 0$ for every $s \in [s_1, b)$ there exists $\lim_{s \to b} w(s) = w_1 \leq M < 0$. Now, by (8), we calculate the limit

$$\lim_{s \to b} w'(s) = \lim_{s \to b} \left[(1 + w^2(s))(1 - h(s)w(s))\right] = +\infty.$$

And this completes the proof. □

The case $\varepsilon = \tilde{\varepsilon} = -1$ can be studied in a similar way. All ideas are already explained, so its proof is left to the reader.

Proposition 3 *Given* $h : [s_0, b) \to \mathbb{R}$, $h \in C^1[s_0, b)$ *and* $b \leq +\infty$ *such that* $h(s) < 0$.

1. *For each* $w_0 \in \mathbb{R}$, *the boundary value problem*

$$w'(s) = -(1 + w^2(s))(1 - h(s)w(s)), \quad w(s_0) = w_0, \tag{9}$$

has a unique C^2*-solution* w *on* $[s_0, b)$.
2. *If* $b = +\infty$, *then* $\lim_{s\to+\infty} h(s)w(s) = 1$.
3. *Assume* $b < +\infty$ *and there exist the limits* $\lim_{s\to b} h(s) = -\infty$ *and* $\lim_{s\to b} \frac{h'(s)}{h^2(s)} = h_1 \in (-\infty, 0]$. *Then, for certain* $w_0 \in \mathbb{R}$, *there exist the limits* $\lim_{s\to b} w(s) = 0$ *and* $\lim_{s\to b} w'(s) = \frac{1}{1+h_1}$.
4. *Assume* $b < +\infty$ *and there exists* $\lim_{s\to b} h(s) = -\infty$. *If for some* $M > 0$ *and* $s_1 \in [s_0, b)$, *it holds* $w(s) \geq M$ *for every* $s \geq s_1$, *then there exist* $\lim_{s\to b} w(s) = w_1 \geq M$ *and* $\lim_{s\to b} w'(s) = -\infty$.
5. *Assume* $b < +\infty$ *and there exist* $M < 0$, $s_1 \in [s_0, b)$ *such that for every* $s \geq s_1$, $w(s) \leq M$, *then there exist* $\lim_{s\to b} w(s) = w_1 \leq M$ *and* $\lim_{s\to b} w'(s) = +\infty$.

5 Constructing Translating Solitons

Our next target is to show the existence of graphical translating solitons which are invariant by the action of a Lie group by isometries, under additional conditions. One simple case is the foliation of the Euclidean plane $\mathbb{R}^2$ by circles centered at the origin, where the Lie group is $SO(2)$, but the origin has to be removed in order to obtain smooth maps. In this case, function $h(s) = 1/s$, because the geodesic curvature of the circles approaches infinity as the radius tends to zero.

Consider (M, g) a semi-Riemannian manifold, and Σ a Lie group acting on M by isometries. We obtain a foliation of M by the orbits of the action of Σ. Assume (1) there is exactly one (singular) orbit O which is a submanifold, but $0 \leq \dim O < \dim M$, and (2) there exists a smooth map $\Phi : (\Sigma/K) \times [0, r) rightarrow M$, for some subgroup K, carrying each $(\Sigma/K) \times \{s\}$, $s \in [0, r)$, into one or the orbits. In other words, we are assuming $O = \Phi((\Sigma/K) \times \{0\})$, and $\Phi : (\Sigma/K) \times (0, r) \to M \backslash O$ is a diffeomorphism. It is possible to reparametrize it to immediately obtain the projection $\pi : M \setminus O \to (0, b)$. When $\nabla\pi$ is never light-like, we can recompute π to obtain a semi-Riemannian submersion, [4]. That is to say, for each $s \in (0, b)$, $\pi^{-1}\{s\}$ is one orbit, which is a hypersurface of constant mean curvature $h(s)$ because Σ acts by isometries, and $\tilde{\varepsilon} = \|\nabla\pi\|^2 = \pm 1$. Note that, $\nabla\pi$ is a unit normal vector field along each non-singular orbit.

Theorem 2 *Under the conditions of this section, assume that the map* $q : (0, r) \to \mathbb{R}$, $q(s) = 1/h(s)$, *can be extended to* $q \in C^1[0, r)$ *and satisfies* $q(0) = 0$, $\tilde{\varepsilon} q'(0) \geq 0$. *Then, there exists a smooth map* $f : [0, \delta) \to \mathbb{R}$ *such that it induces a graphical translating soliton*

$$\Gamma : \Sigma \times [0, \delta) \to (M \times \mathbb{R}, \bar{g} = g + \varepsilon dt^2), \quad \Gamma(\sigma, s) = \big(\Phi(\sigma, s), f(s)\big),$$

whose unit upward normal ν *satisfying* $\bar{g}(\nu, \nu) = \tilde{\varepsilon}$.

In addition, if $\varepsilon = \tilde{\varepsilon}$ *and* $\varepsilon h(s) > 0$ *for any* $s \in (0, r)$*, the translating soliton can be smoothly extended to* $\Gamma : M \to M \times \mathbb{R}$.

Proof By Theorem 1, we just need to define $f(s) = f_1 + \int_0^s w(u)du$, $f_1 \in \mathbb{R}$ being an integration constant.Then, we just need to use Algorithm 1. The boundary condition $f'(0) = 0$ is important to ensure the smoothness of the process.

To extend the translating soliton, we just make use of Sect. 4. □

6 Examples

Example 1 In $\mathbb{R}^n$, with the standard flat metric g_0, we consider the Poincare's Half hyperplane model of $\mathbb{H}^n$, namely

$$\mathbb{H}^n = \left\{(x_1, x_2, ..., x_n) \in \mathbb{R}^n \mid x_n > 0\right\}, g = \frac{1}{x_n^2} g_0.$$

Let Σ be the Lie group $\Sigma = (\mathbb{R}^{n-1}, +)$ acting by isometries on $\mathbb{H}^n$ as usual, namely

$$\Sigma \times \mathbb{H}^n \to \mathbb{H}^n, \quad (w, p) \to (p_1 + w_1, ..., p_{n-1} + w_{n-1}, p_n)$$

where $w = (w_1, w_2, ..., w_{n-1})$ and $p = (p_1, p_2, ..., p_n)$, respectively. Note that the orbits are the well-known *horospheres*.

We define the projection map, with its usual properties:

$$\bar{\tau} : \mathbb{H}^n \to \mathbb{R}, \tau(x_1, x_2, ..., x_n) = \ln(x_n).$$

Consider two local frames $(\partial_{x_1}, \partial_{x_2}, ..., \partial_{x_n})$ and $(E_1 = x_n\partial_{x_1}, E_2 = x_n\partial_{x_2}, ..., E_n = x_n\partial_{x_n})$ of $T\mathbb{H}^n$. A straightforward computation shows

$$\nabla\bar{\tau} = E_n, \quad \text{div}(\nabla\bar{\tau}) = -n + 1. \tag{10}$$

We arrive to the following initial value problem,

$$f''(s) = (1 + (f'(s))^2)(1 + (n - 1)f'(s)), \quad f'(s_0) = f_0, \quad f(s_0) = f_1, \tag{11}$$

where $s_0, f_0, f_1 \in \mathbb{R}$. By the easy change $f'(s) = w(s)$, we transform this problem in

$$w'(s) = (1 + w^2(s))(1 + (n - 1)w(s)), \quad w(s_0) = f_0. \tag{12}$$

The classical change of variable $t = w(s)$ allows to compute a first integral, by the expression $F'(t) = \dfrac{1}{(1+(n-1)t)(1+t^2)}$, so that

$$F : \mathbb{R}\setminus\{-1/(n-1)\} \to \mathbb{R},$$

$$F(t) = \frac{1}{1+(n-1)^2}\left[(n-1)\ln\left(\frac{|1+(n-1)t|}{\sqrt{1+t^2}}\right) + \arctan t\right] + C_0,$$

for some integration constant $C_0 \in \mathbb{R}$. From here we obtain 3 cases.
Case 1: $f_0 = -1/(n-1)$. Then, the function $w(s) = \frac{-1}{n-1}$ is a constant solution to (12). Thus, $f(s) = f_1 - s/(n-1)$ is a solution to (11).

Case 2: $f_0 > -1/(n-1)$. We restrict F, namely $F_1 : \left(\frac{-1}{n-1}, +\infty\right) \to \mathbb{R}$. In this case, $F' > 0$, so that F is injective. To compute its image, we see

$$\begin{aligned}
\lim_{t\to+\infty} F_1(t) &= \lim_{t\to+\infty}\left[\frac{1}{1+(n-1)^2}\left[(n-1)\ln\left(\frac{|1+(n-1)t|}{\sqrt{1+t^2}}\right) + \arctan t\right] + C_0\right] \\
&= \left(\frac{1}{1+(n-1)^2}\right)\left((n-1)\ln(n-1) + \frac{\pi}{2}\right) + C_0 =: K_0, \\
\lim_{t\to\frac{-1}{n-1}^+} F_1(t) &= \lim_{t\to\frac{-1}{n-1}^+}\left[\frac{1}{1+(n-1)^2}(n-1)\ln\left(\frac{|1+(n-1)t|}{\sqrt{1+t^2}} + \arctan t\right) + C_0\right] = -\infty.
\end{aligned}$$

We obtain that $F_1 : \left(\frac{-1}{n-1}, +\infty\right) \to (-\infty, K_0)$ is bijective, and there exists its inverse function

$$F_1^{-1} : (-\infty, K_0) \to \left(\frac{-1}{n-1}, +\infty\right).$$

Now, we recover $w(s) = F_1^{-1}(s)$, $w(s_0) = F_1^{-1}(s_0) = w_0 > \frac{-1}{n-1}$, and $\lim_{s\to-\infty} w(s) = -1/(n-1)$, $\lim_{s\to K_0} w(s) = +\infty$. Finally,

$$f : (-\infty, K_0) \to \mathbb{R}, \quad f(s) = f_1 + \int_{s_0}^{s} w(u)du.$$

Then, we obtain $\lim \lim_{s\to-\infty} \lim f(s) = -\infty$ and $\lim_{s\to K_0} \lim f(s) = +\infty$. Thus, function f has a finite time blow up.

Case 3: $f_0 < -1/(n-1)$. As in the previous case, we restrict F to $F_2 : (-\infty, -1/(n-1)) \to \mathbb{R}$ and compute its image. Indeed,

$$\lim_{t\to-\infty} F_2(t) = \lim_{t\to-\infty}\left[\frac{1}{1+(n-1)^2}\left[(n-1)\ln\left(\frac{|1+(n-1)t|}{\sqrt{1+t^2}}\right)+\arctan t\right]+C_0\right]$$
$$= \left(\frac{1}{1+(n-1)^2}\right)\left((n-1)\ln(n-1)-\frac{\pi}{2}\right)+C_0 =: K_1,$$
$$\lim_{t\to\frac{-1}{n-1}^-} F_2(t) = \lim_{t\to\frac{-1}{n-1}^-}\left[\frac{1}{1+(n-1)^2}\left[(n-1)\ln\left(\frac{|1+(n-1)t|}{\sqrt{1+t^2}}\right)+\arctan t\right]+C_0\right]$$
$$= -\infty.$$

For $t < \frac{-1}{n-1}$, then $F_2'(t) < 0$, so that we obtain the bijection $F_2 : \left(-\infty, \frac{-1}{n-1}\right) \to (-\infty, K_1)$, that is to say,

$$F_2^{-1} : (-\infty, K_1) \to (-\infty, -1/(n-1)).$$

Now, we recover $w(s) = F_2^{-1}(s)$, $w(s_0) = F_2^{-1}(s_0) = f_0 < \frac{-1}{n-1}$, and $\lim_{s\to-\infty} w(s) = -1/(n-1)$, $\lim_{s\to K_1} w(s) = +\infty$. Finally,

$$f : (-\infty, K_1) \to \mathbb{R}, \quad f(s) = f_1 + \int_{s_0}^{s} w(u)du.$$

Then, we obtain $\lim \lim_{s\to-\infty} \lim f(s) = -\infty$ and $\lim_{s\to K_1} \lim f(s) = +\infty$. Therefore, function f has a finite time blow up.

Next, for each case, we resort to Algorithm 1 to obtain our translating solitons. Finally, this example shows that the condition $h > 0$ cannot be removed in Proposition 1.

Example 2 In $\mathbb{R}^{n+1}, n \geq 1$, with its standard flat metric g, consider a round $n-$sphere of radius 1 centered at 0, namely $\mathbb{S}^n$. As usual, we identify the tangent space at $x \in \mathbb{S}^n$,

$$T_x\mathbb{S}^n = \left\{X = (X_1, ..., X_{n+1}) \in \mathbb{R}^{n+1} : g(X, x) = 0\right\}.$$

Now, the Lie group $O(n-1)$ acts by isometries on $\mathbb{S}^n$ as usual:

$$O(n-1) \times \mathbb{S}^n \to \mathbb{S}^n, \ (A, x) \to A.x = \begin{pmatrix} A & 0 \\ 0 & 1 \end{pmatrix} x = \begin{pmatrix} A(x_1, ..., x_n)^t \\ x_{n+1} \end{pmatrix}$$

We restrict our study to $M = \mathbb{S}^n \backslash \{N, S\}$, i.e., we remove the North and South Poles. In this way, the space of orbits can be identified by the following projection map

$$\tau : M \to (-\pi/2, \pi/2), \quad \tau(x) = -\arcsin(x_{n+1}).$$

Then, given $\xi = (0, \ldots, 0, 1)$, simple computations show $x_{n+1} = g(x, \xi)$ and

$$\nabla\tau(x) = -\frac{\xi - g(x,\xi)x}{\sqrt{1-x_{n+1}^2}} = -\frac{\xi - g(x,\xi)x}{\cos(\tau(x))}, \; x \in \mathbb{S}^n, \quad \|\nabla\tau\| = 1, \quad \mathrm{div}(\nabla\tau) = (n-1)\tan(\tau).$$

We obtained that $h(s) = (1-n)\tan(s)$. Thus, we consider the following differential equation:

$$f''(s) = \big(1 + f'(s)^2\big)\big(1 - (n-1)\tan(s)\, f'(s)\big), \tag{13}$$

which we reduce in a first step to ($w = f'$),

$$w'(s) = \big(1 + w^2(s)\big)\big(1 - (n-1)\tan(s) w(s)\big). \tag{14}$$

Needless to say, for each $w_o \in \mathbb{R}$, there exists a solution $w : (-\delta, \delta) \to \mathbb{R}$ such that $w(0) = w_o$. Since $h > 0$ on $(0, \pi/2)$, by Proposition 1, we can extend to $w : (-\delta, \pi/2) \to \mathbb{R}$. Now, by taking $z : (-\pi/2, \delta) \to \mathbb{R}, z(u) = -w(-u)$, it is clear that z is another solution to (14). By Proposition 1, we can extend $z : (-\pi/2, \pi/2) \to \mathbb{R}$. This means that each solution to (14) can be globally defined $w : (-\pi/2, \pi/2) \to \mathbb{R}$. Clearly, for each $f_0 \in \mathbb{R}$, we construct a solution $f(s) = \int w(x)dx + f_0, f : (-\pi/2, \pi/2) \to \mathbb{R}$. Now, by using Algorithm 1, given a solution f, we obtain a translating soliton defined on the sphere except two points, namely $\mathbb{S}^n \backslash \{N, S\}$.

Moreover, a simple computation shows

$$\lim_{s\to\pi/2} \frac{h'(s)}{h^2(s)} = \lim_{s\to\pi/2} \frac{1}{(n-1)\sin^2(s)} = \lim_{s\to-\pi/2} \frac{h'(s)}{h^2(s)} = \frac{1}{n-1} > 0.$$

By Proposition 1, there exist two solutions $\bar{w}$ and $\tilde{w}$ that satisfy the conditions $\lim_{s\to\pi/2} \bar{w}(s) = 0$ and $\lim_{s\to-\pi/2} \tilde{w}(s) = 0$. We take $\bar{f} = \int \bar{w}$ and $\tilde{f} = \int \tilde{w}$, and use Algorithm 1. Then, these translating solitons will admit a tangent plane at points N or S, i. e., they will be smooth. Problem is, we have not been able to show if these two translating solitons coincide.

Example 3 In the standard Euclidean Space $\mathbb{R}^3$, consider a C^∞ plane curve $\alpha : (0, +\infty) \to \mathbb{R}^3, \alpha(s) = (x(s), 0, z(s)), x(s) > 0$ for any $s > 0$, which is arclength, and satisfies $\lim_{s\to 0} \alpha(s) = 0, \lim_{s\to 0} x'(s) = x_0 > 0$. We construct the revolution surface S parametrized by

$$X : \mathbb{S}^1 \times (0, +\infty) \to \mathbb{R}^3, \; X(\theta, s) = (\cos(\theta)x(s), \sin(\theta)y(s), z(s)).$$

This is a smooth surface foliated by circles of radius $1/x(s)$, or rather, invariant by the Lie group $\mathbb{S}^1$ acting by isometries. In this case, consider the map

$$h : (0, \infty) \to \mathbb{R}, \; h(s) = 1/x(s).$$

Since $q(s) := 1/h(s) = x(s)$ can be smoothly C^1-extended to $q : [0, +\infty) \to \mathbb{R}$, with $q'(0) = x_0 > 0$, by our previous results, we can construct a $\mathbb{S}^1$-invariant translating soliton $S \to S \times \mathbb{R} \subset \mathbb{R}^4$. Note that function h is here totally arbitrary.

Acknowledgements M. Ortega has been partially financed by the Spanish Ministry of Economy and Competitiveness and European Regional Development Fund (ERDF), project MTM2016-78807-C2-1-P.

References

1. A.V. Alekseevsky, D.V. Alekseevsky, Riemannian G-Manifold with one-dimensional orbit space. Ann. Global Anal. Geom. **11**(3), 197–211 (1993)
2. J. Clutterbuck, O.C. Schnürer, F. Schulze, Stability of translating solutions to mean curvature flow. Cal. Var. Partial Diff. Eq. **29**(3), 281–293 (2007)
3. E.A. Coddington, N. Levinson, *Theory of Ordinary Differential Equations* (McGraw-Hill Book Company Inc., New York, Toronto, London, 1955)
4. M.A. Lawn, M. Ortega, Translating Solitons from Semi-Riemannian foliations (2016). (preprint)
5. G. Li, D. Tian, C. Wu, Translating solitons of mean curvature flow of noncompact submanifolds. Math. Phys. Anal. Geom. **14**, 83–99 (2011)
6. F. Martin, A. Savas-Halilaj, K. Smoczyk, On the topology of translating solitons of the mean curvature flow, Cal. Var. Partial Diff. Eq. **54**(3), 2853–2882 (2015)
7. X.H. Nguyen, Doubly periodic self-translating surfaces for the mean curvature flow. Geom. Dedicata **174**, 177–185 (2015)
8. B. O'Neill, *Semi Riemannian geometry, With applications to relativity, Pure and Applied Mathematics, 103* (Academic Press, Inc., New York, 1983)
9. S. Wiggins, Introduction to applied nonlinear dynamical systems and chaos. 2nd edn. Texts in Applied Mathematics, 2. Springer-Verlag, New York (2003). ISBN: 0-387-00177-8

Trivalent Maximal Surfaces in Minkowski Space

Wai Yeung Lam and Masashi Yasumoto

Abstract We investigate discretizations of maximal surfaces in Minkowski space, which are surfaces with vanishing mean curvature. The corresponding discrete surfaces admit a Weierstrass-type representation in terms of discrete holomorphic quadratic differentials. There are two particular types of discrete maximal surfaces that are obtained by taking the real part and the imaginary part of the representation formula, and they are deformable to each other by a one-parameter family. We further introduce a compatible notion of vertex normals for general trivalent surfaces to characterize their singularities in Minkowski space as in the smooth theory.

Keywords Discrete differential geometry · Weierstrass-type representation Singularity · Discrete complex analysis · Discrete integrable system

1 Introduction

The study of surfaces in 3-dimensional Lorentzian spaceforms is an ongoing and developing area of research. In particular, the study of constant mean curvature (CMC, for short) surfaces in 3-dimensional Lorentzian spaceforms is a central topic in this subject. Here we focus on spacelike surfaces with vanishing mean curvature, which are called *spacelike maximal surfaces* (maximal surfaces, for short).

Maximal surfaces in 3-dimensional Minkowski space $\mathbb{R}^{2,1}$ maximize their area, in contrast to minimal surfaces in Euclidean space $\mathbb{R}^3$ that minimize the area. Like in

W.Y. Lam
Department of Mathematics, Brown University, Box 1917, Providence, RI 02912, USA
e-mail: lam@math.brown.edu

M. Yasumoto (✉)
Osaka City University Advanced Mathematical Institute, 3-3-138 Sugimoto, Sumiyoshi-ku, Osaka 558-8585, Japan
e-mail: yasumoto@sci.osaka-cu.ac.jp

M.A. Cañadas-Pinedo et al. (eds.), *Lorentzian Geometry and Related Topics*, Springer Proceedings in Mathematics & Statistics 211,
DOI 10.1007/978-3-319-66290-9_10

the case of minimal surfaces in $\mathbb{R}^3$, Kobayashi [11] derived a Weierstrass-type representation for conformal maximal surfaces in $\mathbb{R}^{2,1}$ and constructed several examples. On the other hand, Calabi [6] showed that the only complete maximal surface in $\mathbb{R}^{2,1}$ is the spacelike plane. This fact naturally leads us to the necessity of considering maximal surfaces in $\mathbb{R}^{2,1}$ *with singularities*. In this direction, singularities of maximal surfaces have been well-studied (see [7, 12, 21] for example).

The idea of structure preserving discretizations of differential geometry is rapidly developing. Its goal is establish a discrete theory as rich as the corresponding smooth theory. Various viewpoints of discretizations have been conducted independently.

From the viewpoint of integrable systems, Bobenko, Pinkall [1] described discrete constant negative Gaussian curvature surfaces in $\mathbb{R}^3$, which are compatible with a discrete version of the sine-Gordon equation in [10]. In [2], they further investigated discrete parametrized surfaces in $\mathbb{R}^3$ which are called *discrete isothermic surfaces*. Discrete isothermic surfaces contain interesting classes of discrete surfaces such as discrete minimal surfaces and CMC surfaces. They can be formulated and characterized by employing discrete versions of integrable transformation theory (see [8, 9] for example). Another discretization of surfaces can be found in [4] (see also [16]).

Discretizations of minimal and CMC surfaces in $\mathbb{R}^3$ via the variational approach have also been considered [18, 19]. However, in this approach these surfaces are discretized as triangle meshes, in contrast to quadrilateral meshes in the integrable system approach. It was believed to be difficult to obtain examples that lie in the intersection of the various discretization approaches.

Recently in [14, 15], Pinkall and the first author have launched a new discrete surface theory in $\mathbb{R}^3$ which not only generalizes many previous works to general meshes, but also gives a unified theory that is compatible with both integrable geometric aspects and variational properties. In particular, they introduced discrete minimal surfaces using a Weierstrass representation in terms of discrete holomorphic quadratic differentials, which further generalize the curvature approach [13]. Particular examples are trivalent minimal surfaces that are discrete minimal surfaces where each vertex has three outgoing edges.

On the other hand, a discretization of maximal surfaces has been initiated by the second author [22]. Discrete isothermic surfaces in $\mathbb{R}^{2,1}$ with vanishing mean curvature were introduced, which are called *discrete isothermic maximal surfaces*. A Weierstrass-type representation for discrete isothermic maximal surfaces was derived and several examples were constructed. As in the smooth case, discrete isothermic maximal surfaces generally have configurations of singularities. However, in the realm of discrete differential geometry, it is not possible to detect singularities by means of differentiation with respect to the discrete variables. An important question is how to characterize such singularities of discrete maximal surfaces. Although this phenomenon never occurs in the case of discrete isothermic CMC surfaces in 3-dimensional Riemannian spaceforms, it can be found in the case of Lorentzian spaceforms (see [17, 20, 22] for example).

In this paper, combining the ideas above, we investigate trivalent maximal surfaces in $\mathbb{R}^{2,1}$. Our trivalent maximal surfaces can be locally described by a Weierstrsss-type representation using holomorphic quadratic differentials. We consider two special types of trivalent maximal surfaces in $\mathbb{R}^{2,1}$, which are called *A-maximal* and *C-maximal* surfaces. As will be seen later, an A-maximal surface is obtained by taking the real part of the Weierstrass-type representation and a C-maximal surface is obtained by taking the imaginary part. Moreover, our description enables us to treat associated families of trivalent maximal surfaces.

Further, we characterize singularities of trivalent surfaces in Minkowski space. Like in [22], trivalent maximal surfaces generally have configurations of singularities and so do the surfaces in the associated families. We introduce a notion of vertex normal for trivalent surfaces in Minkowski space. Starting from analyzing singularities of A-maximal and C-maximal surfaces, we characterize singularities of all trivalent maximal surfaces in $\mathbb{R}^{2,1}$. Our results provide not only the unified theory given in [13, 14], but also a better understanding of singular behaviors of trivalent surfaces.

This paper is organized as follows: In Sect. 2, terminologies and related facts are introduced. In Sect. 3, we introduce two types of discrete maximal surfaces in $\mathbb{R}^{2,1}$. In Sect. 4, we investigate trivalent maximal surfaces in $\mathbb{R}^{2,1}$ and derive a Weierstrass-type representation for them. Finally in Sect. 5, we analyze singularities of trivalent maximal surfaces, including their associated families.

2 Preliminaries

Let $\mathbb{R}^{2,1} := (\mathbb{R}^3, \langle \cdot, \cdot \rangle)$ be 3-dimensional Minkowski space with the Lorentz metric

$$\langle (x_1, x_2, x_0)^t, (y_1, y_2, y_0)^t \rangle = x_1 y_1 + x_2 y_2 - x_0 y_0$$

for $(x_1, x_2, x_0)^t, (y_1, y_2, y_0)^t \in \mathbb{R}^{2,1}$ and squared norm

$$\|(x_1, x_2, x_0)^t\|^2 := \langle (x_1, x_2, x_0)^t, (x_1, x_2, x_0)^t \rangle,$$

which can be negative. Note that, for fixed $d \in \mathbb{R}$ and vector $n \in \mathbb{R}^{2,1} \setminus \{0\}$, a plane $\mathcal{P} = \{x \in \mathbb{R}^{2,1} \mid \langle x, n \rangle = d\}$ is *spacelike* or *timelike* or *lightlike* when n is timelike or spacelike or lightlike, respectively. Here we define

$$\mathbb{H}^2_+ := \{X = (x_1, x_2, x_0)^t \in \mathbb{R}^{2,1} | \ \|X\|^2 = -1, \ x_0 > 0\},$$
$$\mathbb{H}^2_- := \{X = (x_1, x_2, x_0)^t \in \mathbb{R}^{2,1} | \ \|X\|^2 = -1, \ x_0 < 0\}.$$

The Minkowski version of the cross product is defined as follows:

$$(x_1, x_2, x_0)^t \times (y_1, y_2, y_0)^t := \begin{vmatrix} \mathbf{e_1} & \mathbf{e_2} & -\mathbf{e_0} \\ x_1 & x_2 & x_0 \\ y_1 & y_2 & y_0 \end{vmatrix} \in \mathbb{R}^{2,1},$$

where $\mathbf{e_1} := (1, 0, 0)^t, \mathbf{e_2} := (0, 1, 0)^t, \mathbf{e_0} := (0, 0, 1)^t$. Note that, for given two vectors $x, y \in \mathbb{R}^{2,1}$, $x \times y$ is perpendicular to both x and y.

2.1 *Smooth Maximal Surfaces in* $\mathbb{R}^{2,1}$

Here we briefly review smooth maximal surfaces in $\mathbb{R}^{2,1}$. Let $F : \mathbb{D}\ (\subset \mathbb{C}) \to \mathbb{R}^{2,1}$ be a spacelike immersion parametrized by complex coordinate $Z \in \mathbb{D}$. Then F is called *maximal* if the mean curvature of F identically vanishes, or equivalently, F locally maximizes its area. In particular, we only consider conformal immersions. For conformal maximal surfaces in $\mathbb{R}^{2,1}$, Kobayshi [11] derived a Weiersrtass-type representation:

Proposition 2.1 *Any conformal maximal surface F can be locally described by*

$$dF = \mathrm{Re}\left((1 + g^2, \sqrt{-1}(1 - g^2), -2g)^t \omega\right),$$

where g is a meromorphic function and $\omega = \hat{\omega} dZ$ is a holomorphic one-form such that $g^2\hat{\omega}$ is holomorphic. This representation formula is called a Weierstrass-type representation *for conformal maximal surfaces and the induced metric is* $(1 - |g|^2)^2|\omega|^2$. *Moreover, the Gauss map ν of F is given by*

$$\nu = \left(\frac{2\mathrm{Re}(g)}{1 - |g|^2}, \frac{2\mathrm{Im}(g)}{1 - |g|^2}, \frac{1 + |g|^2}{1 - |g|^2}\right)^t \in \mathbb{H}^2_+ \cup \mathbb{H}^2_-.$$

So the Gauss map of F can be obtained by the inverse image of the stereographic projection. In this sense we also call g the Gauss map of F.

Remark 2.2 When g takes value in the unit circle $\mathbb{S}^1 \subset \mathbb{C}$, the metric of F degenerates, so F has singularities (see [7, 21] for details).

The Weierstrass-type representation in Proposition 2.1 can be also written as follows:

$$dF = \mathrm{Re}\left((1 + g^2, \sqrt{-1}(1 - g^2), -2g)^t \omega\right) = \mathrm{Re}\left((1 + g^2, \sqrt{-1}(1 - g^2), -2g)^t \frac{\hat{Q} dZ}{g'}\right),$$

where $g' := \dfrac{\partial g}{\partial Z}$, $\hat{Q} := \langle F_{ZZ}, \nu \rangle$ is the coefficient of the Hopf differential $Q = \hat{Q} dZ^2$ of F and encodes its second fundamental form. Note that Q is also called the *holomorphic quadratic differential*. If $\hat{Q}$ is a nonzero real-valued (resp. pure imaginary-valued) function, F is a maximal surface parametrized by conformal curvature line (resp. asymptotic line) coordinates.

3 Discrete Maximal Surfaces in $\mathbb{R}^{2,1}$

In this section we introduce discrete maximal surfaces in $\mathbb{R}^{2,1}$. Throughout this paper, a *discrete surface* is a cell decomposition of a surface $M = (V, E, F)$, where V is a set of points, E is a set of edges, and F is a set of faces. We denote by V_{int} the set of interior vertices and by E_{int} the set of interior edges. For a given discrete surface M, a dual cell decomposition $M^* = (V^*, E^*, F^*)$ of M is a dual surface of M. Then, there are a one-to-one correspondence between $i \in V$ and $i^* \in F^*$, and a one-to-one correspondence between $ij \in E$ and $i^* j^* \in E^*$. For a symbolic notation in the case of trianguated surfaces, see Fig. 1.

In order to derive a Weierstrass-type representation for trivalent maximal surfaces, we set up the notion of discrete holomorphic quadratic differentials.

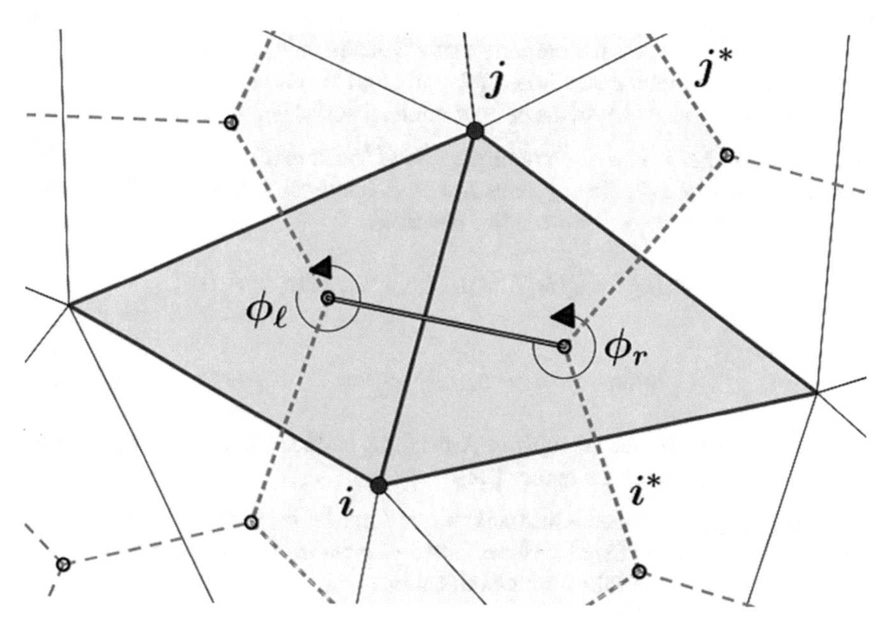

Fig. 1 Two neighboring oriented triangles and the correspondence between $i^* \in F^*$ and $i \in V$. In this picture, we denote the left face ϕ_l with respect to the oriented edge $\overrightarrow{ij}$, and the right face ϕ_r with respect to $\overrightarrow{ij}$

Definition 3.1 *A discrete holomorphic quadratic differential on a nondegenerate planar mesh $z : V \to \mathbb{C}$ is a function $q : E_{int} \to \mathbb{R}$ defined on interior edges such that for every interior vertex i*

$$\sum_j q_{ij} = 0, \ \sum_j \frac{q_{ij}}{z_j - z_i} = 0. \tag{3.1}$$

It was shown in [14] that such a notion is Möbius invariant in the sense that if q is a holomorphic quadratic differential on a planar mesh z and w is a Möbius transform of z, then q is again a holomorphic quadratic on w. Furthermore if the mesh is triangulated, the space of discrete holomorphic quadratic differentials is isomorphic to the space of discrete harmonic functions modulo linear functions. Hence, we can easily obtain a discrete holomorphic quadratic differential on arbitrary planar triangular meshes.

3.1 A-Maximal Surfaces in $\mathbb{R}^{2,1}$

We define two types of discrete maximal surfaces in $\mathbb{R}^{2,1}$. The first type is defined as follows, which reflects the property that any isothermic maximal surface in $\mathbb{R}^{2,1}$ can be obtained by taking its Christoffel dual of an isothermic surface in $\mathbb{H}^2_+ \cup \mathbb{H}^2_-$ (see [22], and see also [13] in the case of minimal surfaces):

Definition 3.2 *Let M be a discrete surface, let M^* be its dual, and let $f : V^* \to \mathbb{R}^{2,1}$ be a realization of M^*. Then f is called an* A-maximal *surface with Gauss map $n : V \to \mathbb{H}^2_+ \cup \mathbb{H}^2_-$ if the following two conditions*

$$dn(e_{ij}) \times df(e^*_{ij}) = 0, \quad \langle n_i + n_j, df(e^*_{ij})\rangle = 0$$

hold, where

$$dn(e_{ij}) := n_j - n_i, \quad df(e^*_{ij}) = f_{\phi_r} - f_{\phi_l}.$$

Here we introduce one example of A-maximal surfaces. The following example was described by the second author [22].

Example 3.3 (Discrete isothermic maximal surfaces in $\mathbb{R}^{2,1}$). We introduce a special class of discrete parametrized surfaces called *discrete isothermic surfaces*.

We first consider the following identification:

$$\mathbb{R}^{2,1} \ni (x_1, x_2, x_0)^t \cong \begin{pmatrix} ix_0 & x_1 - ix_2 \\ x_1 + ix_2 & -ix_0 \end{pmatrix} \in \mathfrak{su}_{1,1},$$

where $\mathfrak{su}_{1,1}$ is the Lie algebra of the Lie group $\mathrm{SU}_{1,1} := \left\{ \begin{pmatrix} a & b \\ \bar{b} & \bar{a} \end{pmatrix} \in \mathrm{SL}_2\mathbb{C} \right\}$. Throughout this example, we write the image of a map M from $\mathbb{Z}^2$ to the target space as

$M(m,n) = M_{(m,n)}$ for all $(m,n) \in \mathbb{Z}^2$, and we call such a map a discrete surface. Like in $\mathbb{R}^3$ ([1]), discrete isothermic surfaces in $\mathbb{R}^{2,1}$ are defined as follows: Let $\hat{F} : \mathbb{Z}^2 \to \mathbb{R}^{2,1}$ be a discrete surface. Then $\hat{F}$ is called a *discrete isothermic surface* if

$$(\hat{F}_q - \hat{F}_p) \cdot (\hat{F}_r - \hat{F}_q)^{-1} \cdot (\hat{F}_r - \hat{F}_s) \cdot (\hat{F}_s - \hat{F}_p)^{-1} = \frac{\alpha_{pq}}{\alpha_{ps}} I, \text{ where}$$
$$p = (m,n),\ q = (m+1,n),\ r = (m+1,n+1),\ s = (m,n+1)$$

for all $(m,n) \in \mathbb{Z}^2$, I is a 2×2 identity matrix, and α_{pq} (resp. α_{ps}) is a real-valued function depending only on the horizontal (resp. vertical) direction satisfying $\frac{\alpha_{pq}}{\alpha_{ps}} < 0$. We call the quantity $\frac{\alpha_{pq}}{\alpha_{ps}}$ a *cross ratio* of $\hat{F}$, and α_{pq}, α_{ps} called the *cross ratio factorizing functions*. In particular, a discrete isothermic surface $g : \mathbb{Z}^2 \to \mathbb{R}^2 (\subset \mathbb{R}^{2,1}) \cong \mathbb{C}$ is called a *discrete holomorphic function*.

A discrete isothermic maximal surface F has the following property:

$$\hat{N}_q - \hat{N}_s = (\alpha_{pq} - \alpha_{ps}) \frac{\hat{F}_r - \hat{F}_p}{\|\hat{F}_r - \hat{F}_p\|^2},\ \hat{N}_r - \hat{N}_p = (\alpha_{pq} - \alpha_{ps}) \frac{\hat{F}_q - \hat{F}_s}{\|\hat{F}_q - \hat{F}_s\|^2},$$

where $\hat{N} := \frac{1}{1 - |\hat{g}|^2} \begin{pmatrix} 2\mathrm{Re}(\hat{g}) \\ 2\mathrm{Im}(\hat{g}) \\ 1 + |\hat{g}|^2 \end{pmatrix}$. Here we denote

$$\mathbb{Z}^2_{even} := \{(m,n) \in \mathbb{Z}^2 \mid m+n \text{ even}\},\ \mathbb{Z}^2_{odd} := \{(m,n) \in \mathbb{Z}^2 \mid m+n \text{ odd}\}.$$

Setting $V := \mathbb{Z}^2_{even},\ V^* := \mathbb{Z}^2_{odd}$, a realization $\hat{F} : V^* \to \mathbb{R}^{2,1}$ is an A-maximal surface with Gauss map $\hat{N} : V \to \mathbb{H}^2_+ \cup \mathbb{H}^2_-$.

3.2 C-Maximal Surfaces in $\mathbb{R}^{2,1}$

Inspired by conical meshes with vanishing mean curvature in $\mathbb{R}^3$ (see [3, 5] for example) next we define another type of discrete maximal surfaces as follows:

Definition 3.4 *Let M be a discrete surface and M^* be its dual. Suppose $f : V^* \to \mathbb{R}^{2,1}$ is a realization of M^* with planar spacelike faces and Gauss map $n : V \to \mathbb{H}^2_+ \cup \mathbb{H}^2_-$. Then f is called a* C-maximal surface *if its integrated mean curvature vanishes on every face $i \in V^*$:*

$$H_i := \sum_j \sigma_{ij} \ell_{ij} \left(\tanh \frac{d_{ij}}{2} \right)^{\sigma_{ij}}$$

Here ℓ denotes the edge length of f while d and σ encodes the geometry of the Guass map:

$$\begin{aligned} d_{ij} &= \cosh^{-1} |\langle N_i, N_j \rangle| \quad \textit{(dihedral angle)} \\ \sigma_{ij} &= \operatorname{sgn}(-\langle N_i, N_j \rangle). \end{aligned}$$

4 Trivalent Maximal Surfaces in $\mathbb{R}^{2,1}$

Here we consider trivalent maximal surfaces in $\mathbb{R}^{2,1}$. A (linear or nonlinear) discrete complex analysis on triangulated surfaces has been developed in [13–15]. Let M be a triangulated surface and M^* be its dual. For a given discrete holomorphic quadratic differential q, we consider a dual one-form $\eta : \overrightarrow{E^*_{int}} \to \mathbb{C}^3$

$$\eta(e^*_{ij}) := \frac{q}{z_j - z_i} \begin{pmatrix} 1 + z_i z_j \\ \sqrt{-1}(1 - z_i z_j) \\ -(z_i + z_j) \end{pmatrix}. \tag{4.1}$$

One can show the following lemma.

Lemma 4.1 *The dual one-form defined by Eq.* (4.1) *is closed, that is, for every interior vertex i,*

$$\sum_j \eta(e^*_{ij}) = 0$$

holds, where j is a neighboring vertex of i.

Proof We can rewrite Eq. (4.1) into

$$\eta(e^*_{ij}) = \frac{q_{ij}}{z_j - z_i} \begin{pmatrix} 1 \\ \sqrt{-1} \\ -2z_i \end{pmatrix} + \frac{q_{ij} z_i z_j}{z_j - z_i} \begin{pmatrix} 1 \\ -\sqrt{-1} \\ 0 \end{pmatrix} - q_{ij} \begin{pmatrix} 0 \\ 0 \\ -1 \end{pmatrix}.$$

The Möbius invariance of q_{ij} implies $\sum_j \frac{q_{ij} z_i z_j}{z_j - z_i} = 0$. Combining this and the definition of discrete holomorphic quadratic differentials, we have the closedness of η. □

Assuming the planar mesh is simply connected, there exists $\mathcal{F} : F \to \mathbb{C}^3$ such that

$$d\mathcal{F}(e^*_{ij}) = \eta(e^*_{ij}).$$

As in the case of discrete minimal surfaces, $\operatorname{Re}(e^{\sqrt{-1}\theta}\mathcal{F})$ defines an S^1-family of discrete maximal surfaces in $\mathbb{R}^{2,1}$. In particular, $f := \operatorname{Re}(\mathcal{F})$ and $\tilde{f} = \operatorname{Re}(\sqrt{-1}\mathcal{F})$

yield a conjugate pair of trivalent maximal surfaces. In the following, we further discuss properties of f and $\tilde{f}$, which suggest two types of discrete maximal surfaces in $\mathbb{R}^{2,1}$.

We define $n : V \to \mathbb{H}^2_+ \cup \mathbb{H}^2_-$

$$n := \frac{1}{|z|^2 - 1} \begin{pmatrix} -z - \bar{z} \\ \sqrt{-1}(z - \bar{z}) \\ (|z|^2 + 1) \end{pmatrix}$$

and $k : E_{int} \to \mathbb{R}$ by $k_{ij} = q_{ij}/|z_j - z_i|^2$. Hence $q/(z_j - z_i) = k_{ij}(\bar{z}_j - \bar{z}_i)$ and

$$\eta(e^*_{ij}) = k_{ij}(\bar{z}_j - \bar{z}_i) \begin{pmatrix} 1 + z_i z_j \\ \sqrt{-1}(1 - z_i z_j) \\ -(z_i + z_j) \end{pmatrix}.$$

From this, we can compute

$$\begin{aligned} \mathrm{Re}(\eta(e^*_{ij})) &= \frac{k_{ij}}{2} \begin{pmatrix} (z_i + \bar{z}_i)(|z_j|^2 - 1) - (z_j + \bar{z}_j)(|z_i|^2 - 1) \\ \sqrt{-1}\left((z_j - \bar{z}_j)(|z_i|^2 - 1) - (z_i - \bar{z}_i)(|z_j|^2 - 1)\right) \\ (|z_j|^2 + 1)(|z_i|^2 - 1) - (|z_i|^2 + 1)(|z_j|^2 - 1) \end{pmatrix} \\ &= \frac{k_{ij}(|z_i|^2 - 1)(|z_j|^2 - 1)}{2}(n_j - n_i). \end{aligned}$$

Since $\|n_j - n_i\|^2 = \dfrac{4|z_j - z_i|^2}{(|z_j|^2 - 1)(|z_i|^2 - 1)}$ and $k_{ij} = \dfrac{q_{ij}}{|z_j - z_i|^2}$, we have

$$\begin{aligned} &df(e^*_{ij}) = \mathrm{Re}(\eta(e^*_{ij})) = 2k_{ij}(n_j - n_i) = 2q_{ij}\frac{n_j - n_i}{\|n_j - n_i\|^2}, \text{ satisfying} \\ &df(e^*_{ij}) \times (n_j - n_i) = 0, \ \langle df(e^*_{ij}), n_i + n_j \rangle = 0. \end{aligned}$$

Furthermore, we have the following lemma.

Lemma 4.2 *The condition* $\sum_j k_{ij}(n_j - n_i) = 0$ *implies* $\sum_j k_{ij}\|n_j - n_i\|^2 = 0$.

Proof One can check that

$$\begin{aligned} &\sum_j k_{ij}\|n_j - n_i\|^2 = \sum_j k_{ij}(-2 - 2\langle n_j, n_i\rangle) = 2\sum_j k_{ij}\{\langle n_i, n_i\rangle - 2\langle n_j, n_i\rangle\} \\ &= -2\sum_j k_{ij}\langle n_j - n_i, n_i\rangle = -2\left\langle \sum_j k_{ij}(n_j - n_i), n_i \right\rangle = 0, \end{aligned}$$

proving the lemma. □

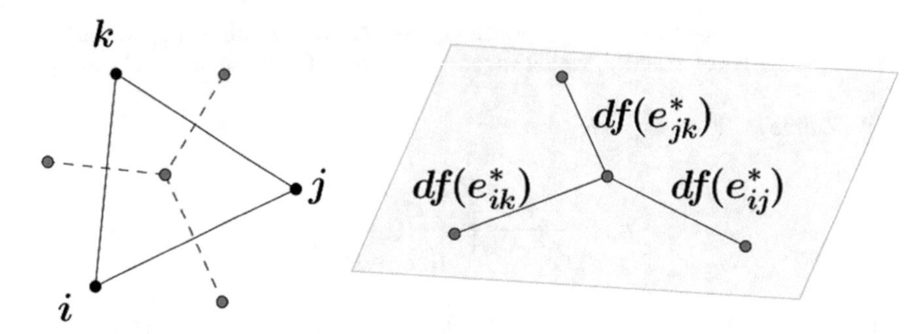

Fig. 2 A planar vertex star condition of an A-maximal surface f

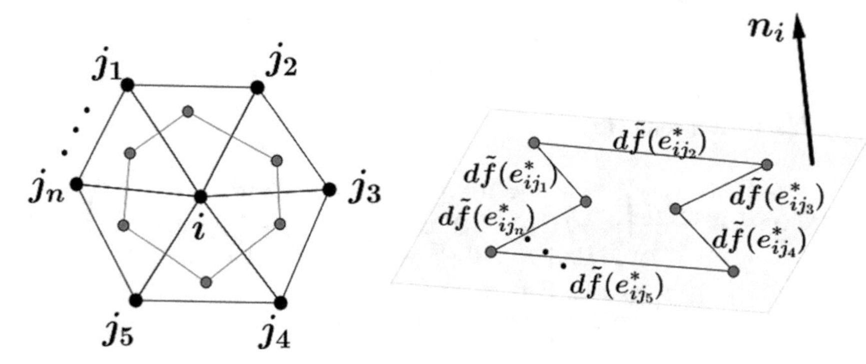

Fig. 3 A planar face of a C-maximal surface $\tilde{f}$. For each dual face $i^* \in F^*$, the resulting face of a C-maximal surface is perpendicular to n_i

Remark 4.3 Because the image of n forms a triangulated mesh, its dual graph f has planar vertex stars, that is, every subgraph obtained by connecting the neighboring three vertices of an image point of f is planar (see Fig. 2).

On the other hand,

$$\begin{aligned}
\mathrm{Re}(\sqrt{-1}\eta(e^*_{ij})) &= \frac{\sqrt{-1}k_{ij}}{2}\begin{pmatrix} (|z_j|^2+1)(z_i-\bar{z}_i)-(|z_i|^2+1)(z_j-\bar{z}_j) \\ (-z_j-\bar{z}_j)(|z_i|^2+1)-(-z_i-\bar{z}_i)(|z_j|^2+1) \\ -(-z_i-\bar{z}_i)(z_j-\bar{z}_j)+(-z_j-\bar{z}_j)(z_i-\bar{z}_i) \end{pmatrix} \\
&= \frac{k_{ij}(|z_i|^2-1)(|z_j|^2-1)}{2}(n_i \times n_j).
\end{aligned}$$

Hence $d\tilde{f}(e^*_{ij}) = 2q_{ij}\dfrac{n_i \times n_j}{\|n_j - n_i\|^2}$ and $\tilde{f}$ has spacelike planar faces (Fig. 3).

Lemma 4.4 *Writing* $df(e^*_{ij}) = k'_{ij}(n_j - n_i)$. *We consider the dihedral angle (hyperbolic distance) between the normals* n_i *and* n_j *as* $d_{ij} \in \mathbb{R}$ *such that*

$$|\langle n_i, n_j\rangle| = \cosh d_{ij}, \ \sinh d_{ij} = \mathrm{sgn}(k'_{ij})|\sinh d_{ij}|.$$

Then we have

$$d\tilde{f}(e_{ij}^*) := n_i \times df(e_{ij}^*) = k'_{ij}(n_i \times n_j)$$

and

$$\langle d\tilde{f}(e_{ij}^*), n_i \rangle = \begin{cases} -\sqrt{\|d\tilde{f}(e_{ij}^*)\|^2} \tanh \frac{d_{ij}}{2} & \text{if } (n_i, n_j \in \mathbb{H}_+^2) \text{ or } (n_i, n_j \in \mathbb{H}_-^2) \\ \sqrt{\|d\tilde{f}(e_{ij}^*)\|^2} \coth \frac{d_{ij}}{2} & \text{if } (n_i \in \mathbb{H}_+^2, n_j \in \mathbb{H}_-^2) \text{ or } (n_j \in \mathbb{H}_+^2, n_i \in \mathbb{H}_-^2). \end{cases}$$

Proof If $n_i, n_j \in \mathbb{H}_+^2$, then $\langle n_i, n_j \rangle = -\cosh d_{ij}$ and we have

$$\langle d\tilde{f}(e_{ij}^*), n_i \rangle = k'_{ij}(1 - \cosh d_{ij}) = -2k'_{ij} \sinh^2 \frac{d_{ij}}{2} = -\sqrt{\|d\tilde{f}(e_{ij}^*)\|^2} \tanh \frac{d_{ij}}{2}.$$

If $n_i \in \mathbb{H}_+^2, n_j \in \mathbb{H}_-^2$, then $\langle n_i, n_j \rangle = \cosh d_{ij}$ and we have

$$\langle d\tilde{f}(e_{ij}^*), n_i \rangle = k'_{ij}(1 + \cosh d_{ij}) = 2k'_{ij} \cosh^2 \frac{d_{ij}}{2} = \sqrt{\|d\tilde{f}(e_{ij}^*)\|^2} \coth \frac{d_{ij}}{2}.$$

□

Lemmas 4.2 and 4.4 imply the condition of vanishing mean curvature of $\tilde{f}$.

Corollary 4.5 *Writing $\sigma_{ij} := \text{sign}(-\langle n_i, n_j \rangle)$. We have*

$$\langle df(e_{ij}^*), n_i \rangle = -\sigma_{ij} \sqrt{\|d\tilde{f}(e_{ij}^*)\|^2} \left(\tanh \frac{d_{ij}}{2} \right)^{\sigma_{ij}}.$$

Furthermore, for each dual face $i^* \in F^*$ we have

$$-\sum_j \sigma_{ij} \sqrt{\|d\tilde{f}(e_{ij}^*)\|^2} \left(\tanh \frac{d_{ij}}{2} \right)^{\sigma_{ij}} = 0.$$

Theorem 4.6 *(*Weierstrass representation for discrete maximal surfaces*) Let M be a simply connected triangulated surface and M^* be its dual. Suppose $z : V \to \mathbb{C}$ is a nondegenerate realization and $q : E_{int} \to \mathbb{R}$ is a discrete holomorphic quadratic differential, then there exists a function $\mathcal{F} : F \to \mathbb{C}^3$ such that*

$$d\mathcal{F}(e_{ij}^*) = \frac{q}{z_j - z_i} \begin{pmatrix} 1 + z_i z_j \\ \sqrt{-1}(1 - z_i z_j) \\ -(z_i + z_j) \end{pmatrix}. \tag{4.2}$$

Furthermore, we denote by $n : V \to \mathbb{H}_+^2 \cup \mathbb{H}_-^2$ the stereographic projection of z given by

$$n := \frac{1}{|z|^2 - 1} \begin{pmatrix} -z - \bar{z} \\ \sqrt{-1}(z - \bar{z}) \\ (|z|^2 + 1) \end{pmatrix}.$$

Then we have $f := \mathrm{Re}(\mathcal{F})$ A-maximal and $\tilde{f} = \mathrm{Re}(\sqrt{-1}\mathcal{F})$ C-maximal. These surfaces yield a conjugate pair of trivalent maximal surfaces with Gauss map n and an associated family of trivalent maximal surfaces

$$f^\theta = \mathrm{Re}(e^{\sqrt{-1}\theta}\mathcal{F}). \tag{4.3}$$

5 Singularities of Trivalent Maximal Surfaces

Finally, in this section, we analyze singularities of trivalent maximal surfaces. We first consider A-maximal surfaces, which are surfaces obtained by taking the real part of Eq. (4.2). Trivalent A-maximal surfaces have planar vertex stars (see in Sect. 3), so each plane at the image of an A-maximal surface passing through its vertex star can be regarded as the "approximation" of the tangent plane of a maximal surface. This observation gives a natural description of singularities of A-maximal surfaces as follows:

Definition 5.1 *A vertex of a trivalent A-maximal surface is singular if the plane containing the vertex and its three neighboring vertices is not spacelike.*

We have the following theorem. The proof is almost the same as the one of Theorem 1.2 in [22], so we omit it.

Theorem 5.2 *Let f be a discrete surface in $\mathbb{R}^{2,1}$ described by the real part of Eq. (4.2). Then f_{ijk^*} is a singular vertex if and only if the circle C_{ijk} passing through z_i, z_j, z_k intersects the unit circle $\mathbb{S}^1 \subset \mathbb{C}$.*

Next, we consider singularities of C-maximal surfaces. We assume that none of the z_i, z_j, z_k is in $\mathbb{S}^1$. First we define a pole and a polar plane in $\mathbb{R}^{2,1}$ (see Fig. 4).

Definition 5.3 *For a given timelike unit normal vector $\tilde{n}$, the plane*

$$\tilde{\mathcal{P}}_{\tilde{n}} := \{X \in \mathbb{R}^{2,1} | \langle X, \tilde{n} \rangle = -1\}$$

is called a polar plane *and $\tilde{n}$ is called the* pole.

Unlike the case of A-maximal surfaces, C-maximal surfaces do not have a planar vertex star. On the other hand, by definition, every oriented edge $df(e^*_{12})$ (resp. $df(e^*_{23}), df(e^*_{13})$) of a C-maximal surface is parallel to $n_1 \times n_2$ (resp. $n_2 \times n_3, n_3 \times n_1$).

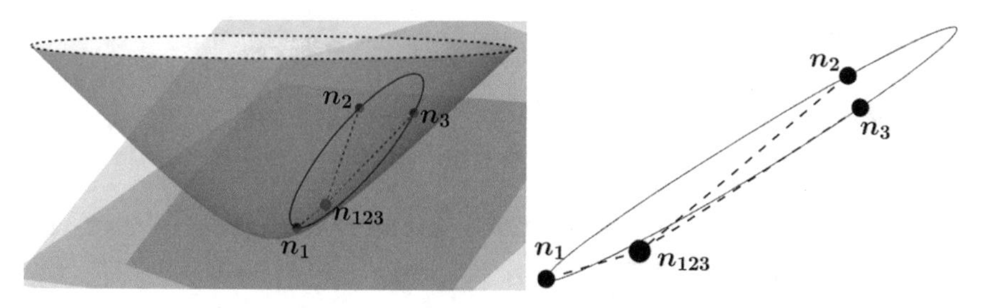

Fig. 4 Poles and polar planes. Three polar planes at three points $n_1, n_2, n_3 \in \mathbb{H}^2$ on the same circle gives rise to the intersection point n_{123}, and the oriented edges $n_{123} - n_i$ $(i = 1, 2, 3)$ is perpendicular to the circle passing through n_1, n_2, n_3

Here we define a notion of vertex normal for trivalent meshes in $\mathbb{R}^{2,1}$. We show that for trivalent maximal surfaces, its direction is invariant in the associated families.

Definition 5.4 *Let $f(v_{123}) \in \mathbb{R}^{2,1}$ be a trivalent vertex and the three outgoing edges $df(e^*_{12})$, $df(e^*_{23})$ and $df(e^*_{31})$ be pairwise linearly independent. We define the* vertex normal *$\mathcal{N}_{123} \in \mathbb{R}^{2,1}$ as follows:*

(1) If $df(e^*_{12})$, $df(e^*_{23})$ and $df(e^*_{31})$ are linearly independent, then there is a unique vector $\mathcal{N}_{123} \in \mathbb{H}^2$, $\mathbb{L}^2 := \{X \in \mathbb{R}^{2,1} | \|X\|^2 = 0\}$, or $\mathbb{S}^{1,1} := \{X \in \mathbb{R}^{2,1} | \|X\|^2 = 1\}$ such that

$$|\langle \mathcal{N}_{123}, \mathcal{N}_1 \rangle| = |\langle \mathcal{N}_{123}, \mathcal{N}_2 \rangle| = |\langle \mathcal{N}_{123}, \mathcal{N}_3 \rangle| = C.$$

where C is a nonnegative constant and $\mathcal{N}_i := \dfrac{df(e^*_{i\,i+1}) \times df(e^*_{i\,i-1})}{\sqrt{|\|df(e^*_{i\,i+1}) \times df(e^*_{i\,i-1})\|^2|}}$.

(2) If $df(e^*_{12})$, $df(e^*_{23})$ and $df(e^*_{31})$ span an affine plane, then

$$\mathcal{N}_{123} := \mathcal{N}_1 = \mathcal{N}_2 = \mathcal{N}_3.$$

Remark 5.5 In Definition 5.4, we assume that the vector $df(e^*_{i\,i+1}) \times df(e^*_{i\,i-1})$ is not lightlike. On the other hand, even if it is lightlike, the face normal $\mathcal{N}$ is still well defined without normalizing $df(e^*_{i\,i+1}) \times df(e^*_{i\,i-1})$.

Then we have the following theorem, which is used to define and characterize singularities of trivalent maximal surfaces.

Theorem 5.6 *Suppose $f^\theta : V^* \to \mathbb{R}^{2,1}$ is an associated family of a maximal surface with a triangular Gauss map $n : V \to \mathbb{H}^2_+ \cup \mathbb{H}^2_-$ related via the Weierstrass representation. Then the face normal $\mathcal{N}_i$ of n is in the same direction as the vertex normal of f^θ, as in Definition 5.4, for all θ.*

Before proving Theorem 5.6, we introduce the following lemma. This is obtained by a direct computation and the result itself is similar to the ordinary one in $\mathbb{R}^3$, up to signs.

Lemma 5.7 *For $a, b, c \in \mathbb{R}^{2,1}$, we have*

$$(a \times b) \times c = \langle b, c\rangle a - \langle a, c\rangle b, \;\; \|a \times b\|^2 = -\|a\|^2\|b\|^2 + \langle a, b\rangle^2,$$
$$\{a \times (b-a)\} \times \{a \times (c-a)\} = -(\langle a \times b, c\rangle)a = -(\langle a, b \times c\rangle)a.$$

Here we prove Theorem 5.6. We denote $\hat{\mathcal{N}}_{123}$ as the face normal of the triangle $n_1n_2n_3$. In particular, we have

$$\langle \hat{\mathcal{N}}_{123}, n_1\rangle = \langle \hat{\mathcal{N}}_{123}, n_2\rangle = \langle \hat{\mathcal{N}}_{123}, n_3\rangle =: d \in \mathbb{R} \text{ and } (n_j - n_i) \times (n_k - n_i)$$
$$= c_1\hat{\mathcal{N}}_{123} \;(c_1 \in \mathbb{R}),$$

where i, j, k are pairwise distinct.

Recall the edge vectors of the trivalent surfaces in the associated family are given by

$$df^\theta(e^*_{ij}) = k_{ij}(\cos\theta \cdot (n_j - n_i) + \sin\theta\, n_i \times n_j) = k_{ij}(\cos\theta(n_j - n_i)$$
$$+ \sin\theta \cdot n_i \times (n_j - n_i)).$$

Using Lemma 5.7, we have

$$\frac{df^\theta(e^*_{12}) \times df^\theta(e^*_{13})}{k_{12}k_{13}} = c_1\{\cos^2\theta \cdot \hat{\mathcal{N}}_{123}$$
$$- \sin^2\theta\langle n_1, \hat{\mathcal{N}}_{123}\rangle \cdot n_1 + \sin\theta\cos\theta \cdot n_1 \times \hat{\mathcal{N}}_{123}\}.$$

So $\mathcal{N}^\theta_1$ is expressed as

$$\mathcal{N}^\theta_1 = \pm\frac{df^\theta(e^*_{12}) \times df^\theta(e^*_{13})}{\sqrt{|\|df^\theta(e^*_{12}) \times df^\theta(e^*_{13})\|^2|}}$$
$$= \pm\frac{\cos^2\theta \cdot \hat{\mathcal{N}}_{123} - d\sin^2\theta \cdot n_1 + \sin\theta\cos\theta \cdot n_1 \times \hat{\mathcal{N}}_{123}}{\sqrt{|\|\hat{\mathcal{N}}_{123}\|^2\cos^2\theta - d^2\sin^2\theta|}}.$$

Thus we have $|\langle\hat{\mathcal{N}}_{123}, \mathcal{N}^\theta_1\rangle| = \sqrt{|\|\hat{\mathcal{N}}_{123}\|^2\cos^2\theta - d^2\sin^2\theta|}$.

Similarly, $|\langle\hat{\mathcal{N}}_{123}, \mathcal{N}^\theta_i\rangle| = \sqrt{|\|\hat{\mathcal{N}}_{123}\|^2\cos^2\theta - d^2\sin^2\theta|}$ for $i = 1, 2, 3$, proving the theorem.

Using the notion of vertex normal, we define singularities of trivalent maximal surfaces as follows:

Definition 5.8 *Let f^θ be a trivalent maximal surface described by Eq. (4.3). Then a vertex (or its image under f^θ) is* singular *if the corresponding vertex normal is not timelike.*

Fig. 5 Examples of trivalent maximal surfaces. The left-hand pictures are A-maximal surfaces, and the right-hand pictures are their conjugate C-maximal surfaces

In conclusion, we have the following theorem, implying a uniform understanding of singularities of trivalent maximal surfaces (Fig. 5).

Theorem 5.9 *Let f^θ be a trivalent maximal surface described by Eq.* (4.3). *Then a vertex $(ijk)^*$ is a singular vertex if and only if the circle C_{ijk} passing through z_i, z_j, z_k intersects $\mathbb{S}^1$, where z is the stereographic projection of the Gauss map of f^θ via the Weierstrass representation.*

Acknowledgements The second author was supported by the JSPS Program for Advancing Strategic International Networks to Accelerate the Circulation of Talented Researchers "Mathematical Science of Symmetry, Topology and Moduli, Evolution of International Research Network based on OCAMI".

References

1. A.I. Bobenko, U. Pinkall, Discrete surfaces with constant negative Gaussian curvature and the Hirota equation. J. Differential Geom. **43**(3), 527–611 (1996)
2. A.I. Bobenko, U. Pinkall, Discrete isothermic surfaces. J. Reine Angew. Math. **475**, 187–208 (1996)
3. A.I. Bobenko, H. Pottmann, J. Wallner, A curvature theory for discrete surfaces based on mesh parallelity. Math. Ann. **348**, 1–24 (2010)
4. A.I. Bobenko, W.K. Schief, Affine spheres: discretization via duality relations. Experiment. Math. **8**(3), 261–280 (1999)
5. A.I. Bobenko, Y. Suris, *Discrete differential geometry: Integrable structure, Graduate Studies in Mathematics, 98* (American Mathematical Society, Providence, RI, 2008)
6. E. Calabi, Examples of Bernstein problems for some non-linear equations. Proc. Sympos. Pure Math. **15**, 223–230 (1970)
7. S. Fujimori, K. Saji, M. Umehara, K. Yamada, Singularities of maximal surfaces. Math. Z. **259**, 827–848 (2008)
8. U. Hertrich-Jeromin, Transformations of discrete isothermic nets and discrete cmc-1 surfaces in hyperbolic space. Manuscripta Math. **102**(4), 465–486 (2000)
9. U. Hertrich-Jeromin, T. Hoffmann, U. Pinkall, A discrete version of the Darboux transform for isothermic surfaces, Discrete integrable geometry and physics (Vienna, 1996), 59–81, Oxford Lecture Ser. Math. Appl., 16, Oxford Univ. Press, New York (1999)
10. R. Hirota, Nonlinear partial difference equations. III. Discrete sine-Gordon equation. J. Phys. Soc. Japan **43**(6), 2079–2086 (1977)
11. O. Kobayashi, Maximal surfaces in the 3-dimensional Minkowski space $\mathbb{L}^3$. Tokyo J. Math. **6**, 297–309 (1983)
12. O. Kobayashi, Maximal surfaces with conelike singularities. J. Math. Soc. Japan **36**(4), 609–617 (1984)
13. W.Y. Lam, *Discrete minimal surfaces: critical points of the area functional from integrable systems*, to appear in IMRN, available from arXiv:1510.08788
14. W.Y. Lam, U. Pinkall, *Isothermic triangulated surfaces*, Math. Ann. **368**, no. 1–2, 165–195 (2017)
15. W.Y. Lam, U. Pinkall, *Holomorphic vector fields and quadratic differentials on planar triangular meshes*, Adv. discrete Differ. Geom. 241–265, Springer, [Berlin] (2016)
16. N. Matsuura, H. Urakawa, Discrete improper affine spheres. J. Geom. Phys. **45**(1–2), 164–183 (2003)
17. Y. Ogata, M. Yasumoto, *Construction of discrete constant mean curvature surfaces in Riemannian spaceforms and applications*, Differ. Geom. Appl. **54**, part A, 264–281 (2017)
18. U. Pinkall, K. Polthier, Computing discrete minimal surfaces and their conjugates. Experiment. Math. **2**(1), 15–36 (1993)
19. K. Polthier, W. Rossman, Discrete constant mean curvature surfaces and their index. J. Reine Angew. Math. **549**, 47–77 (2002)
20. W. Rossman, M. Yasumoto, *Discrete linear Weingarten surfaces and their singularities in Riemannian and Lorentzian spaceforms*, to appear in Advanced Studies in Pure Mathematics
21. M. Umehara, K. Yamada, Maximal surfaces with singularities in Mikowski space. Hokkaido Math. J. **35**, 13–40 (2006)
22. M. Yasumoto, Discrete maximal surfaces with singularities in Minkowski space. Differential Geom. Appl. **43**, 130–154 (2015)

Constant Mean Curvature Hypersurfaces in the Steady State Space: A Survey

Rafael López

Abstract In this survey, we review recent progress in the theory of spacelike hypersurfaces with constant mean curvature in the steady state space. Using the different models of this space, we outline the major concepts, techniques, and results with a special focus on Bernstein-type theorems, hypersurfaces with boundary in a slice, and the Dirichlet problem for the constant mean curvature equation.

Keywords Steady state space · Spacelike hypersurface · Mean curvature
Tangency principle · Omori-Yau maximum principle · Dirichlet problem

MSC 2010: 53A10, 53C42, 53A05

1 Introduction

The *steady state space* $\mathcal{H}^{n+1}$ is the space $\mathbb{R}^{n+1}_+ = \{(x, x_{n+1}) \in \mathbb{R}^n \times \mathbb{R} : x_{n+1} > 0\}$ endowed with the Lorentzian metric

$$g_{(x,x_{n+1})} = \frac{1}{x_{n+1}^2}(|dx|^2 - (dx_{n+1})^2).$$

Thus, $\mathcal{H}^{n+1} = (\mathbb{R}^{n+1}_+, g)$ is the Lorentzian analogue to the hyperbolic space $\mathbb{H}^{n+1}$. From the physical viewpoint, and for $n = 3$, $\mathcal{H}^4$ is a model of the universe proposed by Bondi and Gold [15] and Hoyle [34] under the belief in the "perfect cosmological principle", that is, the space looks the same not only at all points and in all directions (homogeneous and isotropic), but also at all times ([32, Sect. 5.2]). In particular, this model postulates a continuous creation of matter in order to be consistent with the idea of an unchanging universe, existing old and young galaxies in any large

R. López (✉)
Departamento de Geometría y Topología, Instituto de Matemáticas (IEMath-GR), Universidad de Granada, 18071 Granada, Spain
e-mail: rcamino@ugr.es

M.A. Cañadas-Pinedo et al. (eds.), *Lorentzian Geometry and Related Topics*, Springer Proceedings in Mathematics & Statistics 211,
DOI 10.1007/978-3-319-66290-9_11

volume of space which are continuously forming by accretion of new matter. This cosmological model attracted the interest of physicists during part of the twentieth century but nowadays the steady state space has been discarded because it does not predict many physical observations (in contrast to the Big Bang model), as the abundance and the proportion of helium and hydrogen, or the evolution of stars and galaxies. Finally, it was the discovery of the cosmic microwave background (CMB) in late 1964, the most clear evidence against this model: in the initial stages, the universe was denser and hotter than now because it dilutes and cools as it expands. However, the steady state space forbids the existence of CMB because under this model, the density and the temperature are always the same; see [16] and especially [48, Chs. 14, 16].

If we come back to the steady state space viewed as a Lorentzian manifold, it opens up a wide variety of problems in the theory of submanifolds. Surprisingly, it has been until very recently that this space has gained the interest after the work of Montiel [42] in 2003 where, following ideas of [40], it is proved the existence of constant mean curvature spacelike hypersurfaces with boundary in the future infinity of $\mathcal{H}^{n+1}$. This pioneering article was the starting point which many geometers focused in the study of submanifolds in $\mathcal{H}^{n+1}$. Furthermore, this is accompanied by the property that $\mathcal{H}^{n+1}$ can be viewed equivalently in two different coordinates. First, as an open set of the de Sitter space $\mathbb{S}_1^{n+1}$ (such as it appeared in the original works of Bondi and Hoyle). The space $\mathbb{S}_1^{n+1}$ has a high relevance in general relativity that it deserves to study or, as in our case, an open set of $\mathbb{S}_1^{n+1}$. A second model of $\mathcal{H}^{n+1}$ is as a generalized Robertson–Walker (GRW) spacetime and thus forming part of a large family of cosmological models which made that authors working in these spacetimes put their focus in the steady state space.

In this survey, we will study spacelike hypersurfaces with constant mean curvature in $\mathcal{H}^{n+1}$. From a physical viewpoint, the hypersurfaces with constant mean curvature (cmc to abbreviate) are convenient initial data for the Cauchy problem corresponding to the Einstein equations. In spacetimes, there is also an interest to have foliations by means of cmc spacelike hypesurfaces because all points of each leaf of the foliation are instantaneous observers and the timelike unit normal vector of the hypersurfaces measures how the observers get away with respect to the next ones. Our aim in this chapter is twofold. We first want to summarize the most important results in this topic, and second, we are intended to give an overview of the main methods that lie behind the results, especially, the tangency principle, the Omori-Yau maximum principle, or the continuity method for the Dirichlet problem. In this chapter, we also provide a new approach in some results trying to unify, if possible, the techniques of the different models.

This chapter is organized as follows. In Sect. 2 we give a description of the spacelike umbilical hypersurfaces and we present two new models for $\mathcal{H}^{n+1}$, each one will be conveniently employed depending on the problems that we address. Section 3 is devoted to characterize the slices of $\mathcal{H}^{n+1}$ in the class of complete cmc spacelike hypersurfaces obtaining Bernstein-type results and extending some of these results to GRW spacetimes. In the last part of this exposition, we study the shape of a compact cmc hypersurface in relation with its boundary (Sect. 4) and we derive existence results of the Dirichlet problem for the cmc equation (Sect. 5). This chapter ends with a list of open problems.

2 The Steady State Space: A Space and Three Models

We have defined the steady state space as $\mathcal{H}^{n+1} = (\mathbb{R}^{n+1}_+, g)$, and we will say that this is the upper half-space model for $\mathcal{H}^{n+1}$. If $\mathbb{R}^{n+1}_1 = (\mathbb{R}^{n+1}, \langle , \rangle)$ stands for the Lorentz–Minkowski space, where $\langle x, y\rangle = x_1 y_1 + \ldots + x_n y_n - x_{n+1} y_{n+1}$, then $\mathcal{H}^{n+1}$ is nothing but the open set $\mathbb{R}^{n+1}_+ = \mathbb{R}^n \times \mathbb{R}^+$ with the conformal metric $g = \langle , \rangle / x^2_{n+1}$. The time orientation is determined by $e_{n+1} = (0, \ldots, 0, 1)$. As a consequence, the isometries of $\mathcal{H}^{n+1}$ are the conformal transformations of $\mathbb{R}^{n+1}_1$ that preserve the upper half-space $\mathbb{R}^{n+1}_+$, as for example, the rotations about a vertical straight line, the horizontal translations, or the homotheties from a point of $\mathbb{R}^n \times \{0\}$.

We now introduce two equivalent models for $\mathcal{H}^{n+1}$, or to be more precise, we present two types of change of coordinates in $\mathcal{H}^{n+1}$.

1. The de Sitter model. Consider the de Sitter space, that is, the hyperquadric $\mathbb{S}^{n+1}_1 = \{x \in \mathbb{R}^{n+2}_1 : \langle x, x\rangle = 1\}$ of all unit spacelike vectors in $\mathbb{R}^{n+2}_1$, and take $a \in \mathbb{R}^{n+2}_1$ a nonzero null vector in the past half of the null cone. The steady state space is the open region $\mathbb{S}^{n+1}_{1,+} = \{x \in \mathbb{S}^{n+1}_1 : \langle x, a\rangle > 0\}$ with the induced metric. The time orientation is determined by $e_{n+2} = (0, \ldots, 0, 1)$.
2. A GRW spacetime. The steady state space is the vector space $\mathbb{R}^{n+1} = \mathbb{R} \times \mathbb{R}^n = \{(t, x) : t \in \mathbb{R}, x \in \mathbb{R}^n\}$ with the Lorentzian metric $-dt^2 + e^{2t}|dx|^2$. In other words, $\mathcal{H}^{n+1}$ is the generalized Robertson–Walker spacetime $-\mathbb{R} \times_{e^t} \mathbb{R}^n$. Historically, the cosmological model $\mathcal{H}^4$ proposed by Hoyle is $-\mathbb{R} \times_{e^{Ht}} \mathbb{R}^3$ where H is the Hubble constant. The timelike orientation is determined by the vector field ∂_t.

The expressions of the change of coordinates between the three models are the following. The isometry $\Psi : \mathbb{S}^{n+1}_{1,+} \to \mathbb{R}^{n+1}_+$ is

$$\Psi(x) = \frac{1}{\langle x, a\rangle} (x - \langle x, a\rangle b - \langle x, b\rangle a, 1)$$

where $b \in \mathbb{R}^{n+2}_1$ is a null vector such that $\langle a, b\rangle = 1$. This isometry reverses the time orientation. The isometry $\Phi : \mathbb{S}^{n+1}_{1,+} \to -\mathbb{R} \times_{e^t} \mathbb{R}^n$ is

$$\Phi(x) = \left(\log(\langle x, a\rangle), \frac{x - \langle x, a\rangle b - \langle x, b\rangle a}{\langle x, a\rangle} \right)$$

where $b \in \mathbb{R}^{n+2}_1$ is as above. This isometry preserves the time orientation. Finally, the isometry between the GRW model and the upper half-space model is $\Xi : -\mathbb{R} \times \mathbb{R}^n \to \mathbb{R}^{n+1}_+$, $\Xi(t, x) = (x, e^{-t})$ which reverses the time orientation.

Each one of the models has its advantages. For example, it is easy to visualize the isometries in the upper half-space model. In the de Sitter model, the analytic calculations are easier, as for example, when in Sect. 3 we compute the Laplacian of certain functions. Finally, the GRW model allows to see $\mathcal{H}^{n+1}$ as a product manifold with a distinguished role to the fibers of the space and again the analytic calculations in this model are simple (not necessarily easy).

The steady state space has two boundaries at the infinity. The *past infinity* of $\mathcal{H}^{n+1}$ is $\mathcal{J}^- \equiv \{x_{n+1} = \infty\}$ (the nullhypersurface $L_0 = \{x \in \mathbb{S}_1^{n+1} : \langle x, a\rangle = 0\}$ or the vertical hyperplane $\{-\infty\} \times \mathbb{R}^n$ in the GRW model). On the other hand, the *future infinity* $\mathcal{J}^+$ corresponds with the limit hyperplane $\{x_{n+1} = 0\}$ (or $L_\infty = \{x \in \mathbb{S}_1^{n+1} : \langle x, a\rangle = \infty\}$ or $\{\infty\} \times \mathbb{R}^n$ in the GRW model).

Since $\mathcal{H}^{n+1}$ is an open set of $\mathbb{S}_1^{n+1}$, then $\mathcal{H}^{n+1}$ is a non-complete Lorentzian manifold with constant sectional curvature equal to 1. For example, if $\{e_i\}$ is the usual basis of $\mathbb{R}^{n+2}$, and $a = e_1 - e_{n+2}$, the geodesic $\gamma(s) = \cosh(s)e_2 + \sinh(s)e_{n+2}$ is defined only if $s > 0$ with $\lim_{s\to 0}\gamma(s) \in \mathcal{J}^-$ and $\lim_{s\to\infty}\gamma(s) \in \mathcal{J}^+$. Motivated by this example, and following Hawking and Ellis in [32], in the steady state space any fundamental observer has a future event horizon but no past particle horizon. There also exist other geodesics in $\mathcal{H}^{n+1}$ defined for all s, for instance, $\gamma(s) = \cosh(s)e_1 - \sinh(s)e_{n+2}$.

We restrict our interest into *spacelike hypersurfaces* of $\mathcal{H}^{n+1}$. More generally, an immersion $\psi : \Sigma^n \to \mathcal{H}^{n+1}$ of a n-dimensional (connected) manifold Σ is said to be a *spacelike* hypersurface if the induced metric on Σ via ψ is Riemannian. Because the orthogonal subspace $(T_p\Sigma)^\perp$ is timelike and there is a time orientation in the ambient space $\mathcal{H}^{n+1}$, we can define a timelike unit normal vector field N on Σ in such a way that N lies in the future half of the null cone: this concludes that any spacelike hypersurface is orientable. If $\bar{\nabla}$ and ∇ denote the Levi-Civita connections of $\mathcal{H}^{n+1}$ and Σ, respectively, the Gauss equation is $\bar{\nabla}_X Y = \nabla_X Y + \sigma(X, Y)$ for all $X, Y \in \mathfrak{X}(\Sigma)$. Then the mean curvature vector is $\mathbf{H} = \mathrm{tr}(\sigma)/n$. When we write $\mathbf{H} = HN$, then H is called the *mean curvature* of the immersion. In terms of the shape operator A, namely, $AX = -\bar{\nabla}_X N$ for $X \in \mathfrak{X}(M)$, the mean curvature is $H = -\mathrm{tr}(A)/n$. If H is constant, we say that Σ has constant mean curvature and we abbreviated by *H-hypersurface* if we want to emphasize the value of the mean curvature.

Remark 2.1 Throughout the rest of this chapter, all the spacelike hypersurfaces will be oriented with the choice of N pointing to the future. We also keep the convention that the mean curvature H is computed with this choice of N. We know that the isometry Ψ between the de Sitter model and the upper half-space model reverses the time orientation, and the same occurs with the isometry $\Psi \circ \Phi^{-1}$ between the GRW model and the upper half-space model. Thus, a spacelike hypersurface with mean curvature H in the upper half-space model has mean curvature $-H$ in the de Sitter and GRW models.

In the upper half-space model, the mean curvature of a hypersurface Σ can be calculated if we know the mean curvature of Σ viewed as submanifold of $\mathbb{R}_1^{n+1}$. Indeed, since both metrics are conformal, if H' is the mean curvature of $\Sigma \subset \mathbb{R}_1^{n+1}$, then

$$H = x_{n+1}H' - (x_{n+1} \circ N') \tag{1}$$

where $N' = N/x_{n+1}$ is the Gauss map of $\Sigma \subset \mathbb{R}_1^{n+1}$.

An important family of submanifolds in $\mathscr{H}^{n+1}$ are the totally umbilical ones. By the conformality between the metric g and the Lorentzian metric $\langle,\rangle$, these hypersurfaces are the intersection of the umbilical hypersurfaces of $\mathbb{R}_1^{n+1}$ (hyperbolic planes and spacelike planes) with the half-space $\mathbb{R}_+^{n+1}$. First, let us introduce the following notation: for $c \in \mathbb{R}^{n+1}$ and $r > 0$, let

$$\mathbb{H}^n(r; c) = \{x \in \mathbb{R}_+^{n+1} : \langle x - c, x - c\rangle = -r^2\}.$$

Depending on the relation between c and r, this hypersurface has one or two connected components, namely, the upper one $\mathbb{H}_+^n(r; c)$ and the lower one $\mathbb{H}_-^n(r; c)$. We describe the *spacelike umbilical hypersurfaces* of $\mathscr{H}^{n+1}$, including the value of H with respect to the future-directed orientation following our convention of Remark 2.1.

1. A *slice* is a horizontal hyperplane

$$L_\tau = \{x \in \mathbb{R}^{n+1} : x_{n+1} = \tau\}, \quad \tau > 0.$$

   A slice is complete with $H = -1$. After an isometry, a slice is $\mathbb{H}_+^n(r; c)$ where $c_{n+1} = -r$.
2. An *equidistant hypersurface* is a hypersurface of type $\mathbb{H}_-^n(r; c)$. For the existence of $\mathbb{H}_-^n(r; c)$, it is necessary that $c_{n+1} > r$. This hypersurface is not complete and its mean curvature is $H = -c_{n+1}/r$. After an isometry, they are also non-horizontal spacelike hyperplanes or the upper component $\mathbb{H}_+^n(r; c)$ where $c_{n+1} < -r$.
3. A *hyperbolic plane* of center $c \in \mathbb{R}^{n+1}$ and radius $r > 0$ is a hypersurface of type $\mathbb{H}_+^n(r; c)$ where $c_{n+1} > -r$. A hyperbolic plane is complete with $H = c_{n+1}/r$.

Equation (1) allows to write in local coordinates the mean curvature which reveals us the local nature of a cmc hypersurface. Indeed, since a spacelike hypersurface in $\mathscr{H}^{n+1}$ is locally a graph $x_{n+1} = u(x)$ of a function $u \in C^\infty(\Omega)$, $\Omega \subset \mathbb{R}^n \times \{0\}$, by the expression of the mean curvature in $\mathbb{R}_1^{n+1}$ and (1), we obtain

$$Q_H[u] := \operatorname{div}\left(\frac{Du}{\sqrt{1-|Du|^2}}\right) - \frac{n}{u}\left(H + \frac{1}{\sqrt{1-|Du|^2}}\right) = 0. \tag{2}$$

The spacelike condition of the graph is equivalent to $|Du|^2 < 1$. Equation (2) is of elliptic type with the remarkable property that the difference of two solutions of (2) satisfies a linear elliptic PDE, and consequently, we can apply the strong maximum principle of Hopf [31]. This extends the usual tangency principle of Euclidean space for cmc hypersurfaces [37]:

Proposition 2.1 (Tangency principle) *Let Σ_1 and Σ_2 be two spacelike H-hypersurfaces which are tangent at a common interior point p and the unit* *normal vectors coincide at p. If one surface lies on one side of the other in a neighborhood of p, then Σ_1 and Σ_2 coincide in an open set around p. The same holds if $p \in \partial\Sigma_1 \cap \partial\Sigma_2$ provided that the tangent spaces $T_p\partial\Sigma_i$ coincide.*

Remark 2.2 The above notion of Euclidean graph coincides with the one of graph in $\mathscr{H}^{n+1}$. Indeed, if $\Omega \subset L_\tau$ is a (smooth) domain and $f \in C^\infty(\Omega)$, the *graph* of f is the hypersurface $\Sigma_f = \{\gamma(f(x); x) : x \in \Omega\}$, where $\gamma = \gamma(s; x)$ is the geodesic passing by x and orthogonal to L_τ. In the upper half-space model, this geodesic is a vertical line so Σ_f writes as $\{(x, u(x)) : x \in \Omega\}$ for a certain function u. A first example of an entire graph is a slice L_τ which is the graph of the constant function $f(x) = 0$ (or $u(x) = \tau$ in the upper half-space model). In the GRW model, a graph is $\{(u(x), x) : x \in \Omega\}$ where the spacelike condition reads as $|Du|^2 < e^{2u}$.

3 Bernstein-Type Characterizations of Slices

From the above section, we know that there do not exist complete umbilical hypersurfaces of $\mathscr{H}^{n+1}$ with $H < -1$, and that slices (for $H = -1$) are the first such examples. Notice that the steady state space is foliated by means of slices, indeed, $\mathbb{R}^{n+1}_+ = \cup_{\tau>0} L_\tau$ which it is of interest in the cosmological model. Slices also appear, via the tangency principle, as natural barriers for the existence of cmc hypersurfaces: see Sects. 4 and 5. Due to this distinguished role, in this section, we address with the following

> **Problem 1:** Under what geometric assumptions must a complete cmc spacelike hypersurface of $\mathscr{H}^{n+1}$ be a slice?

In this context, the remarkable chapter of Albujer and Alías [3] (part of the Ph. Doctoral Thesis of Albujer [2]) starts a series of works characterizing the slices under certain boundedness assumptions. The purpose of this section is to provide a general view of these results and the techniques employed in their proofs. Here, we use the de Sitter model of $\mathscr{H}^{n+1}$ where a slice corresponds with $L_\tau = \{x \in \mathbb{S}^{n+1}_1 : \langle x, a\rangle = \tau\}$, $\tau > 0$, and its mean curvature is $H = 1$ following the convention of Remark 2.1. First, we need the following definition.

Definition 3.1 A spacelike hypersurface $\psi : \Sigma \to \mathscr{H}^{n+1} = \mathbb{S}^{n+1}_{1,+}$ is said to be *bounded away from the future infinity* (resp. *from the past infinity*) if there exists $\tau > 0$ such that $\psi(\Sigma) \subset \{x \in \mathscr{H}^{n+1} : \langle x, a\rangle \leq \tau\}$ (resp. $\psi(\Sigma) \subset \{x \in \mathscr{H}^{n+1} : \langle x, a\rangle \geq \tau\}$). We say that Σ is *bounded away from the infinity*, or that Σ lies *between two slices*, if Σ is bounded away from the past and from the future infinity.

This definition is coherent with the future and the past infinity of $\mathscr{H}^{n+1}$: since $\mathscr{J}^+$ corresponds with $\langle x, a\rangle = \infty$, if Σ lies "away" from $\mathscr{J}^+$, then $\psi(\Sigma)$ is bounded from above. For a complete spacelike hypersurface $\Sigma \subset \mathscr{H}^{n+1}$, the boundedness of this definition imposes strong restrictions to its topology. First, notice that the spacelike property of Σ implies that the orthogonal projection of Σ on any slice is a local diffeomorphism. If we now suppose that Σ is a complete hypersurface bounded away from the future infinity, then this projection is a covering map on $\mathbb{R}^n$, and because $\mathbb{R}^n$ is simply connected, then it is a diffeomorphism onto $\mathbb{R}^n$, so Σ is an entire graph (in particular, Σ is not compact). This is the reason why results answering to the problem 1 are called of Bernstein-type because in Lorentz–Minkowski space, Cheng and Yau proved that hyperplanes are the only maximal entire graphs in $\mathbb{R}^{n+1}_1$ [23].

Theorem 3.1 ([3]) *If $\Sigma \subset \mathcal{H}^{n+1} = \mathbb{S}^{n+1}_{1,+}$ is a complete spacelike H-hypersurface between two slices, then $H = 1$. Furthermore, if $n = 2$, then Σ is a slice.*

The proof of this theorem involves two ingredients that are the keys in many other results that will appear in this section. First, it is the use of a large family of results known as "maximum principles" in the class of elliptic equations and where the tangency principle (Proposition 2.1) is a first example. For Theorem 3.1, we use he *Omori-Yau maximum principle* which is a type of maximum principle at infinity for complete Riemannian manifolds whose Ricci curvature is bounded from below [43, 49]; see also Remark 3.3.

Lemma 3.1 (Omori-Yau). *Let M be a complete Riemannian manifold with Ricci curvature bounded from below. If $u \in C^\infty(M)$ is a function bounded from above, then there exists a sequence of points $\{p_k\} \subset M$ such that*

$$\lim_{k\to\infty} u(p_k) = \sup_\Sigma u, \quad |\nabla u(p_k)| < \frac{1}{k}, \quad \textit{and} \quad \Delta u(p_k) < \frac{1}{k}. \tag{3}$$

The second ingredient is the use of appropriate functions to which we apply the maximum principles. For a spacelike hypersurface $\Sigma \subset \mathcal{H}^{n+1}$, these functions are the height function $p \mapsto \langle \psi(p), a\rangle$ (abbreviated simply $\langle p, a\rangle$) and the Gauss map $p \mapsto \langle N(p), a\rangle$ (or simply $\langle N, a\rangle$). Following (3), we need to know their Laplacians. Here, the de Sitter model reveals very useful for these calculations, obtaining

$$\Delta\langle p, a\rangle = -n\langle p, a\rangle + nH\langle N, a\rangle, \quad \Delta\langle N, a\rangle = |A|^2\langle N, a\rangle - nH\langle p, a\rangle. \tag{4}$$

See [3, 42]. The expression of $\Delta\langle p, a\rangle$ holds when H is not constant, but for $\Delta\langle N, a\rangle$ is necessary that H is constant. We point out the difference of (4) with formula (8) in [42] by the reverse sign on H according to our convention in Remark 2.1.

Proof (*of Theorem* 3.1) The height function $\langle p, a\rangle$ defined in Σ is bounded because Σ lies between two slices. Since N is future-directed, then $\langle N, a\rangle > 0$. Because $a = \langle p, a\rangle p - \langle N, a\rangle N + a^T$, where a^T is the tangent part of a on Σ, then

$$\begin{aligned} 0 &= \langle a, a\rangle = \langle p, a\rangle^2 - \langle N, a\rangle^2 + |a^T|^2 \geq \langle p, a\rangle^2 - \langle N, a\rangle^2 \\ &= (\langle p, a\rangle - \langle N, a\rangle)(\langle p, a\rangle + \langle N, a\rangle). \end{aligned} \tag{5}$$

We point out that $|\nabla\langle p,a\rangle|^2 = |a^T|^2 = \langle N,a\rangle^2 - \langle p,a\rangle^2$, so $\langle N(p_k),a\rangle \to \sup_\Sigma \langle p,a\rangle$. A computation of the Ricci curvature of Σ gives

$$\mathrm{Ric}_\Sigma(X,X) = n-1+nH\langle AX,X\rangle + \langle AX,AX\rangle \geq n-1-\frac{n^2H^2}{4} \tag{6}$$

for any $X \in \mathfrak{X}(\Sigma)$, in particular, Ric_Σ is bounded from below. Using (3) and (4), we have

$$H < \frac{\langle p_k,a\rangle}{\langle N(p_k),a\rangle} + \frac{1}{nk\langle N(p_k),a\rangle},$$

and letting $k \to \infty$, we conclude $H \leq 1$. The same argument holds with the function $-\langle p,a\rangle$ because the infimum of $\langle p,a\rangle$ is positive since Σ is bounded away from the past infinity. This yields $H \geq 1$, so $H = 1$. When $n = 2$, one can invoke a result of Akutagawa to conclude that Σ is a slice [1]. However, and such as it is rightly pointed in [3], it is better the following argument. From (4) and (5), we get $\Delta\langle p,a\rangle \geq 0$ showing that $\langle p,a\rangle$ is a subharmonic function. As $n = 2$ and $H = 1$, from (6) we deduce $K_\Sigma \geq 0$. Since Σ is complete, a result of Huber asserts that Σ is parabolic [35], so the subharmonic bounded function $\langle p,a\rangle$ is, indeed, constant, showing that Σ is a slice. □

Remark 3.1 If we only assume that Σ is bounded away from the future infinity, the proof yields $H \leq 1$. Since Σ is diffeomorphic to $\mathbb{R}^n$, Σ can not be compact. Taking into account the inequality (6), Bonnet-Myers's theorem implies that Ric_Σ is not bounded from below by 0 and this forces to $H^2 \geq 4(n-1)/n^2$. Thus we have $2\sqrt{n-1}/n \leq H \leq 1$. In the particular case $n = 2$, we conclude $H = 1$, that is, *a complete cmc spacelike hypersurface in $\mathcal{H}^3$ bounded away from the future infinity must be a slice.*

Remark 3.2 We will prove in Remark 5.1 the existence of complete spacelike H-hypersurfaces with $H > 1$ and bounded away from the past infinity.

Here, we present a new approach of Theorem 3.1 using a clever application of the tangency principle in the upper half-space model. Recall our convention on H in Remark 2.1.

Theorem 3.2 *Let $\Sigma \subset \mathcal{H}^{n+1}$ be a spacelike H-hypersurface in the upper half-space model.*

(i) If Σ is an entire graph bounded away from the future infinity, then $H \geq -1$.
(ii) If Σ is complete and bounded away from the past infinity, then $H \leq -1$.

Proof For (i), we know that there exists $\tau_1 > 0$ such that $\Sigma \subset \{x_{n+1} \geq \tau_1\}$. Let $m > 1$, and consider the equidistant hypersurface $P_r = \mathbb{H}^n_-(r; (0,\ldots,0,mr))$ whose mean curvature is $H = -m$. Let us observe that the vertex of P_r is $V_r = (0,\ldots,0,(m-1)r)$. Take $r > 0$ sufficiently small so $(m-1)r < \tau_1$. Then $P_r \cap \Sigma = \emptyset$. Let q be the intersection point of Σ with the x_{n+1}-axis ($q_{n+1} > \tau_1$): this point does

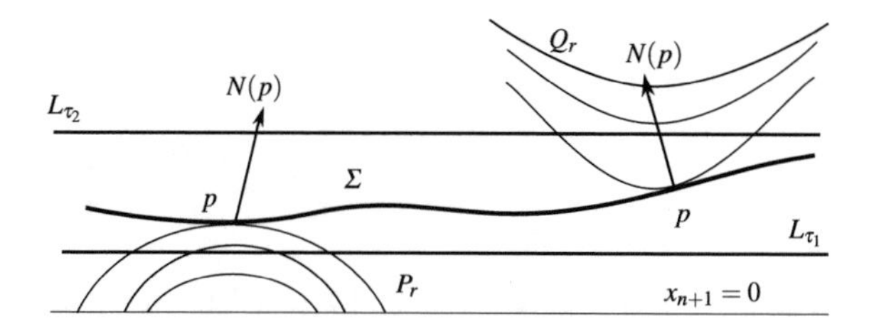

Fig. 1 Proof of Theorem 3.2

exist because Σ is a graph on $\mathbb{R}^n \times \{0\}$. Letting $r \to \infty$, we find a first value r_1, $(m-1)r_1 \leq q_{n+1}$, such that P_{r_1} meets the first time Σ at some point p; see Fig. 1, left. Since $\partial \Sigma = \emptyset$, then p is an interior common point of $\Sigma \cap P_{r_1}$, and the tangency principle says $H > -m$. Because this argument holds for any $m > 1$, we conclude $H \geq -1$.

For (ii), the completeness of Σ together the hypothesis on the boundedness says that Σ is a graph on $x_{n+1} = 0$ and $\Sigma \subset \{x_{n+1} \leq \tau_2\}$ for some $\tau_2 > 0$. The reasoning is similar replacing P_r by hyperbolic planes of type $\mathbb{H}^n_+(r; (0, \ldots, 0, mr))$ coming from $x_{n+1} = \infty$. The vertex of Q_r is $V_r = (0, \ldots, 0, (m+1)r)$ and $(m+1)r > 0$. If r is sufficiently big so $(m+1)r > \tau_2$, then $Q_r \cap \Sigma = \emptyset$. Letting $r \to 0$, we arrive until the first time $r_0 > 0$, $(m+1)r_0 \geq q_{n+1}$, such that Q_{r_0} meets Σ; see Fig. 1, right. The tangency principle gives $m > H$. Since this holds for any $m > -1$, then $H \leq -1$. □

Remark 3.3 (*A contact at "infinity"*) If in the above proof we use slices instead of equidistant hypersurfaces and hyperbolic planes we find with some troubles. For example, and for (i), suppose Σ lies above the slice L_{τ_1} but $\Sigma \not\subset L_t$ for all $t < \tau_1$. We can arrive from below with slices L_τ with $\tau < \tau_1$ without touching Σ, and it could occur that L_{τ_1} touches Σ at some point. Then, the tangency principle would say $\Sigma = L_{\tau_1}$ or $H > -1$ and this would prove the result. But it could happen that L_{τ_1} has a contact "at infinity" with Σ, that is, $L_{\tau_1} \cap \Sigma = \emptyset$ but $L_{\tau_1+\varepsilon} \cap \Sigma \neq \emptyset$ for all $\varepsilon > 0$. In such a case, we can not apply the tangency principle. This illustrates the difference between the tangency principle, which is utilized in local arguments, and the Omori-Yau maximum principle, which considers a touching point "at the infinity" by taking a sequence of points $\{p_k\}$ with the properties of (3).

Theorem 3.2 shows part of Theorem 3.1. It would remain to prove that a spacelike entire graph in $\mathscr{H}^3$ with $H = -1$ between two slices must be a slice. Suppose that Σ lies between the slices L_{τ_1} and L_{τ_2}, $\tau_1 < \tau_2$, and Σ does not lie in another smaller slab. By the tangency principle, it is not possible that Σ has a contact point with L_{τ_1} and L_{τ_2} (see Remark 3.3). Now the statement that we want to prove has the same flavor than the strong half-theorem in Euclidean space for minimal surfaces ([33]; see [44] in hyperbolic space). If in $\mathbb{R}^3$, the proof compares a minimal surface with a family of catenoids, in $\mathscr{H}^3$ a similar idea would be comparing Σ with rotational spacelike surfaces with $H = -1$. In the upper half-space model, and by Eq. (1), a rotational spacelike surface with respect to the x_3-axis has mean curvature H if the profile curve $\alpha(t) = (x(t), 0, z(t))$ satisfies

$$H = \frac{z(t)}{2}\left(\phi'(t) + \frac{\sinh(\phi(t))}{x(t)}\right) - \cosh(\phi(t))$$

where $\alpha'(t) = (\cosh\phi(t), 0, \sinh\phi(t))$. The initial conditions are $x(0) = 0$, $z(0) = z_0 > 0$, and $\phi(0) = \lambda$. Take $H = -1$, and let S_λ denote the rotational surface determined by the parameter λ. If $\lambda = 0$, the unique solution is the slice $x_3 = z_0$. If $\lambda < 0$ (resp. $\lambda > 0$), α is a graph on the x_1-axis with a singularity at $t = 0$, the function z is strictly decreasing (resp. increasing), and there exists $z_\lambda \geq 0$ such that $\lim_{t\to\infty} z(t) = z_\lambda$. This proves that S_λ is asymptotic to a horizontal hyperplane (a slice) and thus the use of S_λ is not adequate because we can not avoid a contact "at infinity" (such as it appeared in Remark 3.3 comparing Σ with slices).

The paper [3] motivated a series of works by different researchers that followed two directions: first, assuming other boundedness of the height function or with other r-mean curvatures, and second, extending to GRW spacetimes.

3.1 Other Assumptions on the Height Function

Caminha and de Lima replaced in [21] the assumption that Σ lies between two slices by some control on the growth of the height function. Let Σ denote a spacelike hypersurface in the GRW model $-\mathbb{R} \times_{e^t} \mathbb{R}^n$ which is oriented according to the future-directed orientation. The height function on Σ is $h = \pi_{\mathbb{R}} \circ \psi$, and since $\langle N, \partial_t\rangle \leq -1$, the *hyperbolic angle* is the function $\theta : \Sigma \to [0, \infty)$ given by $\cosh\theta = -\langle N, \partial_t\rangle$. Let us observe that e^h takes the same value that the function $\langle p, a\rangle$ in $\mathbb{S}^{n+1}_{1,+}$ by the isometry Φ. If $g = -\langle N, \partial_t\rangle$, then $|\nabla h|^2 = g^2 - 1$ and $\Delta h = 1 - n + nHg - g^2$. Hence, we obtain the Laplacian of the functions $v = e^h$ and $\eta = e^h\langle N, \partial_t\rangle$:

$$\Delta v = -nv + nH\eta, \quad \Delta\eta = nHv + |A|^2\eta, \tag{7}$$

where in the computation of $\Delta\eta$ we use that H is constant. Both equations are the analogous ones to (4) in the GRW model.

Theorem 3.3 ([21]) *Let $\Sigma \subset \mathscr{H}^{n+1} = -\mathbb{R} \times_{e^t} \mathbb{R}^n$ be a complete spacelike H-graph with $H \geq 1$. If $h \leq -\log(-\langle N, \partial_t\rangle - 1)$, then $H = 1$.*

Here, the growth of the height function h is bounded, in some sense, by the hyperbolic angle g. The proof uses (7) to conclude that the function $-v - \eta$ is subharmonic and the relation between h and g says exactly that this function is bounded, which allows to use the Omori-Yau maximum principle. In fact, we can replace the hypothesis on h by $h \leq -c\log(g - 1)$ for some $c > 0$. We give here the proof in the de Sitter model.

Proof Consider the function $\varphi = \langle N, a\rangle - \langle p, a\rangle$. The hypothesis says that φ is bounded from above so there exists $\sup_\Sigma \varphi$. Note that $\varphi \geq 0$ by (5). Then (4) yields

$$\begin{aligned}\Delta\varphi &= (|A|^2 - nH)\langle N, a\rangle + n(H-1)\langle p, a\rangle \\ &\geq (nH^2 - nH)\langle N, a\rangle + n(H-1)\langle p, a\rangle \\ &= n(H-1)(H\langle N, a\rangle - \langle p, a\rangle) \geq n(H-1)\varphi \geq 0\end{aligned}$$

where we have used $|A|^2 \geq nH^2$ and $H \geq 1$. This proves that φ is a subharmonic function, and taking the sequence $\{p_k\}$ of (3), the above inequality of $\Delta\varphi$ implies

$$0 \leq n(H-1)\varphi(p_k) \leq \Delta\varphi(p_k) < \frac{1}{k}.$$

By contradiction, suppose $H > 1$. Letting $k \to \infty$, we conclude $\sup_\Sigma \varphi = 0$, so $\varphi = 0$ on Σ. Then $\langle N, a\rangle = \langle p, a\rangle$ and $|\nabla\langle p, a\rangle|^2 = 0$, proving that Σ is a slice, that is, $H = 1$, a contradiction. This proves the theorem. □

In the following result, Camargo, Caminha, and de Lima replace the constancy of H by $H \geq 1$ and the integrability of the gradient of the height function. Now we do not conclude $H = 1$ (as in Theorems 3.1 and 3.3), but that Σ is a slice.

Theorem 3.4 ([18]) *Let $\Sigma \subset \mathscr{H}^{n+1} = -\mathbb{R} \times_{e^t} \mathbb{R}^n$ be a complete spacelike hypersurface between two slices with (not necessarily constant) mean curvature $H \geq 1$. If $|\nabla\langle p, a\rangle| = |a^T|$ is Lebesgue integrable, then Σ is a slice.*

Proof By contradiction, suppose that $H > 1$ on Σ. From (4), $\Delta\langle p, a\rangle = n\langle HN - p, a\rangle)$. As $\langle HN - p, HN - p\rangle = 1 - H^2 < 0$, then $HN - p$ is a timelike vector on Σ, so $\langle HN - p, a\rangle$ is positive on Σ or its negative on Σ. Then $\Delta\langle p, a\rangle \geq 0$ or $\Delta\langle p, a\rangle \leq 0$. Up to a change of a sign, the function $\langle p, a\rangle$ is a subharmonic function which is bounded from above because Σ lies between two slices. Since Σ is complete and $|\nabla\langle p, a\rangle| \in \mathscr{L}^1(\Sigma)$, then $\langle p, a\rangle$ is harmonic by a result of Yau in [50]. Then $\Delta\langle p, a\rangle = 0$ and we derive that the timelike vector $HN - p$ satisfies $\langle HN - p, a\rangle = 0$, a contradiction. This proves that $H = 1$ on Σ. From (5), we have $\langle N, a\rangle - \langle p, a\rangle \geq 0$, and thus $\Delta\langle p, a\rangle \geq 0$ and $\langle p, a\rangle$ is bounded from above. The same argument as before proves that $\Delta\langle p, a\rangle = 0$ so $\langle N, a\rangle - \langle p, a\rangle = 0$, $|a^T| = 0$, and consequently, Σ is a slice. □

When $n = 2$, the same conclusion holds if we replace $|a^T| \in \mathscr{L}^1(\Sigma)$ by $K_\Sigma \geq 0$. This was proved by Aquino et al. in [14] and the argument is the same as in Theorem 3.1: now Σ is a parabolic surface, so the subharmonic function $\langle p, a\rangle$ must be constant, proving that Σ is a slice.

Finally, subsequent results in the literature replace the conditions on H by others about the higher order mean curvatures H_r. In order to use similar arguments, one needs, among other things, to extend the Laplacian operator and the Omori-Yau maximum principle. For the Laplacian, we take the so-called *r-th Newton transformations* P_r which are self-adjoint linear transformations on Σ involving H_r and the shape operator A. Then we define the second-order linear differential operator $L_r = \mathrm{tr}(P_r \circ \mathrm{Hess})$: for instance, $L_0 = \mathrm{tr}(\mathrm{Hess})$ is nothing that the Laplacian operator Δ. For $r = 1$, we have the known Yau's square operator $\square = \mathrm{tr}(P_1 \circ \mathrm{Hess})$. It is also necessary to generalize the Omori-Yau maximum principle (this was done for the operator $\square$ by Caminha and de Lima in [20, 22]). Besides that, the ellipticity of L_r is not assured and it depends on bounds on H_r. All these considerations make a bit difficult to give a clear statement of the results: we refer the interested reader to [4, 7, 12, 13, 18, 25], where much more of the results hold in GRW spacetimes. We refer also the reader to [9] for a recent account on Omori-Yau-type maximum principles for more general operators and its geometric applications.

3.2 Extension to GRW Spacetimes

A second scenario to extend Theorem 3.1 is by considering GRW spacetimes because of the third model for $\mathscr{H}^{n+1}$. We first review some basics of these spaces (we refer the article of Alías, Romero and Sánchez [11]). A *generalized Robertson–Walker* (GRW) spacetime is the product manifold $\bar{M} = I \times M$ endowed with the Lorentzian metric $-dt^2 + f^2\langle,\rangle_M$: here $(M^n, \langle,\rangle_M)$ is a n-dimensional Riemannian manifold, I is a 1-dimensional manifold (either a circle or an open interval of $\mathbb{R}$), and $f : I \to \mathbb{R}$ is a positive smooth function, called the *warping function*. We denote this space as $-I \times_f M$. Among examples of GRW spacetimes, we have the Lorentz–Minkowski space $\mathbb{R}^{n+1}_1 = -\mathbb{R} \times_1 \mathbb{R}^n$, de Sitter space $\mathbb{S}^{n+1}_1 \equiv -\mathbb{R} \times_{\cosh(t)} \mathbb{S}^n$, and the steady state space $\mathscr{H}^{n+1} \equiv -\mathbb{R} \times_{e^t} \mathbb{R}^n$. If $f(t) = e^t$, we say that $\bar{M}$ is a *steady state type space*, where besides $\mathscr{H}^{n+1}$, we point out the remarkable *de Sitter cusp space* $-\mathbb{R} \times_{e^t} \times \mathbb{T}^n$ in terminology of Galloway [29], where $\mathbb{T}^n$ is a flat n-torus.

A slice in $-I \times_f M$ is a hyperplane $L_\tau = \{\tau\} \times M$, for some $\tau \in I$. Then L_τ is an umbilical spacelike hypersurface with $H = f'(\tau)/f(\tau)$ computed with respect to $N = \partial_t$. Again, the question that we address is under what geometric assumption must a complete spacelike cmc hypersurface be a slice. The functions $\langle p, a\rangle$, and $\langle N, a\rangle$ used for the steady state space $\mathscr{H}^{n+1}$ correspond now with the height function $h = \pi_{|I} \circ \psi$ and the hyperbolic angle $\langle N, \partial_t\rangle$. If one wants to use the same techniques done in the previous results, it is necessary to consider the following remarks:

1. In $\mathbb{S}^{n+1}_{1,+}$, we computed the Laplacian of $\langle p, a\rangle$ in order to use the Omori-Yau maximum principle. In the GRW model, $\langle p, a\rangle$ corresponds with the function e^h which is nothing that $f(h)$ for the warping function $f(t) = e^t$. Thus, it is natural to compute the Laplacian of $f(h)$ in a GRW spacetime obtaining (see [26]):

$$\Delta f(h) = \left(\frac{f''f - f'^2}{f}(h)\right)|\nabla h|^2 - n\left(\frac{f'^2}{f}(h) + f'(h)H\langle N, \partial_t\rangle\right).$$

   If we want f to be subharmonic, then it is enough that both summands are not negative. For the first one is equivalent to say that $\log(f)$ is a convex function.

2. For the Omori-Yau maximum principle, we need that Ric_Σ is bounded from below. Here, we recall that a spacetime obeys the *null convergence condition* (NCC) if the Ricci tensor $\overline{\text{Ric}}$ of $\bar{M}$ satisfies $\overline{\text{Ric}}(Z, Z) \geq 0$ for any null vector Z. In a GRW spacetime, this inequality is expressed in terms of the warping function as

$$\text{Ric}_M \geq (n-1)\sup_I (ff'' - f'^2)\langle,\rangle_M = (n-1)\sup_I f^2(\log f)''\langle,\rangle_M. \quad (8)$$

3. The condition $H \geq 1$ in Theorems 3.3 and 3.4 can be viewed as a comparison between H and the mean curvature of each slice. In a GRW spacetime, we would need to relate H with the quotient f'/f.

An example of generalization of Theorem 3.1 that indicates the type of results that we are referring is the following:

Theorem 3.5 *Let Σ be a complete spacelike hypersurface between two slices in a GRW spacetime. Suppose $|\nabla h| \in \mathscr{L}^1(\Sigma)$.*

1. If $f'(h)H \geq f'^2/f(h) > 0$ ([26]), or
2. If H is bounded and $f(h)H_2 \geq f'(h)H \geq 0$ ([6]),

then Σ is a slice.

In both statements we see again a comparison criterion between mean curvature quantities without being constant. For example, in item 2, we have $H_2^2/H^2 \leq f'^2/f^2$ where the right-hand side is the square of the mean curvature of the slice L_τ. The reader can see other Bernstein-type results for complete hypersurfaces in GRW spacetimes in [5, 17, 19, 27, 28, 30, 45].

4 Compact Spacelike Hypersurfaces with Boundary

In this section, we study how the boundary of a compact cmc hypersurface affects on the shape of the whole hypersurface. First, we precise our setting. Let $\psi : \Sigma \to \mathcal{H}^{n+1}$ be a spacelike immersion of a compact hypersurface Σ, in particular, $\partial\Sigma \neq \emptyset$ and let $\Gamma \subset \mathcal{H}^{n+1}$ be a $(n-1)$-submanifold. We say that Σ is a *hypersurface with boundary* Γ if the restriction of the immersion ψ to the boundary $\partial\Sigma$ is a diffeomorphism onto Γ. We abbreviate by saying that Γ is the boundary of Σ, or that $\partial\Sigma = \Gamma$, or that Σ spans Γ. We address the following

Problem 2: Given a compact $(n-1)$-submanifold Γ included in a slice, does the geometry of Γ impose restrictions to the shape of a compact spacelike cmc hypersurface spanning Γ?

Related to the above problem, we study three specific questions:

(i) Whether the boundary Γ determines the position and the height of the hypersurface that spans with respect to the slice containing Γ.

(ii) Whether the geometry of Γ imposes restrictions to the possible values H of mean curvatures of H-hypersurfaces spanning Γ.
(iii) Whether the symmetries of Γ are inherited by the whole hypersurface. In other words, suppose that Γ is invariant by a rigid motion $M : \mathscr{H}^{n+1} \to \mathscr{H}^{n+1}$, that is, $M(\Gamma) = \Gamma$. If Σ is a cmc spacelike hypersurface with $\partial\Sigma = \Gamma$, we ask if Σ is also invariant by M. The simplest case is when Γ is a geodesic sphere and whether Σ is a hypersurface of revolution.

In this section, we will use the notation $\mathbb{H}^n_+(r; c)$ or $\mathbb{H}^n_-(r; c)$, assuming that $c = (0, \dots, 0, c)$.

4.1 Height Estimates

Let Σ be a compact spacelike H-hypersurface in the upper half-space model with $\partial\Sigma \subset L_\tau$. Notice that if $\partial\Sigma$ is a simple closed curve, then Σ is a graph because the spacelike condition says that the orthogonal projection from Σ in L_τ is a covering map onto a simply connected domain. A first result gives us the position of Σ with respect to L_τ depending whether $H < -1$ or $H > -1$, that is, comparing H with the mean curvature of L_τ.

Proposition 4.1 *Let Σ be a compact spacelike hypersurface with $\partial\Sigma \subset L_\tau$. If $H < -1$ (resp. $H > -1$, $H = -1$), then $x_{n+1} \geq \tau$ in Σ (resp. $x_{n+1} \leq \tau$, $\Sigma \subset L_\tau$).*

Proof It is enough to consider the case $H < -1$. By contradiction, suppose that there are points strictly below L_τ. Let $q \in \Sigma$ be the lowest point with respect to L_τ, and let $\bar{\tau} = q_{n+1}$. We place the slice $L_{\bar{\tau}}$ at q. The orientation of Σ and $L_{\bar{\tau}}$ coincide at q (both ones are pointing to the future, so pointing upward). Since Σ lies above $L_{\bar{\tau}}$ around q, a comparison of the mean curvatures between Σ and $L_{\bar{\tau}}$ yields $H \geq -1$, a contradiction. This proves that $x_{n+1} \geq \tau$ in Σ. Following with the same argument, and if $\bar{\tau} < \tau$ and $H = -1$, the tangency principle would say that Σ lies contained in $L_{\bar{\tau}}$, a contradiction again because $\partial\Sigma \subset L_\tau$ and $\tau \neq \bar{\tau}$. □

Other approach to Proposition 4.1 is studying $\Delta\langle p, a\rangle$ in the de Sitter model. In these coordinates, Proposition 4.1 says that if $H > 1$ (resp. $H < 1$, $H = 1$), then $\langle p, a\rangle \leq \tau$ (resp. $\langle p, a\rangle \geq \tau$, $\Sigma \subset L_\tau$). We know from (5) that $\langle p, a\rangle - \langle N, a\rangle \leq 0$. Let $H > 1$. Since $\langle N, a\rangle > 0$, then (4) yields

$$\Delta\langle p, a\rangle \geq -n\langle p, a\rangle + n\langle N, a\rangle \geq 0$$

and the maximum principle asserts that $\langle p, a\rangle \leq \max_{\partial\Sigma}\langle p, a\rangle = \tau$, obtaining Proposition 4.1. When $H \leq 0$, then (4) gives directly $\Delta\langle p, a\rangle \leq 0$ and the maximum principle concludes $\langle p, a\rangle \geq \tau$. It would remain the case $0 < H < 1$ which is not deduced directly from $\Delta\langle p, a\rangle$, and the reasoning is the following. As $0 < H < 1$, $\langle HN - p, HN - p\rangle = H^2 - 1 < 0$, and consequently, the vector $HN - p$ is timelike so $\langle HN - p, a\rangle > 0$ on Σ or $\langle HN - p, a\rangle < 0$ on Σ. At the farest point $q \in$

$\mathrm{int}(\Sigma)$ from L_τ, $|\nabla\langle p, a\rangle|(q) = 0$ and (5) gives $\langle N(q), a\rangle = \langle q, a\rangle$. Then $\langle HN - p, a\rangle(q) = (H-1)\langle q, a\rangle < 0$. This proves definitively that $\langle HN - p, a\rangle < 0$ on Σ yielding $\Delta\langle p, a\rangle \leq 0$ and consequently, $\langle p, a\rangle \geq \tau$ by the maximum principle.

Once we know that Σ lies on one side of L_τ, we want to estimate, if possible, how far Σ rises up from L_τ. In Euclidean space, this height is less than $1/|H|$ for H-graphs whose boundary Γ lies in a hyperplane, and thus this estimate is independent on the size of Γ (here and for a general reference in Euclidean space, we refer [37]). Following ideas of Meeks, the usual manner is by considering a linear combination of the height function and the Gauss map, then prove that this function is subharmonic, and finally apply the maximum principle to get the desired estimates. From (4) and because $|A|^2 - nH^2 \geq 0$, we have

$$\Delta(-H\langle p, a\rangle + \langle N, a\rangle) = (|A|^2 - nH^2)\langle N, a\rangle \geq 0 \tag{9}$$

proving that $-H\langle p, a\rangle + \langle N, a\rangle$ is subharmonic. The maximum principle asserts

$$-H\langle p, a\rangle + \langle N, a\rangle \leq \max_{\partial\Sigma}(-H\langle p, a\rangle + \langle N, a\rangle) = -H\tau + \max_{\partial\Sigma}\langle N, a\rangle. \tag{10}$$

However, and in contrast to the Euclidean case, we can not get a similar estimate. For example, from (10) and for $H > 0$, we have

$$\tau + \frac{\min_\Sigma\langle N, a\rangle - \max_{\partial\Sigma}\langle N, a\rangle}{H} \leq \langle p, a\rangle \tag{11}$$

but this estimate is not given only in terms of the boundary and H: inequality (11) is a bizarre estimate because involves the function $\langle N, a\rangle$ in Σ.

Proposition 4.1 generalizes to GRW spacetimes employing similar arguments. We replace $\Delta\langle p, a\rangle$ by the Laplacian of any primitive F of the warping function f. If h is the height function and $g = \langle N, \partial_t\rangle \leq -1$, the computation of ΔF in [30] yields $\Delta F = -nh((\log f)'(h) + Hg)$. Then it is immediate following the result proved by García-Martínez and Impera.

Theorem 4.1 ([30]) *Let Σ be a compact spacelike hypesurface in a GRW spacetime $-I \times_f M$ with $\partial\Sigma \subset L_\tau$.*

1. *If $H \geq \max\{0, \sup_I(\log f)'\}$, then $h \leq \tau$.*
2. *If $H \leq \min\{0, \inf_I(\log f)'\}$, then $h \geq \tau$.*

Let us observe that in $\mathscr{H}^{n+1}$, where $f(t) = e^t$, the above theorem says that if $H \geq 1$ (resp. $H \leq 0$), then $h \leq \tau$ (resp. $h \geq \tau$) but no information is obtained when $0 < H < 1$. If we proceed with the same arguments as in Eq. (9), and when H is constant, we consider the function $HF + fg$. Then

$$\Delta(HF + fg) = fg\left(|A|^2 - nH^2 + \mathrm{Ric}_M(N^*, N^*) - (n-1)(\log f)''|\nabla h|^2\right) \tag{12}$$

where $N^* = (\pi_M)_*(N)$. Because $\langle N^*, N^* \rangle_M = |\nabla h|^2/f^2$ and $|A|^2 \geq nH^2$, if we want to bound from below the parenthesis in (12), we need to estimate the Ricci curvature of M. Using (8) and the maximum principle, we have

Theorem 4.2 ([30]) *Let Σ be a compact spacelike H-hypersurface in a GRW spacetime satisfying NCC. Suppose $\partial\Sigma \subset L_\tau$. If f is nondecreasing function and $H \geq \max\{0, \sup_I (\log f)'\}$, then*

$$\tau - \alpha \leq h_{|\Sigma} \leq \tau, \quad \alpha = \frac{\frac{f(\tau)}{f(\min_\Sigma h)} \max_{\partial\Sigma}(-g) - 1}{H} \geq 0.$$

When we particularize to $\mathcal{H}^{n+1}$, then $H \geq 1$ and this estimate corresponds, up to a change of the models, with (11).

In the de Sitter model, and for $H > 1$, there exist compact H-hypersurfaces with boundary in a given slice and with arbitrary height as it is shown in the next example (we point out a gap in [24, Theorem 3.1] related with Remark 2.1).

Example 4.1 Consider the upper half-space model and H-hypersurfaces with $H < -1$. Fix the slice L_1. Consider the equidistant hypersurfaces $\mathbb{H}_-^{n+1}(r; c_r)$, $c_r = (0, \ldots, 0, -Hr)$, whose mean curvature is H. For $r > -1/(1+H)$, let $\Sigma_r = \mathbb{H}_-^n(r; c_r) \cap \{x_{n+1} \geq 1\}$, which is a compact H-hypersurface with $\partial\Sigma_r \subset L_1$. The height of Σ_r about L_1 is given by its vertex and this height is $\log|r(1+H)|$ which tends to ∞ as $r \to \infty$.

Following the above example, we observe that the boundary of Σ_r is a sphere of arbitrary large radius, namely, $\sqrt{(H^2-1)r^2 + 2Hr + 1}$. We now give an estimate of the height of a H-hypersurface with $H < -1$ depending only on H and the size of Γ.

Theorem 4.3 ([42]) *Let $\Gamma \subset L_\tau$ be a $(n-1)$-submanifold and let $\Omega \subset L_\tau$ denote the domain bounded by Γ. If $\Sigma \subset \mathcal{H}^{n+1}$ is a compact spacelike H-hypersurface with $\partial\Sigma = \Gamma$ and $H < -1$ in the upper half-space model, then the height h of Σ with respect to L_τ satisfies*

$$h \leq \log(h_0), \quad h_0 = \frac{-H + \sqrt{(H^2-1)R^2 + 1}}{1 - H}, \quad 2R = \operatorname{diam} \Omega. \tag{13}$$

Proof After an isometry of $\mathcal{H}^{n+1}$, suppose that $\tau = 1$ and let $B_R \subset L_1$ be the ball of radius $R > 0$ containing Ω inside: here R coincides with the Euclidean radius because the induced metric in L_1 is the Euclidean one in $\mathbb{R}^n \times \{1\}$. After a horizontal translation, if necessary, we assume the $(0, \ldots, 0, 1)$ is the center of B_R. Consider the equidistant hypersurfaces $\mathbb{H}_-^n(r; c_r)$ where $c_r = (0, \ldots, 0, -rH)$. For r sufficiently big, we can trap Σ in the convex domain of $\mathcal{H}^{n+1}$ determined between $\mathbb{H}_-^n(r; c_r)$ and $x_{n+1} \geq 1$. In particular, the disk determined by the sphere $L_1 \cap \mathbb{H}_-^n(r; c_r)$ contains B_R inside. Letting $t \searrow 0$, we arrive until the first value

$r = r_0$ such that $L_1 \cap \mathbb{H}^n_-(r_0; c_{r_0}) = \partial B_R$, just when $\mathbb{H}^n_-(r_0, c_{r_0})$ touches $\partial\Sigma$. During this process decreasing r, the tangency principle forbids the existence of an interior contact point between Σ and $\mathbb{H}^n_-(r; c_r)$ because both hypersurfaces have the same constant mean curvature. Thus, $r = r_0$ is the first time that $\mathbb{H}^n_-(r; c_r)$ meet Σ. This proves that the height of Σ is less than the height of $\mathbb{H}^n_-(r_0; c_{r_0})$, and this concludes (13). □

4.2 A Flux Formula

For the question (ii), we work in the de Sitter model following de Lima [36]. Let $\Gamma \subset L_\tau$ be a $(n-1)$-submanifold and let Ω denote the bounded domain that bounds Γ in L_τ. Let Σ be a compact spacelike H-hypersurface spanning Γ. The following argument follows the same steps as in Euclidean space. Consider the n-cycle $\Sigma \cup \Omega$ and define the Killing vector field in $\mathscr{H}^{n+1}$

$$Y_p = (\langle p, b\rangle a - \langle p, a\rangle b)/\langle a, b\rangle$$

where $b \in L_\tau$ with $\langle a, b\rangle \neq 0$. By the divergence theorem, we have

$$n|H||\Omega| = \left|\int_{\partial\Sigma} \langle Y_p, \nu\rangle\right| \tag{14}$$

where ν is the unit conormal vector field pointing to Σ and $|\Omega|$ is the volume of Ω. Identity (14) is usually called a *flux formula*. Let us observe that the left-hand side of (14) does not depend on Σ. Since $\langle p, a\rangle = \tau$ in L_τ, we have

$$n|H||\Omega| \leq \int_{\partial\Sigma} |\langle Y, \nu\rangle| = \int_{\partial\Sigma} \left|\frac{\langle p, b\rangle\langle \nu, a\rangle}{\tau} - \langle \nu, b\rangle\right|. \tag{15}$$

Although p, b, and ν are unit spacelike vectors, we cannot bound $\langle p, b\rangle$ and $\langle \nu, b\rangle$ by 1 (here we observe a gap in [36, p. 975]).

In Euclidean space $\mathbb{R}^{n+1}$, if $\partial\Sigma$ lies in a hyperplane, then (15) gives $|H| \leq |\Gamma|/((n-1)|\Omega|)$, obtaining an upper estimate for H depending *only on* Γ. For example, if Γ is a sphere of radius $r > 0$, then $|H| \leq 1/r$. However, in Lorentz–Minkowski space $\mathbb{R}^{n+1}_1$, and when the boundary lies in a spacelike hyperplane $< a >^\perp$, $|a| = 1$, the flux formula (14) gives $n|H||\Omega \leq \int_{\partial\Sigma} |\langle \nu, a\rangle|$ but we have the same problem than in (15) (see [8] for $n = 2$ and [10] for the general n-dimensional case).

We analyze the case when Γ is a sphere $\mathbb{S}^{n-1} \subset L_\tau$. Since L_τ is isometric to $\mathbb{R}^n$, in the upper half-space model, $\mathbb{S}^{n-1} \subset L_\tau$ is an Euclidean sphere in the hyperplane $x_{n+1} = \tau$. We have explicit examples of compact cmc hypersurfaces spanning a sphere obtained as follows. When we meet an umbilical hypersurface with a slice L_τ, we obtain a sphere $\mathbb{S}^{n-1}$ separating the umbilical hypersurface in two connected components. Depending on each case, there is at most one compact component, and called a *hyperbolic cap*. In fact, hyperbolic caps do exist always for equidistant hypersurfaces $\mathbb{H}^n_-(r;c)$, and only do exist for hyperbolic planes $\mathbb{H}^n_+(r;c)$ when $0 < c_{n+1} + r < \tau$. Furthermore, $\mathbb{S}^{n-1}$ determines a round disk called a *planar disk* in the very slice L_τ. Then we have:

Proposition 4.2 *Let $\Gamma = \mathbb{S}^{n-1}$ be a sphere of radius $\rho > 0$.*

1. *If $H < -1$, there exists a unique hyperbolic H-cap spanning Γ.*
2. *If $H = -1$, there exists a unique planar disk spanning Γ.*
3. *If $H \geq 0$, then there exists a hyperbolic H-cap spanning Γ if and only if $\rho < 1$. Moreover, this cap is unique.*
4. *Let $-1 < H < 0$. If $\rho \leq 1$, there exists a unique hyperbolic H-cap spanning Γ. If $\rho > 1$, then there exists a hyperbolic H-cap spanning Γ if and only if $H \leq H_0$, where $H_0 = -\sqrt{\rho^2 - 1}/\rho$. Moreover, the hyperbolic cap is unique if $H = H_0$ and there exactly two hyperbolic H-caps if $H < H_0$.*

Proof After an isometry of $\mathcal{H}^{n+1}$, suppose $\tau = 1$. Then the radius ρ of Γ coincides with the Euclidean radius.

1. For $H < -1$, we take the equidistant hypersurfaces $\mathbb{H}^n_-(r; -Hr)$ for all $r > 0$. When we meet with the slice L_1, we obtain spheres of arbitrary radius ρ.
2. Immediate.
3. Take a hyperbolic plane $\mathbb{H}^n_+(r; Hr)$. When we intersect with L_1, the sphere, if exists, has radius ρ such that $Hr + \sqrt{\rho^2 + r^2} = 1$. Thus, we study the solutions of $q(r) = 1$, where $q(r) = Hr + \sqrt{\rho^2 + r^2}$. Since $H \geq 0$, $q(r)$ is strictly increasing with $\lim_{r\to 0} q(r) = \rho$ and $\lim_{r\to\infty} q(r) = \infty$; see Fig. 2, left. This proves the result.
4. The above function $q(r)$ is now decreasing around $r = 0$, $q(r)$ has a unique minimum at $r_0 = -H\rho/\sqrt{1 - H^2}$ and $\lim_{r\to\infty} q(r) = \infty$; see Fig. 2, right. If $\rho \leq 1$, the result is immediate. If $\rho \geq 1$, it suffices to prove that $q(r_0) \leq 1$, which holds if and only if $H \leq H_0$. □

Example 4.2 Proposition 4.2 allows to give an example of two compact spacelike H-hypersurfaces spanning the same boundary. Let Γ be a sphere of radius $\rho > 1$. Let H be satisfying $-1 < H < -\sqrt{\rho^2 - 1}/\rho$, and denote by r_1 and r_2 the two roots of $Hr + \sqrt{\rho^2 + r^2} = 1$. Then the hyperbolic H-caps of radius ρ determined by $\mathbb{H}^n_+(r_i; Hr_i)$ and L_1 span Γ: see Fig. 3.

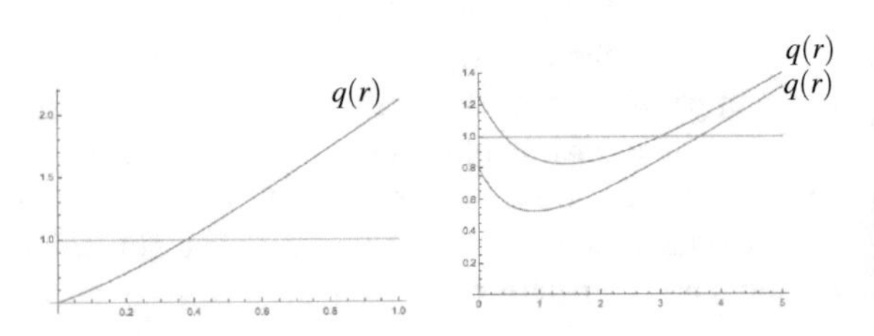

Fig. 2 Solutions of $q(r) = 1$ in the proof of Proposition 4.2. Left: $H = 1$ and $\rho = 1/2$; Right: $H = -3/4$ and $\rho = 0.8$ and $\rho = 1.25$

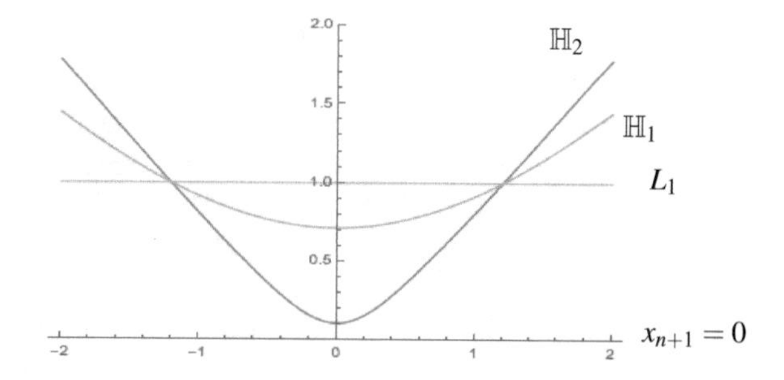

Fig. 3 Two hyperbolic H-caps $\mathbb{H}_1$ and $\mathbb{H}_2$ with the same boundary. Here, $H = -0.7$ and the boundary is a sphere of radius $\rho = 1.2$ in the slice L_1

Example 4.3 The above non-uniqueness result can be also proved as follows. For $H \in (-1, 0)$, let r be sufficiently small, so $\mathbb{H}^n_+(r; Hr)$ intersects L_1 in a sphere Γ of radius $\rho > 1$ and defining a hyperbolic cap C below L_1. The Euclidean cone with vertex the origin and containing Γ intersects $\mathbb{H}^n_+(r; Hr)$ again in other sphere of radius $\rho' \neq \rho$ and determining other hyperbolic cap C'. Without loss of generality, suppose $\rho' > \rho$, and thus, $C \subset C'$. If h denotes the homothety from the origin of $\mathbb{R}^{n+1}$ of ratio ρ/ρ', then $h(C')$ is a H-hypersurface spanning Γ and $h(C') \neq C$.

From Proposition 4.2, we deduce that the radius ρ of a sphere $\mathbb{S}^{n-1}$ imposes restrictions to the values H for hyperbolic H-caps. Indeed, we have:

1. If $\rho < 1$, then any real number H is a value of a hyperbolic H-cap spanning $\mathbb{S}^{n-1}$.
2. If $\rho \geq 1$, then $H \in (-\infty, -\sqrt{\rho^2 - 1}/\rho)$.

In general, we prove that the shape and the size of $\Gamma \subset L_\tau$ imposes restrictions to the value H for a compact H-hypersurface spanning Γ.

Theorem 4.4 ([38]) *Let $\Gamma \subset L_\tau$ be a closed submanifold and let $\Omega \subset L_\tau$ denote the domain bounded by Γ. Let $\Sigma \subset \mathscr{H}^{n+1}$ be a compact H-hypersurface spanning Γ.*

1. *If Ω contains a ball of radius* 1, *then $H < 0$.*
2. *If Ω contains a ball of radius*

$$\rho_0 = \sqrt{\frac{1-H}{1+H}}, \tag{16}$$

then $H \notin (-1, 0)$.

Proof We only prove the item 1. By contradiction, suppose $H \geq 0$. After an isometry, assume $\tau = 1$ and that Ω contains a sphere of radius 1 centered at $(0, \ldots, 0, 1)$. We know from Theorem 4.1 that Σ lies below L_1. Consider the hyperbolic H-caps of type $\mathbb{H}^n_+(r; Hr)$ with boundary in L_1. The radius of the boundaries of these caps is ρ, $0 < \rho < 1$, with $Hr + \sqrt{\rho^2 + r^2} = 1$, and its vertex is the point $V_r = (0, \ldots, 0, (H + 1)r)$. When ρ is very small (and r is close to $1/(H + 1)$), the cap lies below L_1 but it does not meet Σ. Letting $r \to 0$, the vertex goes to $(0, \ldots, 0)$ but $\rho < 1$. Thus, there will be a first value r_0 such that the hyperbolic plane $\mathbb{H}^n_+(r_0; Hr_0)$ intersects Σ at an interior contact point, a contradiction with the tangency principle. □

4.3 The Spherical Boundary Case

For the question (iii), we utilize the upper half-space model because we will work with the isometries of $\mathscr{H}^{n+1}$. As it was announced in Sect. 2, the foliation by slices of $\mathscr{H}^{n+1}$ allows a certain control on the position of a compact cmc hypersurface such as it appeared in Proposition 4.1 when $H = -1$. Assuming that the boundary of the surface lies in other type of umbilical hypersurface, Proposition 4.1 extends straightforward provided that we have a foliation of the ambient space by H-hypersurfaces for a given value of H (in Proposition 4.1 this value was $H = -1$).

Theorem 4.5 ([38]) *Let $H_0 \in (-\infty, -1] \cup [0, \infty)$. Let Σ be a compact spacelike H-hypersurface whose boundary is contained in an umbilical H_0-hypersurface P. Then we have one of the following three possibilities: either $H = H_0$ and $\Sigma \subset P$; or $H < H_0$ and Σ lies above P; or $H > H_0$ and Σ lies below P.*

Here, we revise a gap in [38] where it asserted that this result holds for any value H_0.

Proof 1. Case $H_0 < -1$. Then P is an equidistant hypersurface $\mathbb{H}^n_-(r; c)$ and take all Euclidean homotheties of P from a fix point of $x_{n+1} = 0$, obtaining the desired foliation of $\mathscr{H}^{n+1}$; see Fig. 4, left.

2. Case $H_0 \geq 0$. Without loss of generality, suppose that P is the hyperbolic plane $\mathbb{H}^n_+(r; c_r)$, where $c_r = (0, \ldots, 0, H_0 r)$. Let us observe that P is asymptotic to the upper light cone $\mathscr{C}^+ = \{x \in \mathbb{R}^{n+1}_+ : \langle x - c_r, x - c_r\rangle = 0, x_{n+1} \geq H_0 r\}$, and denote W the convex domain of $\mathbb{R}^{n+1}_+$ determined by $\mathscr{C}^+$. Let Σ be a compact H-hypersurface with $\partial\Sigma \subset P$. Take Q a lightlike hyperplane tangent to $\mathscr{C}^+$. Move parallely Q sufficiently far from Σ, and then come back to its initial position. By using that $\partial\Sigma \subset P$ and that P is asymptotic to $\mathscr{C}^+$, it is not possible that Q meets Σ because it should be at an interior point $p \in \Sigma$ and Q would be the tangent space of Σ at p violating the spacelike condition of Σ. If we do the same argument with all these hyperplanes Q, we have that Σ is included in W. Consider now $P_t = h_t(P)$ the homothety of P centered at the origin and ratio $t \in (0, \infty)$. Then $\{P_t\}$ is a foliation of the convex domain determined by the light cone $\langle x, x\rangle = 0$: note that this domain contains W which lies Σ. The argument follows the same steps as for Theorem 4.5. Suppose that $\text{int}(\Sigma)$ has points in both sides of P. Coming from hyperbolic planes P_t with $t = \infty$ until the first contact point with Σ, the tangency principle implies that $H_0 > H$. A similar reasoning with planes P_t coming from $t = 0$ gives $H_0 < H$; see Fig. 4, right. This contradiction proves

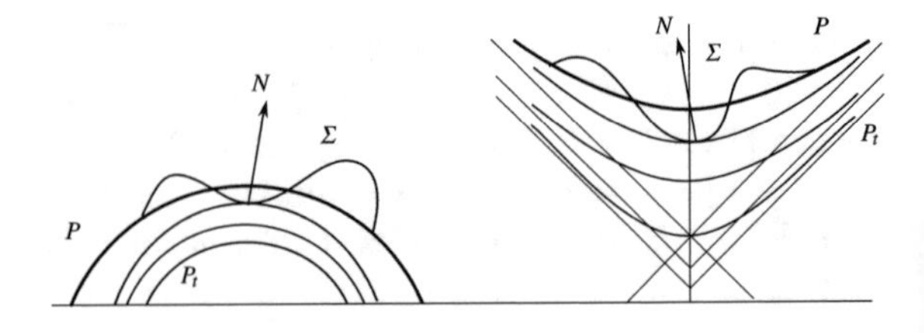

Fig. 4 Proof of Theorem 4.5: case $H_0 < -1$ (*left*) and case $H_0 \geq 0$ (*right*)

that Σ lies only on one side of P, but precisely the above argument gives that if $H < H_0$ (resp. $H > H_0$), then Σ lies above (resp. below) P. □

When the mean curvature H_0 lies in the range of the interval $(-1, 0)$, the family of umbilical H-hypersurfaces does not provide a foliation of the ambient space because any two members of this family meet each other: this appeared in Example 4.3.

Finally, we answer to the question (iii) when Γ is a sphere.

Theorem 4.6 ([38]) *Planar disks and hyperbolic caps are the only compact spacelike H-hypersurfaces in $\mathcal{H}^{n+1}$ spanning a sphere.*

Proof The proof uses the classical Alexandrov Reflection Method (see details [37] in the Euclidean space). The idea is to use the same hypersurface to compare with its reflection about a vertical hyperplane which preserves the constancy of the mean curvature, and finally, use the tangency principle. Let L_τ denote the slice containing Γ and $\Omega \subset L_\tau$ the bounded domain by Γ. We know by Theorem 4.5 that Σ is Ω when $H = -1$ or Σ lies on one side of L_τ if $H \neq -1$. In this case, $\Sigma \cup \Omega$ defines a domain in $\mathbb{R}^{n+1}_+$ and we can use the Alexandrov reflection method with reflections about vertical hyperplanes. This proves that Σ is a hypersurface of revolution with respect to a straight line orthogonal to L_τ. Finally, the hyperbolic caps are the only compact rotational cmc hypersurfaces spanning a sphere. □

Remark 4.1 When $H \in (-\infty, -1] \cup [0, \infty)$, we can do a proof of this result based on Theorem 4.5. For this argument, it is enough to prove that there exists an umbilical H-hypersurface containing Γ. If $H \leq -1$, the result follows from Proposition 4.2 (i, ii). If $H \geq 0$, then Theorem 4.4 says that $\rho < 1$, and now we use Proposition 4.2 (iii). Let us observe that we cannot complete the proof when $H \in (-1, 0)$.

We compare Theorem 4.6 to what happens in other ambient spaces. In Lorentz–Minkowski space $\mathbb{R}^{n+1}_1$ and in hyperbolic space when $|H| \leq 1$, the umbilical hypersurfaces are the only compact cmc hypersurfaces spanning a sphere [8, 10, 37]. However, it is an open question nowadays if spherical caps and planar disks are the only embedded compact H-hypersurfaces of $\mathbb{R}^{n+1}$ spanning a sphere (or in $\mathcal{H}^{n+1}$ when $|H| > 1$).

5 The Dirichlet Problem for the Mean Curvature Equation

In this section, we will prove the existence of complete H-hypersurfaces of $\mathcal{H}^{n+1}$, $H < -1$, whose boundary lies in the future infinity $\mathcal{J}^+$. This was proved by Montiel in [42] and was motivated by the Goddard conjecture: "the only complete spacelike cmc hypersurfaces in $\mathbb{S}^{n+1}_1$ must be umbilical." The hypersurfaces obtained in [42] (see Corollary 5.1 below) illustrate that this conjecture is not true in the steady state space. Recall that in $\mathbb{S}^{n+1}_1$, there was a great work of answering to this conjecture which is false, although in some cases is true, for example, when $|H| \leq 1$ and $n = 2$, when $0 \leq H^2 < 4(n-1)/n^2$ and $n \geq 3$, or when the hypersurface is compact [1, 41].

In order to prove the existence of Montiel's examples, and since the boundary lies in $\mathscr{I}^+$, the strategy is solving the Dirichlet problem in domains of slices, then take a sequence of such domains going to $\mathscr{I}^+$ and their corresponding solutions and having a suitable control of the solutions that ensures its convergence in the limiting process. Thus, in this section, we will study the existence of compact spacelike H-graphs on a domain of a slice.

The Dirichlet problem for the mean curvature equation is the following:

Problem 3: Given $\Omega \subset L_\tau$ a bounded domain, $H \in \mathbb{R}$ and $\tau > 0$, find a solution of

$$Q_H[u] = 0 \text{ on } \Omega, \quad u = \tau > 0 \text{ along } \partial\Omega. \tag{17}$$

The uniqueness of solutions of (17) is not assured by the standard theory because the term on u in the expression of $Q_H[u]$ in (2) is not necessarily nondecreasing. Recall that we showed in Example 4.2 two spacelike H-graphs on a disk of L_1 with the same boundary curve and $-1 < H < 0$.

The solvability of the Dirichlet problem (17) strongly depends whether $H < -1$ or $H > -1$, just the value of the mean curvature of a slice and, depending on each case, the hypersurface lies on one side of L_τ by Theorem 4.1.

Theorem 5.1 ([42]) *Let $\Omega \subset L_\tau$ be a compact bounded domain which has mean convex boundary. If $H < -1$, then there exists a unique solution of (17).*

We solve the Dirichlet problem using the method of continuity. For this technique, we refer the reader to [40] in the context of cmc hypersurfaces in hyperbolic space (the general reference for elliptic equations is Gilbarg and Trudinger [31]). Without loss of generality, suppose that $\tau = 1$. The method of continuity considers the uniparametric family of Dirichlet problems

$$\begin{cases} Q_{H(t)}[u] = 0 & \text{in} \Omega, \\ u = 1 & \text{along} \partial\Omega \end{cases} \tag{18}$$

where $H(t) = t(1 + H) - 1$, $t \in [0, 1]$. Let us observe that for $t = 1$, a solution of (18) is the solution that we are looking for (17). We show that the subset of $[0, 1]$ defined by

$$\mathscr{A} = \{t \in [0,1] : \exists u_t \in C^{2,\alpha}(\overline{\Omega}), \text{ such that } Q_{H(t)}(u_t) = 0 \text{ and } u_{t|\partial\Omega} = 1\}$$

is nonempty, open, and closed in $[0, 1]$. In such a case, $1 \in \mathscr{A}$ proving that there exists a solution $u \in C^{2,\alpha}(\overline{\Omega})$ of (17). Finally, as H is constant and Ω is smooth, the regularity theory for the cmc equation proves that any $C^{2,\alpha}$ solution of (17) will be smooth, proving Theorem 5.1. We observe that $\mathscr{A} \neq \emptyset$ because $0 \in \mathscr{A}$ since $H(0) = -1$ and $u = 1$ are a solution in the domain Ω. The proof that $\mathscr{A}$ is open of [0, 1] is a consequence of the implicit function theorem in Banach spaces and it follows standard arguments.

The difficulties lie in proving that $\mathscr{A}$ is a closed set of [0, 1]. This follows once we establish a priori C^1 estimates of the prospective solutions of (18), that is, estimates of $|u|$ and $|Du|$ depending only on H and Ω.

The estimate for $|u|$ was proved in Theorem 4.3 where the estimate (13) depends only on H and Ω. For the estimate of $|Du|$, we need to work in the de Sitter model. The graph Σ_u of u corresponds with a graph $\Sigma_f \subset \mathbb{S}_1^{n+1}$ and the slice $L_1 \subset \mathbb{R}_+^{n+1}$ with the slice L_1 in $\mathbb{S}_1^{n+1}$. By the isometry Ψ, we have $u = e^f$ and

$$\langle p, a\rangle = \frac{1}{u} = e^{-f}, \quad \langle N, a\rangle = \frac{1}{u\sqrt{1-|Du|^2}}. \tag{19}$$

Thus, we will obtain bounds for $|Du|$ provided that we have a certain control of the functions $\langle p, a\rangle$ and $\langle N, a\rangle$. We proved in (10) that $-H\langle p, a\rangle + \langle N, a\rangle \leq -H + \langle N(q), a\rangle$ for some $q \in \partial\Sigma_u$. At q, the maximum principle gives again

$$H\langle \nu_q, a\rangle + \langle (dN)_q(\nu_q), a\rangle = \langle \nu_q, a\rangle(-H + \langle (dN)_q(\nu_q), \nu_q\rangle \leq 0$$

where ν is the unit conormal pointing along $\partial\Sigma_u$ toward Σ. Since $\langle \nu_q, a\rangle > 0$, then $-H + \langle (dN)_q(\nu_q), \nu_q\rangle \leq 0$. Using the mean convexity of Ω, we deduce $-H + \langle N(q), a\rangle \leq 0$, and thus

$$-H\langle p, a\rangle + \langle N, a\rangle \leq -H + \langle N(q), a\rangle \leq 0.$$

Hence, we deduce $\langle N, a\rangle \leq H\langle p, a\rangle \leq H$ because $\langle p, a\rangle \leq 1 (= \tau)$ by Proposition 4.1. Finally, using (19) we conclude

$$|Du|^2 \leq \frac{H^2 - 1}{H^2} \tag{20}$$

obtaining the desired uniform estimate for $|Du|$ and proving Theorem 5.1. The uniqueness is a consequence of the standard theory [31] and the inequality (20).

The estimates (13) and (20) are independent on the slice where lies Ω. This extends Theorem 5.1 allowing $u = 0$ along $\partial\Omega \subset \mathbb{R}^n \times \{0\}$. Indeed, we go by considering the solutions u_τ of (17) and by letting $\tau \to 0$, the domains $\Omega \times \{\tau\}$ converge to $\Omega \times \{0\} \subset \mathscr{J}^+$, and the funtions u_τ have a priori C^1-estimates independent on τ.

Theorem 5.2 ([39]) *Let $\Omega \subset \mathbb{R}^n$ be a compact domain which is mean convex boundary. If $H < -1$, then there exists a solution of (17) with $u = 0$ along $\partial\Omega$.*

If we introduce the concept of *asymptotic future boundary* of Σ as $\partial_\infty\Sigma = \overline{\Sigma} \cap \mathscr{J}^+$, the above result can be re-phrased as follows:

Corollary 5.1 ([39]) *Let $\Omega \subset \mathscr{J}^+$ be a compact domain which is mean convex boundary. If $H < -1$, then there exists a complete embedded spacelike H-hypersurface with $\partial_\infty\Sigma = \partial\Omega$.*

Proof It only remains to prove that the induced metric ds^2 on Σ_u is complete. From (20) we have

$$ds^2 = \frac{1}{u^2}(|dx|^2 - \langle Du, dx\rangle^2) \geq \frac{1}{H^2-1}\frac{\langle Du_x, dx\rangle^2}{u^2} \geq \frac{1}{H^2-1}|d\log u|^2.$$

This says that the length of any curve in Σ_u reaching $\partial_\infty \Sigma_u$ must be infinity. □

Remark 5.1 Let us observe that (13) says that $u \leq ct$ and this implies that the complete H-hypersurface obtained in Corollary 5.1 has $H < -1$ and it is bounded away from the past infinity of $\mathscr{H}^{n+1}$.

Consider now the Dirichlet problem (17) for values $H > -1$, so the graph lies below L_τ by Theorem 4.5. From Theorem 4.4, it is expectable some kind of smallness assumption on the domain Ω. If in Theorem 5.1 it was assumed to be mean convex, now we need strong convexity assumptions. For $\kappa > 0$, a domain $\Omega \subset L_\tau$ is said to be *κ-convex* if all the principal curvatures κ_i of $\partial\Omega$ with respect to the inward normal vector satisfy $\kappa_i \geq \kappa$.

Theorem 5.3 *Let $-1 \leq H < 0$ and let $\Omega \subset L_\tau$ be a κ-convex domain strictly contained in a ball of radius 1 in L_τ. If*

$$\kappa \geq \sqrt{1-H^2}, \tag{21}$$

then there exists a solution of the Dirichlet problem (17).

We utilize the method of continuity again. Suppose $\tau = 1$ and $(0, \ldots, 0, 1) \in \Omega$. Since the induced metric in L_1 is the Euclidean one, Ω is included in a ball Ω_1 of Euclidean radius 1. As $0 < u < 1$ in Ω and $\overline{\Omega} \subset \Omega_1$, then the spacelike condition, the convexity of Ω, and a similar argument as for Theorem 4.5, proves the existence of $0 < c < 1$ depending only on Σ_u such that $c < x_{n+1} \leq 1$ in Σ_u: this provides a priori estimates for $|u|$ in (17).

In Theorem 5.1, the gradient estimates were obtained using the subharmonicity of the function $-H\langle p, a\rangle + \langle N, a\rangle$. Now this is not enough and we will prove first the existence of an apriori estimate of $|Du|$ along $\partial\Omega$, and once obtained this estimate, we will get an estimate of $|Du|$ in the domain Ω.

Proposition 5.1 *Under the assumptions of Theorem 5.3, we have*

$$\sup_{\Omega} |Du| < \sqrt{1-H^2} \tag{22}$$

for every solution u of the Dirichlet problem (17).

Proof 1. Estimates along $\partial\Omega$. Because Ω is κ-convex and inequality (21), we can trap Σ_u in the domain determined by a hyperbolic H-hyperplane $\mathbb{H}^n_+(r; Hr)$ and the slice L_1 such the intersection $L_1 \cap \mathbb{H}^n_+(r; Hr)$ is the boundary of a round ball

B_ρ of radius $\rho = 1/\sqrt{1-H^2}$. Denoted by $\mathbb{H}^+_\rho \subset \mathbb{H}^n_+(r; Hr)$, the hyperbolic cap is determined by L_1. Then $\mathbb{H}^+_\rho$ lies below L_1, and Σ_u lies between $\mathbb{H}^+_\rho$ and B_ρ. The assumption on the κ-convexity of Ω says $\kappa \geq 1/\rho$, so the domain Ω has the following Blaschke's outer rolling sphere property: for every $p \in \partial\Omega$, there exists a ball in L_1 of radius ρ touching p and leaving the domain Ω inside the ball. By moving horizontally $\mathbb{H}^+_\rho$, we go touching every point of $\partial\Omega$ in such a way that during these translations (isometries of $\mathcal{H}^{n+1}$), the tangency principle forbids a contact between $\mathbb{H}^+_\rho$ and Σ_u. Therefore, we can reach any point of $\partial\Omega$ leaving Σ_u sandwiched by $\mathbb{H}^+_\rho$ and L_1, in particular, the slope of Σ_u along $\partial\Omega$ is bounded by the one of $\mathbb{H}^+_\rho$. This estimate is written precisely as $\sup_{\partial\Omega} |Du| < \sqrt{1-H^2}$, proving (22) for boundary points of Ω.
2. Estimates on Ω. In the de Sitter model, the function $-H\langle p, a\rangle + \langle N, a\rangle$ is subharmonic and the maximum principle gives

$$-H\langle p, a\rangle + \langle N, a\rangle \leq \sup_{\partial\Omega}\left(-H + \frac{1}{\sqrt{1-|Du|^2}}\right) < \frac{1-H^2}{H}$$

where we have used that $|Du| < \sqrt{1-H^2}$ along $\partial\Omega$. Hence, we obtain $\langle N, a\rangle$, and using (19), we conclude $\sup_\Omega |Du| < \sqrt{1-H^2}$. □

Finally, Proposition 5.1 proves definitively Theorem 5.3.

6 Oulook and Open Problems

The interest of the steady state space comes because it is the Lorentzian analogous of hyperbolic space. In this chapter, we have focused on spacelike hypersurfaces with constant mean curvature but the theory of submanifolds is much more: for example, one can study spacelike hypersurfaces with constant Gaussian curvature as in [46, 47]. The spacelike condition on the hypersurface is a strong difference between $\mathcal{H}^{n+1}$ and $\mathbb{H}^{n+1}$. The three models of $\mathcal{H}^{n+1}$ allow different approaches for a given problem, but we have observed in the literature similar results written in different coordinates. This is the reason why we have strived to make a common line of the progress in $\mathcal{H}^{n+1}$ after the paper of Montiel [42] and later of Albujer and Alías [3]. We have not fully studied the extension of results in other GRW spacetimes because we believe that a generalization in these spaces would bring long statements in the conditions on the warping function as well as the fiber manifold that could go away from our initial aim. Finally, we have emphasized on the techniques employed, where the maximum principles (tangency principle, Omori-Yau) play an important role.

In the literature, there has been a great interest for characterizing the slices, in connection with the Bernstein problem but we find missing more efforts in other directions. We have collected a list of open problems some of which could be tackled in the near future.

1. Obtain other types of characterizations for slices. The condition to be included between two slices in Theorem 3.1 is too strong. Extend the characterizations of Bernstein-type for hyperbolic planes, for example assuming that the hypersurface lies in the convex side of another hyperbolic plane.
2. Study spacelike H-hypersurfaces with $-1 < H < 0$ because this range of values for H appears remarkably in some results: Example 4.2 and Theorem 4.4 and 5.3.
3. Investigate the family of spacelike cmc hypersurfaces invariant by a uniparametric group of isometries. We are thinking not only in the rotational examples but also hypersurfaces invariant by a parabolic or a hyperbolic group of rotations.
4. Characterize cmc hypersurfaces whose asymptotic future boundary is known. In the upper half-space model, we can assume that $\partial_\infty \Sigma$ is one point, an Euclidean sphere or two concentric Euclidean spheres. In the literature, there are similar results in the hyperbolic space.
5. Study complete spacelike hypersurfaces in $\mathscr{H}^{n+1}$ with constant mean curvature $|H| = 1$. We know that if $n = 2$, then the hypersurface is a slice, but we do not know the existence of other examples in arbitrary dimension.
6. Solve the Dirichlet problem for the case $H > 0$. Following the discussion in Sect. 5, consider the case $H < -1$ but dropping the mean convex boundary assumption in Theorem 5.1.
7. We have not discussed on timelike cmc hypersurfaces in $\mathscr{H}^{n+1}$. Recall that there is a great activity in recent years in this topic in Lorentz–Minkowski space $\mathbb{R}^3_1$. The mean curvature equation for timelike surfaces is not elliptic so we cannot make use of the maximum principle. However, we have timelike slices which could be characterized with results of Bernstein-type.

Acknowledgements The author has been partially supported by the MINECO/FEDER grant MTM2014-52368-P.

References

1. K. Akutagawa, On spacelike hypersurfaces with constant mean curvature in the de Sitter space. Math. Z. **196**, 13–19 (1987)
2. A. Albujer, Geometría global de superficies maximales en espacios producto Lorentzianos. Ph.D. Thesis, Universidad de Murcia, Murcia, Spain (2008)
3. A. Albujer, L.J. Alías, Spacelike hypersurfaces with constant mean curvature in the steady state space. Proc. Amer. Math. Soc. **137**, 711–721 (2009)
4. J.A. Aledo, R.M. Rubio, Constant mean curvature spacelike surfaces in Lorentzian warped products. Adv. Math. Phys. **2015**(5) (2015). Article ID 761302
5. L.J. Alías, A.G. Colares, Uniqueness of spacelike hypersurfaces with constant higher order mean curvature in generalized Robertson-Walker spacetimes. Math. Proc. Camb. Philos. Soc. **143**, 703–729 (2007)

6. L.J. Alías, A.G. Colares, H.F. de Lima, On the rigidity of complete spacelike hypersurfaces immersed in a generalized Robertson-Walker spacetime. Bull. Braz. Math. Soc. (N.S.) **44**, 195–217 (2013)
7. L.J. Alías, M. Dajzcer, Uniqueness of constant mean curvature surfaces properly immersed in a slab. Comment. Math. Helv. **81**, 653–663 (2006)
8. L.J. Alías, R. López, J.A. Pastor, Compact spacelike surfaces with constant mean curvature in the Lorentz-Minkowski 3-space. Tohoku Math. J. **50**, 491–501 (1998)
9. L.J. Alías, P. Mastrolia, M. Rigoli, *Maximum Princ. Geom. Appl.*, Springer Monographs in Mathematics (Springer Verlag, Berlin, Heidelberg, Nueva York, 2016)
10. L.J. Alías, J.A. Pastor, Constant mean curvature spacelike hypersurfaces with spherical boundary in the Lorentz-Minkowski space. J. Geom. Phys. **28**, 85–93 (1998)
11. L.J. Alías, A. Romero, M. Sánchez, Uniqueness of complete spacelike hypersurfaces of constant mean curvature in generalized Robertson-Walker spacetimes. Gen. Relativ. Gravit. **27**, 71–84 (1995)
12. C.P. Aquino, H.F. de Lima, Uniqueness of complete hypersurfaces with bounded higher order mean curvatures in semi-Riemannian warped products. Glasg. Math. J. **54**, 201–212 (2012)
13. C.P. Aquino, H.F. de Lima, F.R. dos Santos, M.A.L. Velásquez, Spacelike hypersurfaces with constant rth mean curvature in steady state type spacetimes. J. Geom. **106**, 85–96 (2015)
14. C.P. Aquino, H.F. de Lima, F.R. dos Santos, M.A. Velásquez, Characterizations of spacelike hyperplanes in the steady state space via generalized maximum principles. Milan J. Math. **83**, 199–209 (2015)
15. H. Bondi, T. Gold, On the generation of magnetism by fluid motion. Mon. Not. R. Astron. Soc. **110**, 607–611 (1950)
16. D.G. Brush, Prediction and theory evaluation: cosmic microwaves and the revival of the Big Bang. Persp. Sci. **1**, 565–602 (1993)
17. M. Caballero, A. Romero, R.M. Rubio, Constant mean curvature spacelike surfaces in three-dimensional generalized Robertson-Walker spacetimes. Lett. Math. Phys. **93**, 85–105 (2010)
18. F. Camargo, A. Caminha, H.F. de Lima, Bernstein-type theorems in semi-Riemannian warped products. Proc. Amer. Math. Soc. **139**, 1841–1850 (2011)
19. F. Camargo, A. Caminha, H.F. de Lima, U. Parente, Generalized maximum principles and the rigidity of complete spacelike hypersurfaces. Math. Proc. Cambridge Philos. Soc. **153**, 541–556 (2012)
20. A. Caminha, H.F. de Lima, A generalized maximum principle for Yau's square operator, with applications to the steady state space. Advances in Lorentzian geometry, 7–19, Shaker Verlag, Aachen (2008)
21. A. Caminha, H.F. de Lima, Complete vertical graphs with constant mean curvature in semi-Riemannian warped products. Bull. Belg. Math. Soc. Simon Stevin **16**, 91–105 (2009)
22. A. Caminha, H.F. de Lima, Complete spacelike hypersurfaces in conformally stationary Lorentz manifolds. Gen. Relativ. Gravit. **41**, 173–189 (2009)
23. S.Y. Cheng, S.T. Yau, Maximal spacelike hypersurfaces in the Lorentz- Minkowski space. Ann. Math. **104**, 407–419 (1976)
24. A.G. Colares, H.F. de Lima, Spacelike hypersurfaces with constant mean curvature in the steady state space. Bull. Belg. Math. Soc. Simon Stevin **17**, 287–302 (2010)
25. A.G. Colares, H.F. de Lima, On the rigidity of spacelike hypersurfaces immersed in the steady state space H^{n+1}. Publ. Math. Debr. **81**, 103–119 (2012)
26. A.G. Colares, H.F. de Lima, Some rigidity theorems in semi-Riemannian warped products. Kodai Math. J. **35**, 268–282 (2012)
27. J. Dong, X. Liu, Uniqueness of complete spacelike hypersurfaces in generalized Robertson-Walker spacetimes. Balkan J. Geom. App. **20**, 38–48 (2015)
28. D. de la Fuente, A. Romero, P.J. Torres, Radial solutions of the Dirichlet problem for the prescribed mean curvature equation in a Robertson-Walker spacetime. Adv. Nonlinear Stud. **15**, 171–181 (2015)
29. G. Galloway: Cosmological spacetimes with $\Lambda > 0$. Advances in Differential Geometry and General Relativity, eds. S. Dostoglou, P. Ehrlich, Contemp. Math. **359**, Amer. Math. Soc. (2004)

30. S. García-Martínez, D. Impera, Height estimates and half-space theorems for spacelike hypersurfaces in generalized Robertson-Walker spacetimes. Differ. Geom. Appl. **32**, 46–67 (2014)
31. D. Gilbarg, N.S. Trudinger, *Elliptic Partial Differential Equations of Second Order*. Reprint of the 1998 edition. Classics in Mathematics. Springer-Verlag, Berlin (2001)
32. S.W. Hawking, G.F.R. Ellis, *The Large Scale Structure of Space-Time* (Cambridge University Press, Cambridge, 1973)
33. D. Hoffman, W. Meeks, The strong half-space theorem for minimal surfaces. Invent. Math. **101**, 373–377 (1990)
34. F. Hoyle, A new model for the expanding universe. Mon. Not. R. Astron. Soc. **108**, 372–382 (1948)
35. A. Huber, On subharmonic functions and differential geometry in the large. Comment. Math. Helv. **32**, 13–72 (1957)
36. H.F. de Lima, Spacelike hypersurfaces with constant higher order mean curvature in de Sitter space. J. Geom. Phys. **57**, 967–975 (2007)
37. R. López, *Constant Mean Curvature Surfaces with Boundary*, Springer Monographs in Mathematics (Springer Verlag. Berlin, Heidelberg, Nueva York, 2013)
38. R. López, A characterization of hyperbolic caps in the steady state space. J. Geom. Phys. **98**, 214–226 (2015)
39. R. López, Spacelike graphs of prescribed mean curvature in the steady state space. Adv. Nonlinear Stud. **16**, 807–819 (2016)
40. R. López, S. Montiel, Existence of constant mean curvature graphs in hyperbolic space. Calc. Var. Partial Differ. Equ. **8**, 177–190 (1999)
41. S. Montiel, An integral inequality for compact spacelike hypersurfaces in de Sitter space and applications to the case of constant mean curvature. Indiana Univ. Math. J. **37**, 909–917 (1988)
42. S. Montiel, Complete non-compact spacelike hypersurfaces of constant mean curvature in de Sitter spaces. J. Math. Soc. Japan **55**, 915–938 (2003)
43. H. Omori, Isometric immersions of Riemannian manifolds. J. Math. Soc. Japan **19**, 205–214 (1967)
44. L. Rodriguez, H. Rosenberg, Half-space theorems for mean curvature one surfaces in hyperbolic space. Proc. Amer. Math. Soc. **126**, 2755–2762 (1998)
45. A. Romero, R.M. Rubio, J.J. Salamanca, Spacelike graphs of finite total curvature in certain 3-dimensional generalized Robertson-Walker spacetime. Rep. Math. Phys. **73**, 241–254 (2014)
46. J. Spruck, The asymptotic Plateau problem for convex hypersurfaces of constant curvature in hyperbolic space. Mat. Contemp. **43**, 247–280 (2012)
47. J. Spruck, L. Xiao, Convex spacelike hypersurfaces of constant curvature in de Sitter space. Discrete Contin. Dyn. Syst. Ser. B **17**, 2225–2242 (2012)
48. S. Weinberg, *Gravitation and Cosmology: Principles and Applications of the General Theory of Relativity*. Wiley, New York (1972)
49. S.T. Yau, Harmonic functions on complete Riemannian manifolds. Comm. Pure Appl. Math. **28**, 201–228 (1975)
50. S.T. Yau, Some function-theoretic properties of complete Riemannian manifolds and their applications to geometry. Indiana Univ. Math. J. **25**, 659–670 (1976)

Calabi–Bernstein-Type Problems in Lorentzian Geometry

Rafael M. Rubio

Abstract The study of maximal hypersurfaces in Lorentzian manifolds is an interesting mathematical problem, which connects differential geometry, nonlinear partial differential equations, and certain problems in mathematical relativity. One of the more celebrated results in the context of global geometry of maximal hypersurfaces is the Calabi–Bernstein theorem in the Lorentz–Minkowski spacetime. The nonparametric version of this theorem states that the only entire solutions to the maximal hypersurface equation in the Lorentz–Minkowski spacetime are spacelike affine hyperplanes. The present work reviews some of the classical and recent proofs of the theorem for the two-dimensional case, as well as several extensions for Lorentzian-warped products and other relevant spacetimes. On the other hand, the problem of uniqueness of complete maximal hypersurfaces is analyzed under the perspective of some new results.

Keywords Maximal hypersurfaces in spacetimes · Calabi–Bernstein type problems · Lorentzian geometry

1 Introduction

We begin with two examples of nonlinear partial differential equations, which arise in the context of some differential geometric problems.

First, we recall the well-known minimal hypersurface equation in the Euclidean space $\mathbb{R}^{n+1}$. So, for a smooth function $u : \Omega \longrightarrow \mathbb{R}$ on a domain Ω in $\mathbb{R}^n$, the problem is given by the following nonlinear differential equation in divergence form,

---

The author is partially supported Spanish MINECO and ERDF Project MTM2016-78807-C2-1-P.

R.M. Rubio (✉)
Departamento de Matemáticas, Campus de Rabanales, Universidad de Córdoba,
14071 Córdoba, Spain
e-mail: rmrubio@uco.es

M.A. Cañadas-Pinedo et al. (eds.), *Lorentzian Geometry and Related Topics*,
Springer Proceedings in Mathematics & Statistics 211,
DOI 10.1007/978-3-319-66290-9_12

$$\operatorname{div}\left(\frac{Du}{\sqrt{1+|Du|^2}}\right)=0, \tag{1}$$

where D and div denote the gradient and divergence operators in the Euclidean n-plane $\mathbb{R}^n$ respectively. This equation is elliptic and it is easy to see that the affine functions are trivial solutions.

Second, the maximal spacelike hypersurface equation in the Lorentz–Minkowski spacetime $\mathbb{L}^{n+1}$. With coordinates $(t, x_1, ..., x_n)$ (and Lorentzian form $g = -dt^2 + \sum_{j=1}^{n} dx_j^2$), the problem is given for $t = u(x_1, ..., x_n)$ by

$$\operatorname{div}\left(\frac{Du}{\sqrt{1-|Du|^2}}\right)=0, \quad |\; Du \;|^2 < 1. \tag{2}$$

where D and div denote the gradient and divergence operators in the Euclidean n-plane $\mathbb{R}^n$ respectively.

The condition $|\; Du \;|^2 < 1$ assures that the graph of every solution is spacelike, this is, the induced fundamental form on the graph is definite positive. Moreover, the problem is elliptic thanks to this extra constraint.

Note that if we take a unitary normal vector field on the graph $t = u(x_1, ..., x_n)$ in the same time-orientation of the timelike coordinate vector field $\frac{\partial}{\partial t} := \partial_t$, then its mean curvature is given by

$$nH = \operatorname{div}\left(\frac{Du}{\sqrt{1-|Du|^2}}\right).$$

On the other hand, the graph of u is extremal, among functions satisfying the spacelike condition under interior variations (with compact support) for the volume integral,

$$V = \int \sqrt{1- |\; Du \;|^2}\, dx_1 \wedge .. \wedge dx_n.$$

Again, trivial solutions of Eq. (2) are affine functions (with spacelike graph).

1.1 Bernstein Theorem

The early seminal result of Bernstein [13], amended by Hopf [35], is the well-known following uniqueness theorem,

> The only entire solutions to the Eq. (1) on the Euclidean plane $\mathbb{R}^2$ are the affine functions.

This result is known as the classical Bernstein theorem. In 1968, Simons [61] proved a result which in combination with theorems of De Giorgi [32] and Fleming [30] yields a proof of the Bernstein theorem for $n \leq 7$. Moreover, there is a counterexample $u \in C^\infty(\mathbb{R}^n)$ to the Bernstein conjecture for each $n \geq 8$.

1.2 Calabi–Bernstein Theorem

One of the most relevant results in the context of global geometry of spacelike surfaces is the classical Calabi–Bernstein theorem. This result was established in 1970 by Calabi [21] inspired in the classical Bernstein theorem, via a duality between solutions to Eqs. (1) and (2).

In its non-parametric version, it asserts that the only entire solutions to the maximal surface equation

$$\operatorname{div}\left(\frac{Du}{\sqrt{1-|Du|^2}}\right)=0, \quad |Du|<1$$

on the Euclidean plane $\mathbb{R}^2$ are affine functions.

In fact, Calabi also shows that the result holds for the case of maximal hypersurfaces in $\mathbb{L}^4$. Later on, Cheng and Yau [22] extended the Calabi–Bernstein theorem to the general $n+1$-dimensional case. Another important achievement in [22] was the introduction of a new procedure, the so-called Omori–Yau generalized maximum principle [44, 63].

The Calabi–Bernstein Theorem can also be formulated in a parametric way. In this case, it states that the only complete maximal hypersurfaces in $\mathbb{L}^{n+1}$ are the spacelike hyperplanes. In their proof of the parametric version, Cheng and Yau obtain a Simons-type formula, that is, the authors compute the Laplacian of the trace of the square of the associate shape operator to the unitary normal vector field on the maximal hypersurface. Subsequently, assuming completeness and making use of a consequence of their new maximum principle, the authors obtain the result in parametric version. Nevertheless, both versions (parametric and nonparametric ones) are not equivalent a priori, since there exist examples of spacelike entire graphs in $\mathbb{L}^{n+1}$ which are not complete (see for instance [6]). This fact, is a notable difference and difficulty with respect to the Riemannian case, where thank to the Hopf–Rinow theorem all entire graph in $\mathbb{R}^{n+1}$ must be complete. So, Cheng and Yau prove that in the case where the mean curvature is constant, a embedded spacelike hypersurface in $\mathbb{L}^{n+1}$ must be complete, which allows to obtain the non-parametric version.

2 Some Approaches to the Classical Calabi–Bernstein Theorem

After the general proof by Cheng and Yau, several authors have approached to the classical version of Calabi–Bernstein theorem from different perspectives, providing diverse extensions and new proofs of the result in $\mathbb{L}^3$.

In 1983, Kobayashi [38] derived the Calabi–Bernstein Theorem as a consequence of the corresponding Weierstrass–Enneper parameterization for maximal surfaces in $\mathbb{L}^3$. Below, we briefly describe the proof of Kobayashi.

2.1 Kobayashi Approach

Consider the Lorentz–Minkowski spacetime $\mathbb{L}^3$, which is given by the Lorentzian manifold $(\mathbb{R}^3, \langle\,,\rangle)$, where

$$\langle\,,\rangle = -dt^2 + dx^2 + dy^2.$$

Let S be a connected maximal surface in $\mathbb{L}^3$. The surface S must be orientable and let N be the unitary normal vector field on S such that $\langle N, \frac{\partial}{\partial t}\rangle < 0$. Let $N : S \longrightarrow \mathbb{H}^2_+$ be the Gauss map of S and define a stereographic projection $\sigma : \mathbb{D} \longrightarrow \mathbb{H}^2_+$, from $\mathbb{D} = \{z \in \mathbb{C}/ \mid z \mid < 1\}$ onto $\mathbb{H}^2_+$ as follow

$$\sigma(z) = \Big(\frac{2\,\mathrm{Re}(z)}{1-\mid z\mid^2}, \frac{2\,\mathrm{Im}(z)}{1-\mid z\mid^2}, \frac{1+\mid z\mid^2}{1-\mid z\mid^2}\Big). \tag{3}$$

The map σ is conformal and bijective assigning to each point $z \in \mathbb{D}$, the point in $\mathbb{H}^2_+$ obtained as the intersection of the straight line determined by $(0, 0-1)$ and $(z, 0)$ with $\mathbb{H}^2_+$.

On the other hand, the Gauss map is also conformal. Taking this into account the author shows the following result (Weierstrass–Enneper formula):

Theorem 2.1 *Any maximal surface S in $\mathbb{L}^3$ is represented as*

$$\phi(z) = \mathrm{Re}\int \Big(\frac{1}{2}f(1+g^2), \frac{i}{2}f(1-g^2), -fg\Big)dz, \quad z \in D, \tag{4}$$

where D is a domain in $\mathbb{C}$, and f (resp. g) is a holomorphic (resp. meromorphic) function on D such that fg^2 is holomorphic in D and $\mid g(z)\mid \neq 1$ for $z \in D$. Moreover,

(i) The Gauss map N is given by $N(z) = \sigma(g(z))$, where σ is the map defined in (3).

(ii) The induced metric is given by $ds = \frac{1}{2}\mid f\mid (\mid 1-\mid g\mid^2\mid)\mid dz\mid$.

(iii) The Gauss curvature of the surface is given by $K = \big(\frac{4|\partial_z g|}{|f|(|1-|g|^2|)^2}\big)^2$.

Thus, we consider the immersion $\phi : D \longrightarrow \mathbb{L}^3$ with the induced metric from $\mathbb{L}^3$.

Assume that the maximal surface is complete. Since the surface is not compact, the uniformization theorem allow us to affirm that D must be conformal to $\mathbb{C}$ or the unit disc $\mathbb{D}$. Suppose that D is conformal to $\mathbb{D}$. Now, the author makes use of a result of Osserman [46, p. 67] to show that there is a divergent curve γ in D such that $\int_\gamma \mid f(z)\mid\mid dz\mid < \infty$. Thus using (ii) of Theorem 2.1 we will conclude that the surface cannot be complete. Hence, D must be conformal to $\mathbb{C}$. Finally, taking into account (i) in Theorem 2.1 we have $\mid g\mid < 1$ and g is holomorphic. Now, it is enough to call Liouville's theorem to obtain that g is a constant function and as a direct consequence the immersion is a spacelike plane.

2.2 About Other Approaches

In 1994, a new proof of the Classical Calabi–Bernstein theorem is given by Estudillo and Romero [29] making use again of the Weierstrass–Enneper representation. The authors find an adequate local upper bound for the Gaussian curvature of a maximal surface. Estudillo and Romero inspired in a paper by Osserman [47] obtain the following inequality of the Gauss curvature at any point p, $K(p)$, of a maximal surface S with boundary in $\mathbb{L}^3$,

$$K(p) \leq \frac{4}{d(p, \partial S)^2}, \quad \text{for any} \quad p \in S,$$

where d is the Riemannian distance on S.

As a consequence, if we consider a complete maximal surface S and an arbitrary point $p \in S$, we can take a geodesic disc with center at p and radius r. Now, it is enough to choose r as large as we desire to conclude that S must be totally geodesic.

On the other hand, using a conformal metric, the authors get a new proof of the nonparametric version.

In the real field, a simple proof of the nonparametric version, which only requires the Liouville theorem for harmonic functions on the Euclidean plane $\mathbb{R}^2$, was given in 1994 by Romero [52]. As the author says, the proof is inspired in a proof of the classical Bernstein theorem given by Chern [24]. So, the author obtains a conformal metric on the entire graph, which is complete and flat. Thus, via Cartan's theorem, the graph endows with the conformal metric is isometric to Euclidean plane. On the other hand, Romero shows that the function $\frac{1}{\langle N,a\rangle}$, where N is the unitary (future directed) normal vector field on the graph and a is an suitable constant lightlike vector, is a positive harmonic function globally defined on the graph. Finally, taking into account, the invariance of harmonic functions by conformal changes of metric we have that $\langle N, a\rangle$ is constant.

Via a local integral inequality for the Gaussian curvature of a maximal surface, in 2000, Alías and Palmer [7] provided another new proof for the parametric case. The authors obtain an upper bound for the total curvature of geodesic discs in a maximal surface in terms of the local geometry of the surface and its hyperbolic image. Specifically, the authors show

Theorem 2.2 *Let $x : S \longrightarrow \mathbb{L}^3$ be a maximal surface in the three-dimensional Lorentz–Minkowski spacetime. Let $p \in S$ and $R > 0$ be a positive real number such that the geodesic disc of radius R about p satisfies $D(p, R) \subset\subset S$. Then for all $0 < r < R$, the total curvature of a geodesic disc $D(p, r)$ satisfies*

$$0 \leq \int_{D(p,r)} K\,dA \leq c_r \frac{L(r)}{r \log(R/r)}, \tag{5}$$

where dA is the area element, $L(r)$ denotes the lenght of $\partial D(p, r)$ and $c_r = c_r(p, r)$ is a constant.

Making use of this integral inequality, Alías and Palmer get a new proof of the Calabi–Bernstein theorem. Indeed, if S is complete, R can approach to infinity for any fixed point p and fixed radius r, now from (5) $K \equiv 0$.

These authors get also a new proof of the nonparametric version based on a duality result with minimal surface equation in the Euclidean case [9]. Recently (2010), yet another short proof of both versions has been given by Romero and Rubio [53] making use of the interface between the parabolicity of a Riemannian surface and the capacity of geodesic annuli. Finally, a more recent (2015) original new proof has been given by Aledo, Romero and Rubio by using a development inspired by the well-known Bochner's technique.

We must emphasize that for several of the proof of the classical results, it is essential in a way or another that any complete maximal surface in $\mathbb{L}^3$ must be parabolic.

2.3 Romero–Rubio's Proof of the Classical Result

In this section, we will describe the new approach given by Romero and Rubio [53] to the two-dimensional version of the Calabi–Bernstein theorem.

Consider the Lorentz–Minkowski spacetime $\mathbb{L}^3$, which is given by the Lorentzian manifold $(\mathbb{R}^3, \langle\, ,\, \rangle)$, where

$$\langle\, ,\, \rangle = -dt^2 + dx^2 + dy^2.$$

Let $x : S \longrightarrow \mathbb{L}^3$ be a (connected) immersed spacelike surface in $\mathbb{L}^3$. Observe that S must be orientable and let N be the unitary normal vector field on S such that $\langle N, \partial_t \rangle > 0$, where ∂_t denotes the coordinate vector field $\frac{\partial}{\partial t}$. If $\theta(p)$ denotes the hyperbolic angle between N and $-\partial_t$ at $p \in S$, then $\cosh\theta = \langle N, \partial_t \rangle$.

We will denote by $\overline{\nabla}$ and ∇ the Levi–Civita connections of $\mathbb{L}^3$ and S, respectively. Then the Gauss and Weingarten formulas for S in $\mathbb{L}^3$ are given, respectively, by

$$\overline{\nabla}_X Y = \nabla_X Y - \langle A(X), Y \rangle N \tag{6}$$

and

$$A(X) = -\overline{\nabla}_X N, \tag{7}$$

for all tangent vector fields $X, Y \in \mathfrak{X}(S)$, where $A : \mathfrak{X}(S) \longrightarrow \mathfrak{X}(S)$ stands for the *shape operator* associated to N.

On the other hand, the tangential component of ∂_t at any point of S is given by $\partial_t^\top = \partial_t + \cosh\theta N$.

We suppose that S is maximal. It is immediate to see that

$$\nabla \cosh\theta = -A\partial_t^\top$$

being A the shape operator associated to N. It is not difficult to obtain by standard computation the following formulas:

$$| \nabla \cosh \theta |^2 = \frac{1}{2} \text{trace}\,(A^2) \sinh^2 \theta \quad \text{and} \quad \Delta \cosh \theta = \text{trace}\,(A^2) \cosh \theta$$

where ∇ and Δ are respectively the gradient and Laplacian relative to the induced Riemannian metric on S.

The following technical result, which is a reformulation by Romero and Rubio [53] of a Lemma by Alías and Palmer [8], is a key piece of the this new approach. Previously, we need to recall some general preliminaries.

Let S be an $n(\geq 2)$-dimensional Riemannian manifold and let B_r denote the geodesic ball of radius r around a fixed point $p \in S$. For $0 < r < R$ let $A_{r,R}$ be the geodesic annulus $A_{r,R} := B_R \setminus \bar{B}_r$. Denote by $\omega_{r,R}$ the harmonic measure of ∂B_R with respect to $A_{r,R}$, that is the solution of the elliptic problem

$$\Delta \omega = 0 \ \text{ in } \ A_{r,R}, \quad \omega \equiv 0 \ \text{ on } \ \partial B_r, \quad \text{and} \quad \omega \equiv 1 \ \text{ on } \ \partial B_R.$$

The capacity of the annulus is defined to be

$$\frac{1}{\mu_{r,R}} := \int_{A_{r,R}} |\nabla \omega_{r,R}|^2 \, dV.$$

It is well known that S is parabolic if and only if

$$\lim_{R \to \infty} \frac{1}{\mu_{r,R}} = 0.$$

Now, we can enunciate the technical lemma.

Lemma 2.3 *Let S be an $n(\geq 2)$-dimensional Riemannian manifold and let $v \in C^2(S)$ which satisfies $v \Delta v \geq 0$. Let B_R be a geodesic ball of radius R in S. For any r such that $0 < r < R$ we have*

$$\int_{B_r} |\nabla v|^2 \, dV \leq \frac{4\, \text{Sup}_{B_R} v^2}{\mu_{r,R}},$$

where B_r denote the geodesic ball of radius r around p in S and $\frac{1}{\mu_{r,R}}$ is the capacity of the annulus $B_R \setminus \bar{B}_r$.

Proof For any $\zeta \in C^\infty(B_R)$ with $\text{supp}(\zeta) \subset B_R$, from the divergence theorem we have

$$\int_{B_R} (\zeta^2 |\nabla v|^2 + \zeta^2 v \Delta v + 2\zeta v \langle \nabla \zeta, \nabla v \rangle) dV = 0,$$

and as a consequence,

$$\int_{B_R} \zeta^2 |\nabla v|^2 dV \leq 2 \int_{B_R} |\zeta v \langle \nabla \zeta, \nabla v \rangle| dV \leq a^2 \int_{B_R} \zeta^2 |\nabla v|^2 dV + \frac{1}{a^2} \int_{B_R} v^2 |\nabla \zeta|^2 dV,$$

for all $a > 0$, and hence taking $a = 1/\sqrt{2}$ we obtain

$$\int_{B_R} \zeta^2 |\nabla v|^2 dV \leq 4 \int_{B_R} v^2 |\nabla \zeta|^2 dV \leq 4\, \mathrm{Sup}_{B_R} v^2 \int_{B_R} |\nabla \zeta|^2 dV.$$

Define ζ by

$$\zeta(x) = \begin{cases} 1 & \text{if } x \in \bar{B}_r \\ 1 - \omega_{r,R} & \text{if } x \in A_{r,R} \end{cases}$$

Finally, although ζ is not smooth it can be approximated by smooth function, and so we obtain

$$\int_{B_r} |\nabla v|^2 dV \leq \frac{4\, \mathrm{Sup}_{B_R} u^2}{\mu_{r,R}}.$$

□

We are now in a position to describe the proof of the parametric version. So, consider the auxiliary function $v : S \longrightarrow (\frac{\pi}{2}, \frac{3\pi}{2})$, $v(p) = \arctan(\cosh\theta(p))$, which has an advantage on the original $\cosh\theta$, that is, v is bounded.

It is immediate to verify that $v\Delta v \geq 0$. From the previous Lemma, and taking into account that

$$\nabla v = \frac{1}{1 + \cosh^2\theta} \nabla \cosh\theta,$$

we have

$$\int_{B_r} |\nabla v|^2\, dV \leq \frac{9\pi^2}{\mu_{r,R}},$$

for $0 < r < R$, which easily gives

$$\int_{B_r} |\nabla(\cosh\theta)|^2\, dV \leq \frac{C}{\mu_{r,R}},$$

where B_r denote the geodesic disc of radius r around p in S, $\frac{1}{\mu_{r,R}}$ is the capacity of annulus $B_R \setminus \bar{B}_r$ and $C = C(p, r) > 0$ is constant.

Now, the surface S is necessarily non compact and from the Gauss formula it has curvature $K \geq 0$. If we assume that S is complete, a classical result by Ahlfors and Blanc–Fiala–Huber (see for instance, [37]), affirms that a complete two-dimensional Riemannian manifold with nonnegative Gauss curvature is parabolic.

On other hand, it is well known that S will be parabolic if and only if $\lim_{R\to\infty} \frac{1}{\mu_{r,R}} = 0$. We get that R can approach to infinity for a fixed arbitrary point p and a fixed r, obtaining that $\cosh\theta$ is constant on S.

2.4 The Nonparametric Case

We finish the approach by Romero and Rubio with a sketch of the proof given by the authors for the nonparametric version of the classical Calabi–Bernstein theorem.

For each $u \in C^\infty(\Omega)$, note that the induced metric on $\Omega \subset \mathbb{R}^2$, via the graph $\{(u(x, y), x, y) : (x, y) \in \Omega\} \subset \mathbb{L}^3$, is $g_u := -du^2 + g_0$, where g_0 is the usual Riemannian metric of $\mathbb{R}^2$. The metric g_u is positive definite, if and only if u satisfies $|Du| < 1$, where Du denote the gradient of u in $(\mathbb{R}^2, g_0)$.

On the other hand, the graph of u is spacelike and has zero mean curvature if and only if u is a solution to the maximal surface Eq. (2) in the Lorentz–Minkowski space.

We consider on $\mathbb{R}^2$ the function $\cosh\theta = \frac{1}{\sqrt{1-|Du|^2}}$ and the conformal metric $g' = (\cosh\theta + 1)^2 g_u$, which taking into account the relation between curvatures for conformal changes (see for instance, [14]) is flat.

If the graph is entire, then g' is complete, because $L' \geq L_0$ where L' and L_0 denote the lengths of a curve on $\mathbb{R}^2$ with respect to g' and the usual metric of $\mathbb{R}^2$. Taking into account the invariance of subharmonic functions by conformal changes of metric, we are in position to use the same argument as in the parametric case on the Riemannian surface (F, g') to get the result.

2.5 A New Proof Using the Bochner Technique

Recently, yet another proof of the classical Calabi–Bernstein theorem has been given by Aledo Romero and Rubio [4]. In this paper, the authors make use of the Bochner technique. By mean of the Bochner–Lichnerowicz's Formula and a well-known Liouville-type result, the authors show the parametric version of the aforementioned theorem.

Next, we will explain the main steps of the Aledo–Romero–Rubio's proof.

Consider $x : S \longrightarrow \mathbb{L}^3$ a (connected) immersed maximal surface in the Lorentz–Minkowski spacetime $\mathbb{L}^3$. We choose a unit timelike normal vector field N globally defined on S in the same time-orientation of $\frac{\partial}{\partial t}$.

Making use of the Gauss equation for a surface in $\mathbb{L}^3$, it is easy to verify that

$$\operatorname{trace}(A^2) = 2K, \tag{8}$$

where A denotes the shape operator associated to the normal vector field N and K is the Gaussian curvature of the surface.

The idea of the proof is to choose a suitable function on the maximal surface and to apply the Bochner–Lichnerowicz's Formula.

Recall that the well-known Bochner–Lichnerowicz's Formula (see, for instance [23]) states that

$$\frac{1}{2}\Delta\left(|\nabla u|^2\right) = |\mathrm{Hess}\, u|^2 + \mathrm{Ric}(\nabla u, \nabla u) + \langle \nabla u, \nabla(\Delta u)\rangle \tag{9}$$

for $u \in C^\infty(S)$. Here Ric stands for the Ricci tensor of S and $|\mathrm{Hess}\, u|^2$ is the square algebraic trace-norm of the Hessian of u, namely $|\mathrm{Hess}\, u|^2 := \mathrm{trace}(H_u \circ H_u)$ where H_u denotes the operator defined by $\langle H_u(X), Y\rangle := \mathrm{Hess}\,(u)(X, Y)$ for all $X, Y \in \mathfrak{X}(S)$.

Let us choose $a \in \mathbb{L}^3$ a null vector, i.e., a nonzero vector such that $\langle a, a\rangle = 0$, and consider the function $\langle N, a\rangle$ on S.

Now, applying Schwarz's inequality (for symmetric square matrix), we have,

$$|\mathrm{Hess}\,\langle N, a\rangle|^2 \geq \frac{1}{2}(\Delta\langle N, a\rangle)^2. \tag{10}$$

On the other hand, from the Weingarten formula (7) it is easy to obtain the gradient of the function $\langle N, a\rangle$ on S,

$$\nabla\langle N, a\rangle = -A(a^\top), \tag{11}$$

where $a^\top = a + \langle N, a\rangle N$ is tangent to S and standard computations allow us to obtain

$$\Delta\langle N, a\rangle = \langle N, a\rangle \mathrm{trace}(A^2). \tag{12}$$

From (11) and taking into account that $|a^\top|^2 = \langle N, a\rangle^2$ and that S is maximal, we get

$$|\nabla\langle N, a\rangle|^2 = K\langle N, a\rangle^2 \tag{13}$$

and so

$$\mathrm{Ric}(\nabla\langle N, a\rangle, \nabla\langle N, a\rangle) = K|\nabla\langle N, a\rangle|^2 = K^2\langle N, a\rangle^2. \tag{14}$$

With the previous computations, we can to apply the Bochner–Lichnerowicz's Formula to the chosen function $\langle N, a\rangle$ on S and so, to obtain the following inequality

$$\Delta K \geq 4K^2. \tag{15}$$

Since the Gauss curvature of S is nonnegative and if we assume that S is complete, then we can use the following known result (see, for instance [62]),

Lemma 2.4 *Let S be a complete Riemannian surface whose Gaussian curvature is bounded from below and $u \in \mathcal{C}^\infty(S)$ a nonnegative function such that $\Delta u \geq cu^2$ for a positive constant c. Then u vanishes identically on S.*

As a consequence, $K \equiv 0$ and so S is totally geodesic.

Observe that totally geodesic spacelike surfaces in Minkowski spacetime $\mathbb{L}^3$ are spacelike planes. Nevertheless, in Lorentzian warped products this is not necessarily true, and this is why some additional hypotheses are sometimes needed in theses spaces.

3 Some Extension of the Classical Result

In this section, we will describe some recent extensions of the Calabi–Bernstein theorem in the two-dimensional case, as well as, others Calabi–Bernstein-type problems. The three-dimensional Lorentzian ambient will be given by a Lorentzian product or in a more general case for a Lorentzian-warped product.

Consider (F, g) a Riemannian manifold, let $(I, -dt^2)$ be a real interval with negative metric, and $f : I \longrightarrow \mathbb{R}$ a smooth positive function. Recall that the warped product $I \times_f F$ is given by the Lorentzian manifold $(M = I \times F, \langle\, ,\rangle)$, where

$$\langle\, ,\, \rangle = -\pi_I^*(dt^2) + f(\pi_I)^2 \pi_F^*(g), \tag{16}$$

and π_I, π_F denote the projections from M onto the base I and the fiber F, respectively.

In particular, when $f \equiv 1$ we have the Lorentzian product of $(I, -dt^2)$ and (F, g).

Recall that any warped product $I \times_f F$ possesses an infinitesimal timelike conformal symmetry which is an important tool. Indeed, the vector field

$$\xi := f(\pi_I)\, \partial_t, \tag{17}$$

which is timelike and, from the relationship between the Levi–Civita connections of M and those of the base and the fiber, satisfies

$$\overline{\nabla}_X \xi = f'(\pi_I)\, X \tag{18}$$

for any $X \in \mathfrak{X}(M)$, where $\overline{\nabla}$ is the Levi–Civita connection of the warped metric. Thus, ξ is conformal with $\mathcal{L}_\xi \langle\, ,\, \rangle = 2\, f'(\pi_I)\, \langle\, ,\, \rangle$ and its metrically equivalent 1-form is closed.

Spacetimes given as a Lorentzian warped product $I \times_f F$ are introduced in General Relativity literature in [10] and they are called generalized Robert–Walker spacetimes (GRW).

In any GRW spacetime $M = I \times_f F$, the level hypersurfaces of the function $\pi_I : M \longrightarrow I$ constitute a distinguished family of spacelike hypersurfaces: the so-called *spacelike slices*. Along this work, we will represent by $t = t_0$ the spacelike slice $\{t_0\} \times F$. For a given spacelike hypersurface $x : S \longrightarrow M$, we have that $x(S)$

is contained in $t = t_0$ if and only if $\pi_I \circ x = t_0$ on S. We will say that S is a spacelike slice if $x(S)$ equals to $t = t_0$, for some $t_0 \in I$, and that S is contained between two slices if there exist $t_1, t_2 \in I$, $t_1 < t_2$, such that $x(S) \subset [t_1, t_2] \times F$.

If we take the unitary normal vector field to every spacelike slice given by $-\partial_t$, then the shape operator and the mean curvature of the spacelike slice $t = t_0$ are respectively $A = f'(t_0)/f(t_0)\, I$, where I denotes the identity transformation, and the constant $H = -f'(t_0)/f(t_0)$. Thus, a spacelike slice $t = t_0$ is maximal if and only if $f'(t_0) = 0$ (and hence, totally geodesic).

The following nonlinear elliptic differential equation, in divergence form represents the maximal surface equation in a three-dimensional Lorentzian warped product $I \times_f F$ (dim $F = 2$, in this case),

$$\operatorname{div}\left(\frac{Du}{f(u)\sqrt{f(u)^2 - \mid Du \mid^2}}\right) = -\frac{f'(u)}{\sqrt{f(u)^2 - \mid Du \mid^2}}\left(2 + \frac{\mid Du \mid^2}{f(u)^2}\right) \tag{E.1}$$

$$\mid Du \mid < f(u) \tag{E.2}$$

where f is the warping function defined on the open interval I of the real line $\mathbb{R}$, the unknown u is a function defined on a domain Ω of the Riemannian surface (F, g), $u(\Omega) \subseteq I$, D and div denote the gradient and the divergence of (F, g) and $\mid Du \mid^2 := g(Du, Du)$.

The constraint (E.2) assures the spatiality of the graph $\{(u(p), p)/p \in \Omega\}$ and it is the ellipticity condition. From now on, we will refer to the nonlinear problem E.1+E.2 as equation (E):

On the other hand, the solutions of (E) are the extremals under interior variations for the functional

$$u \longmapsto \int f(u)\sqrt{f(u)^2 - \mid Du \mid^2}\, dA,$$

where dA is the area element for the Riemannian metric g, which acts on functions u such that $u(\Omega) \subseteq I$ and $\mid Du \mid < f(u)$.

Observe that when $I = \mathbb{R}$, $F = \mathbb{R}^2$ and $f = 1$, the equation (E) is the maximal surface equation in $\mathbb{L}^3$.

Note that a constant function $u = c$ is a solution to the equation (E), if and only if $f'(c) = 0$.

We will begin with a new example of non-parametric Calabi–Bernstein-type problem given by Latorre and Romero [39]. We have to say that this paper is the first one dealing with the maximal surface equation for warped Lorentzian products, whose fiber is a complete (non-compact) 2-Riemannian manifold, in particular, the Euclidean plane $\mathbb{R}^2$. In this work, the authors assume that the sectional curvature of the Lorentzian manifold is not zero on any proper open subset, i.e., the warping function is not locally constant, although the curvature of the ambient satisfies a natural geometric assumption arising from Relativity theory, the *null convergence condition* (NCC), which says that the Ricci quadratic form on null tangent vectors is nonnegative. Obviously the Calabi–Bernstein theorem is not included in this case.

In their proof, Latorre and Romero introduce a new conformal metric on the graph. With this metric, the authors prove that the entire maximal graph is complete and parabolic, which allows to conclude that the warping function restricted to the graph is superharmonic and consequently constant. Since, the warping function is not locally constant, the only entire solutions are given by the constant functions $u = c$ with $f'(c) = 0$.

Another approach to the previous problem have been obtained by Romero and Rubio [54]. Following the ideas in Alías-Palmer's papers [7, 8], the authors obtain a local integral estimate of the squared length for the gradient of a distinguished function on the maximal surface, which is constant if and only if the surface is contained in a spacelike slice $t = t_0$, with $f'(t_0) = 0$ in the warped product.

A new extension of nonparametric Calabi–Bernstein theorem in the case of a Lorentzian product $\mathbb{R} \times F$, where F denotes a Riemannian 2-manifold, with non-negative curvature, has been given by Albujer and Alías, [2, 3]. Moreover, the authors find examples of complete and nontrivial entire maximal graphs in $\mathbb{H}^2 \times \mathbb{R}$, where $\mathbb{H}^2$ denotes the hyperbolic plane [2, Example 5.2], (see also [1, 31]), which show the need for certain curvature assumption on the fiber for possible extensions of the Calabi–Bernstein theorem.

Recently, another Calabi–Bernstein-type results in the more general ambient of a warped Lorentzian product are given by Caballero, Romero and Rubio [18, 19]. So, the authors obtain several extensions of the classical Calabi–Bernstein theorem to three-dimensional warped products satisfying suitable energy conditions and whose fiber can be unnecessary of nonnegative Gaussian curvature in some cases. Moreover, in the particular case where the warping function is constant, the authors recover the non-parametric extension of the classical result in Lorentzian product spaces given by Albujer and Alíias.

In a different direction, yet another extension of the classical result has been given by Pelegrín, Romero and Rubio. This time, the ambient is a three-dimensional space-time which admits a parallel lightlike vector field and obeying the energy condition known as *timelike energy condition* (TCC) [48]. Note that, it is normally argued that TCC is the mathematical translation that gravity, on average, attracts. More precisely, the authors show that in a three-dimensional Lorentzian manifold, which admits a parallel global lightlike vector field and obeys the timelike energy condition, then every complete isometrically immersed maximal surface must be totally geodesic.

Finally, we will describe with more detail a new extension of the classical Calabi–Bernstein theorem by Rubio–Salamanca [59]. In this last work, the authors study entire solutions to the maximal surface equation in a Lorentzian three-dimensional warped product, whose fiber is given by a Riemannian surface with finite total curvature.

Recall that a complete Riemannian surface has finite total curvature if the integral of the absolute value of its Gaussian curvature is finite. Of course, the Euclidean plane has finite total curvature, but note that any complete surface, whose curvature is nonnegative outside a compact subset has finite total curvature. Also, it is well-known that a complete Riemannian surface has finite total curvature if the negative part of its Gaussian curvature is integrable (see, for instance [40, Sect. 10]).

On the other hand, examples of complete minimal surfaces in $\mathbb{R}^3$ with finite total curvature are known [34]. Examples in a different ambient space can be found in [50].

3.1 *On the Proof of Rubio–Salamanca's Extension*

The authors deal with the maximal surface equation for Lorentzian-warped product (E), when the fiber (F, g) is a complete (non-compact) Riemannian surface with finite total curvature.

In their work, the authors are mainly interested in uniqueness and nonexistence results for entire solutions (i.e. defined on all F) of equation (E).

Let Ω be a domain of the Riemannian surface F, for each $u \in C^\infty(\Omega)$, $u(\Omega) \subseteq I$, the induced metric on Ω from the Lorentzian metric (16), via its graph $\Sigma_u = \{(u(p), p) : p \in \Omega\}$ in M, is written as follows

$$g_u = -du^2 + f(u)^2 g,$$

and it is positive definite, i.e. Riemannian, if and only if u satisfies $\mid Du \mid < f(u)$ everywhere on Ω.

When g_u is Riemannian, $f(u)\sqrt{f(u)^2 - \mid Du \mid^2}\, dA$ is the area element of (Ω, g_u). Therefore (E.1) of (E) is the Euler–Lagrange equation for the area functional, its solutions are spacelike graphs of zero mean curvature in $M = I \times_f F$, and this equation is called the maximal surface equation in M.

If we denote by N the unit normal vector field N on a spacelike graph Σ_u such that $\langle N, \partial_t \rangle \geq 1$ on Σ_u, where $\partial_t := \partial/\partial t \in \mathfrak{X}(M)$, then

$$N = \frac{-f(u)}{\sqrt{f(u)^2 - \mid Du \mid^2}} \left(1, \frac{1}{f(u)^2} Du\right),$$

and the hyperbolic angle θ between $-\partial_t$ and N is given by

$$\langle N, \partial_t \rangle = \cosh\theta = \frac{f(u)}{\sqrt{f(u)^2 - \mid Du \mid^2}}.$$

On the other hand, the Lorentzian-warped product spaces considered by the authors must satisfy certain natural energy condition, which turns out to have an expression in terms of the curvature of its fiber (F, g) and the warping function f. So, recall that a Lorentzian manifold obeys NCC if its Ricci tensor $\overline{\mathrm{Ric}}$ satisfies

$$\overline{\mathrm{Ric}}(Z, Z) \geq 0,$$

for any null vector Z, i.e., $Z \neq 0$ such that $\langle Z, Z \rangle = 0$.

Taking into account how the Ricci tensor of M is obtained from the Gaussian curvature of the fiber K^F and the warping function f (see for instance [45, Corollary 7.43]) it is easy to check that a Lorentzian-warped product space $I \times_f F$ with a 2-dimensional fiber obeys NCC if and only if

$$\frac{K^F(\pi_F)}{f^2} - (\log f)'' \geq 0. \tag{19}$$

From now on, let's consider spacelike entire graphs. So, let $\Sigma_u = \{(u(p), p) : p \in F\}$ be, the graph of $u \in C^\infty(F)$ such that $u(F) \subseteq I$ in the Lorentzian warped product $M = I \times_f F$. Suppose that the graph is spacelike.

Note that $\pi_I(u(p), p) = u(p)$ for any $p \in F$, and so π_I on the graph and u can be naturally identified by the isometry between $(\Sigma_u, \langle\, ,\, \rangle)$ and (F, g_u). Analogously, the differential operators ∇ and Δ in $(\Sigma_u, \langle\, ,\, \rangle)$ can be identified with those ones ∇_u and Δ_u in (F, g_u).

If we denote $\partial_t^\top = \partial_t + \langle N, \partial_t \rangle N$ the tangential component of ∂_t on Σ_u, then it is not difficult to see

$$\nabla \pi_I|_{\Sigma_u} := \nabla u = -\partial_t^\top.$$

Now, suppose the graph maximal and consider the distinguished function

$$\langle N, \xi \rangle = f(u) \cosh \theta,$$

defined on the graph. It is immediate to see that

$$\nabla \langle N, \xi \rangle = -A\xi^\top,$$

where A denotes the shape operator of the graph. Taking a orthonormal frame consisting of the eigenvectors of the shape operator, we obtain

$$\mid \nabla \langle N, \xi \rangle \mid^2 = \frac{1}{2} \text{trace}\,(A^2)\{\langle N, \xi \rangle^2 - f(\pi_I)^2\}. \tag{20}$$

Moreover, using the Gauss and Codazzi equations, as well as, the expression for the Ricci tensor of M (see for instance [45, Chap. 7]), it is a standard computation to obtain (via the isometry)

$$\begin{aligned} \Delta_u(f(u)\cosh\theta) &= \left\{ \frac{K^F}{f(u)^2} - (\log f)''(u) \right\} |\nabla_u u|^2 f(u) \cosh\theta \\ &\quad + \frac{1}{2} \text{trace}(A^2) f(u) \cosh\theta \end{aligned} \tag{21}$$

On the other hand, taking into account the Gauss equation and using again the expression for the Ricci tensor of M, then the Gauss curvature of a maximal graph is

$$K = \frac{f'(u)^2}{f(u)^2} + \left\{ \frac{K^F}{f(u)^2} - (\log f)''(u) \right\} \mid \partial_t^{\top} \mid^2 + \frac{K^F}{f(u)^2} + \frac{1}{2}\mathrm{trace}(A^2). \quad (22)$$

As a direct consequence, from (21) we have the following alternative expression,

$$\Delta_u(f(u)\cosh\theta) = \left\{ K_u - \frac{f'(u)^2}{f(u)^2} - \frac{K^F}{f(u)^2} + \frac{1}{2}\mathrm{trace}(A^2) \right\} f(u)\cosh\theta. \quad (23)$$

One of the fundamental tools in the work of Rubio and Salamanca is to introduce a conformal metric on the graph, which allows certain control on its Gaussian curvature. So, on the manifold F we consider the following Riemannian metric

$$g'_u := f(u)^2 \cosh^2\theta \, g_u, \quad (24)$$

where

$$f(u)\cosh\theta = \frac{f(u)^2}{\sqrt{f(u)^2 - \mid Du \mid^2}}$$

and $\mid Du \mid^2 := g(Du, Du)$. Therefore, if $\epsilon := \mathrm{Inf}(f) > 0$ we get the following inequality

$$L' \geq \epsilon^2 L,$$

where L' and L denote the lengths of a curve in F with respect to g'_u and g, respectively. Consequently, g'_u is complete whenever g is complete.

Now, suppose that $\mathrm{Sup} f(u) < \infty$. Put $\lambda = \mathrm{Sup} f(u)$ and consider the new Riemannian metric

$$g^*_u := (f(u)\cosh\theta + \lambda)^2 g_u \quad (25)$$

on F. The completeness of the metric (24) assures that g^*_u is also complete. Moreover, it has the advantage over g'_u that we can control its Gaussian curvature under reasonable assumptions. In order to concrete this assertion, denote by K^*_u and K_u the Gaussian curvatures of the Riemannian metrics g^*_u and g_u, respectively. From (25) and using the relation between Gaussian curvatures for conformal changes (see for instance, [13]), we have

$$K_u - (f(u)\cosh\theta + \lambda)^2 K^*_u = \Delta_u \log(f(u)\cosh\theta + \lambda). \quad (26)$$

The following lemma is key to the achievement of the principal result, since it allows to assure that the graph endowed with the appropriate conformal metric will has finite total curvature.

Lemma 3.1 *Suppose that (F, g) is complete, with finite total curvature. If* $\mathrm{Inf} f > 0$, $\mathrm{Sup} f < \infty$ *and the inequality* $\frac{K^F}{f(u)^2} - (\log f)''(u) \geq 0$ *holds on F, then the complete Riemannian surface (F, g^*_u) has finite total curvature.*

Proof From the previous expressions (22) and (21) we get,

$$\Delta_u \log(f(u)\cosh\theta + \lambda) \leq \frac{1}{f(u)\cosh\theta + \lambda}\left\{\left(K_u - \frac{K^F}{f(u)^2}\right)f(u)\cosh\theta + \left(K_u - \frac{K^F}{f(u)^2}\right)\lambda\right\}$$

$$\leq K_u - \frac{K^F}{f(u)^2}.$$

Since the Riemannian area elements of the metrics g and g_u^* satisfy

$$dA_u^* = \frac{(f(u)\cosh\theta + \lambda)^2 f(u)^2}{\cosh\theta}\, dA,$$

making use of (26), we obtain

$$\int_F \max(-K_u^*, 0)\, dA_u^* \leq \int_F \max(-K^F, 0)\frac{1}{\cosh\theta}\, dA < \int_F \max(-K^F, 0)\, dA < \infty.$$

Therefore, the Riemannian surface (F, g_u^*) is complete and it has finite total curvature. □

Now, we can state one of main results of Rubio–Salamanca's work.

Theorem 3.2 *Let $M = I \times_f F$ a Lorentzian warped product, with fiber (F, g) a complete Riemannian surface, which has finite total curvature and whose warping function satisfies* $\operatorname{Inf} f > 0$ *and* $\operatorname{Sup} f < \infty$. *If M obeys the NCC, then any entire maximal graph $(\Sigma_u, \langle\, ,\, \rangle)$ must be totally geodesic. Moreover, if there exists a point $p \in F$ such that $\frac{K^F(p)}{f(u(p))^2} - (\log f)''(u(p)) > 0$, then u is constant.*

Proof From previous Lemma, we have that (F, g_u^*) is complete with finite total curvature. Consider the function $\frac{1}{f(u)\cosh\theta}$ on (F, g_u). Then, some computations allow to show that the Laplacian

$$\Delta_u\left(\frac{1}{f(u)\cosh\theta}\right) = -\frac{1}{f(u)^2\cosh^2\theta}\Delta_u(f(u)\cosh\theta) + 2\frac{|\nabla_u(f(u)\cosh\theta)|^2}{(f(u)^3\cosh^3\theta)}$$

is non-positive

Taking into account the invariance of superharmonic functions by conformal changes of metric, we get a positive superharmonic function on the complete parabolic Riemannian surface (F, g_u^*) and as a consequence the function $f(u)\cosh\theta$ must be constant. Thus, from the second term of (21), whose expression we will recall,

$$\Delta_u(f(u)\cosh\theta) = \left\{\frac{K^F}{f(u)^2} - (\log f)''(u)\right\}|\nabla_u u|^2 f(u)\cosh\theta$$
$$+ \frac{1}{2}\mathrm{trace}(A^2) f(u)\cosh\theta,$$

we obtain that the graph $(\Sigma_u, \langle\, , \,\rangle)$ is totally geodesic.

On the other hand, if in addition there exists a point $p \in F$ such that $\frac{K^F(p)}{f(u(p))^2} - (\log f)''(u(p)) > 0$, taking into account the first addend of (21), then there exists an open neighbourhood of $(p, u(p))$ in Σ_u which is contained in the complete maximal graph $u = u_0$, with $f'(u_0) = 0$. As $(\Sigma_u, \langle\, , \,\rangle)$ is entire and totally geodesic, it must coincide with the totally geodesic spacelike slice $t = u_0$. □

Of course, this last result extends the classical Calabi–Bernstein theorem in its nonparametric version. Moreover, the extension given by Alías and Albujer [2, 3] for Lorentzian products is also included. On the other hand, the Theorem 3.2 is independent of those given for Lorentzian warped product by Caballero, Romero and Rubio [18, 19].

4 Uniqueness of Complete Maximal Hypersurfaces in Spacetimes

In this new section we will make a brief review on some uniqueness results about maximal hypersurfaces in spacetimes. These results can be considered parametric versions of Calabi–Bernstein-type problems. On the other hand, these types of problems have both mathematical and physical interest due to their relevance in Mathematical Relativity

The importance in General Relativity of maximal and constant mean curvature spacelike hypersurfaces in spacetimes is well-known; a summary of several reasons justifying it can be found in the paper of Marsden and Tipler [42].

Recall that each maximal hypersurface can describe, in some relevant cases, the transition between the expanding and contracting phases of a relativistic universe. Moreover, they can constitute an initial set for the Cauchy problem [51]. Specifically, Lichnerowicz proved that a Cauchy problem with initial conditions on a maximal hypersurface is reduced to a second-order nonlinear elliptic differential equation and a first-order linear differential system [41]. Also, the deep understanding of this kind of hypersurfaces is essential to prove the positivity of the gravitational mass.

On the other hand, they are also interesting for Numerical Relativity, where maximal hypersurfaces are used for integrating forward in time [36]. From a mathematical point of view, it is necessary to study the maximal hypersurfaces of a spacetime in order to understand its structure. Especially, for some asymptotically flat spacetimes, the existence of a foliation by maximal hypersurfaces is established (see for instance, [12] and references therein).

Thus, the existence results and, consequently, uniqueness appear as kernel topics.

Let us remark that the completeness of a spacelike hypersurface is required whenever we study its global properties, and also, from a physical viewpoint, completeness implies that the whole physical space is taken into consideration.

On the other hand, recall that a maximal hypersurface is (locally) a critical point for a natural variational problem, namely of the volume functional (see, for instance [16]). After the relevant result of the Bernstein–Calabi conjecture [21] for the n-dimensional Lorentz–Minkowski spacetime given by Cheng and Yau [22], classical papers dealing with uniqueness results are [17, 27, 42].

In their work [17], Brill and Flaherty replaced the Lorentz–Minkowski spacetime by a spatially closed universe, and proved uniqueness results for CMC hypersurfaces in the large by assuming $\overline{\mathrm{Ric}}(z, z) > 0$ for every timelike vector z. This assumption may be interpreted as the fact that there is real present matter at every point of the spacetime. It is known as the Ubiquitous Energy Condition.

This energy condition was relaxed by Marsden and Tipler [42] to include, for instance, non-flat vacuum spacetimes.

More recently, Bartnik [12], proved very general existence theorems and consequently, he claimed that it would be useful to find new satisfactory uniqueness results.

Later, Alías, Romero and Sánchez [10], proved new uniqueness results in the class of spacetimes that they called spatially closed Generalized Robertson–Walker (GRW) spacetimes under TCC. Generalized Robertson–Walker spacetimes extend classical Robertson–Walker ones to include the cases in which the fiber has not constant sectional curvature, i.e., they are given as a Lorentzian warped product as we have already described in Sect. 3. Although to be spatially homogeneous is reasonable as a first approximation of the large scale structure of the universe, this assumption could not be appropriate when we consider a more accurate scale. On the other hand, small deformations of the metric on the fiber of classical Robertson–Walker spacetimes fit into the class of generalized Robertson–Walker spacetimes.

Recall that a spacetime is said spatially closed if there exists a compact spacelike hypersurface in the spacetime. In this work, the authors show that a GRW spacetime is spatially closed if and only its fiber is compact.

Alías, Romero and Sánchez [10], introduce a new technique based on Minkowski-type integral formulas, applying the divergence theorem to the tangent part of the conformal vector field ξ (see, formula 17) on the spacelike hypersurface, as well as, on its image for the shape operator. So, the authors can show that in a spatially closed GRW spacetime obeying the TCC, every compact spacelike hypersurface of constant mean curvature is totally umbilical. In the case of a GRW spacetime $(I \times_f F, \langle\, ,\rangle)$, this energy condition is equivalent to following inequalities,

$$f'' \leq 0 \tag{27}$$

and

$$\mathrm{Ric} \geq (n-1)(ff'' - f'^2)\langle\, ,\rangle, \tag{28}$$

where Ric denote the Ricci tensor of the n-dimensional fiber (F, g). Taking this equivalence into account, the authors show that in a GRW spacetime satisfying TCC with strict inequality in (28) the only compact spacelike hypersurfaces of constant mean curvature are level spacelike hypersurfaces of the timelike function $t := \pi_I$ (spacelike slices).

In [5], Alías and Montiel using a well-known generalized maximum principle improve the last result aforementioned and prove that in a GRW spacetime whose warping function satisfies the convexity condition $(\log f)'' \leq 0$, the spacelike slices are the only constant mean curvature compact spacelike hypersurfaces.

In 2011, this result was generalized by Caballero, Romero, and Rubio [20] for a larger class of spatially closed spacetimes, those who have a gradient conformal timelike vector field. In addition to this, the global structure of this class of spacetimes is analyzed and the relation with its well-known subfamily of generalized Robertson–Walker spacetimes is exposed in detail.

Up to this point, except the Cheng–Yau theorem, all the uniqueness results aforementioned are shown in spatially closed spacetimes. In spite of the historical importance of spatially closed GRW spacetimes, a number of observational and theoretical arguments about the total mass balance of the universe [25] suggest the convenience of taking into consideration open cosmological models. Even more, a spatially closed GRW spacetime violates the holographic principle [15] whereas a GRW spacetime with non-compact fiber could be a suitable model that follows that principle [11]. More precisely, the entropy contained in any spacelike region cannot exceed the area of the regions boundary. That is, if Ω is a compact region of a spacelike hypersurface, and $S(\Omega)$ denotes the entropy of all matter systems in Ω, then

$$S(\Omega) \leq \frac{\text{Area}(\partial\Omega)}{4}.$$

The previous inequality cannot be satisfied by some physical spatially closed spacetimes. For instance, let us consider that a spacetime contains a compact spacelike hypersurface such that it contains a matter system that does not occupy the whole of it. Then, consider a sequence of compact sets contained in the region where no matter system exists, having a point as limit. Using the previous inequality in the outside part of this sequence, we have that the entropy of the whole matter system becomes arbitrarily small, and then it has to be zero. We found a contradiction.

Recently, Romero, Rubio, and Salamanca [55] introduce a new class of spatially open GRW spacetimes, which is called spatially parabolic GRW spacetimes. This new notion of spatially parabolic GRW spacetime is a natural counterpart of the spatially closed GRW spacetime. So, a GRW spacetime is spatially parabolic if its fiber is a parabolic Riemannian manifold.

Recall that a complete (non-compact) Riemannian manifold is said to be parabolic if the only positive superharmonic functions are the constants.

On the other hand, the parabolicity of the fiber of a GRW spacetime could also be supported by some physical reasons. For instance, galaxies can be understood as molecules (see, for instance, [45, Chap. 12]), if a sonde is sent to the space, its motion may be approached by a Brownian motion, [33]. In fact, the distribution of galaxies and their velocities are not completely known. Parabolicity may favor that the sonde could be observed in any region, since the Brownian motion is recurrent in any parabolic Riemannian manifold [33].

The authors show that under reasonable assumptions on the restriction of the warping function to the spacelike hypersurface and on the boundedness of the hyperbolic angle between the unit normal vector field and the timelike coordinate vector field ∂_t, a complete spacelike hypersurface in a spatially parabolic GRW spacetime is shown to be parabolic, and the existence of a simply connected parabolic spacelike hypersurface in a GRW spacetime also leads to the parabolicity of its fiber. Note that the assumption on the hyperbolic angle of the maximal hypersurface has a physical consequence, this is, relative speed between normal and comoving observers do not approach the speed of light at every point of the hypersurface (see, [60, pp. 45–67]). Then, all the complete maximal hypersurfaces in spatially parabolic GRW spacetimes are determined in several cases, extending, in particular, to this family of open cosmological models several well-known uniqueness results for the case of spatially closed GRW spacetimes (see also [56]).

For arbitrary dimension, parabolicity has no clear relationship with sectional curvature. Indeed, the Euclidean space $\mathbb{R}^n$ is parabolic if and only if $n \leq 2$. Moreover, there exist parabolic Riemannian manifolds whose sectional curvature is not bounded from below.

The family of spatially parabolic GRW spacetimes is very large, although some other interesting GRW spacetimes do not belong to this family. For instance, those Robertson–Walker spacetimes whose fiber is the hyperbolic space $\mathbb{H}^n$ are excluded.

Making use of two maximum principles: the strong Liouville property and the Omori–Yau generalized maximum principle, Romero, Rubio, and Salamanca [57] obtain new uniqueness results in other relevant spatially open GRW spacetimes for complete maximal hypersurfaces which are between two spacelike slices (time bounded) and/or have a bounded hyperbolic angle. In contrast to parabolicity, some curvature assumptions should be imposed here.

On the other hand, in the case of the Einstein–de Sitter spacetime, which is a spatially open model, which shows a reasonable fit to recent observations [64], a new uniqueness result for complete maximal (and constant mean curvature spacelike) hypersurfaces is given [58]. The result is obtained applying to the sine of the hyperbolic angle of the hypersurface, a Liouville-type theorem (see, [43, 62]), which is a consequence of the Omori–Yau generalized maximum principle.

Finally, focusing on the problems of uniqueness and nonexistence of complete maximal hypersurfaces immersed in a spatially open Robertson–Walker spacetime with flat fiber, Pelegrín, Romero and Rubio [49] give new nonexistence and uniqueness results on complete maximal hypersurfaces. Note that these models have aroused a great deal of interest, since recent observations have shown that the current universe is very close to a spatially flat geometry [28].

It is important to say that the authors do not need the hyperbolic angle of the hypersurface to be bounded, which was an assumption used in some previous works studying the spatially open case. Thus, they are able to deal with spacelike hypersurfaces approaching the null boundary at infinity, such as hyperboloids in Minkowski spacetime.

Acknowledgements The author would like to thank the referee for his deep reading and valuable suggestions.

References

1. A.L. Albujer, New examples of entire maximal graphs in $\mathbb{H}^2 \times \mathbb{R}_1$. Differ. Geom. Appl. **26**, 456–462 (2008)
2. A.L. Albujer, L.J. Alías, Calabi-Bernstein results for maximal surfaces in Lorentzian product spaces. J. Geom. Phys. **59**, 620–631 (2009)
3. A.L. Albujer, L.J. Alías, Parabolicity of maximal surfaces in Lorentzian product spaces. Math. Z. **267**, 453–464 (2011)
4. J.A. Aledo, A. Romero, R.M. Rubio, The classical Calabi-Bernstein theorem revisited. J. Math. Anal. Appl. **431**, 1172–1177 (2015)
5. L.J. Alías, S. Montiel, Uniqueness of spacelike hypersurfaces with constant mean curvature in generalized Robertson-Walker spacetimes. *Differential geometry, Valencia*, vol. 5969 (Publ., River Edge, NJ, World Sci, 2001), p. 2002
6. L.J. Alías, P. Mira, On the Calabi-Bernstein theorem for maximal hypersurfaces in the Lorentz-Minkowski space, in *Proceedings of the Meeting Lorentzian Geometry-Benalmádena*, Spain, vol. 5(2003) (Pub. RSME, 2001), pp. 23–55
7. L.J. Alías, B. Palmer, On the Gaussian curvature of maximal surfaces and the Calabi-Bernstein theorem. Bull. Lond. Math. Soc. **33**(4), 454–458 (2001)
8. L.J. Alías, B. Palmer, Zero mean curvature surfaces with non-negative curvature in flat Lorentzian 4-spaces. Proc. R. Soc. Lond. A **455**, 631–636 (1999)
9. L.J. Alías, B. Palmer, A duality result between the minimal surface equation and the maximal surface equation. An. Acad. Bras. Cienc. **73**, 161–164 (2001)
10. L.J. Alías, A. Romero, M. Sánchez, Uniqueness of complete spacelike hypersurfaces of constant mean curvature in Generalized Robertson-Walker spacetimes. Gen. Relat. Gravit. **27**, 71–84 (1995)
11. D. Bak, S.-J. Rey, Cosmic holography. Class. Quantum Grav. **17**, L83–L89 (2000)
12. R. Bartnik, Existence of maximal surfaces in asymptotically flat spacetimes. Commun. Math. Phys. **94**, 155–175 (1984)
13. S. Bernstein, Sur une théorème de géometrie et ses applications aux équations dérivées partielles du type elliptique. Commun. Soc. Math. Kharkov **15**, 38–45 (1914)
14. A.L. Besse, *Einstein Manifolds* (Springer, 1987)
15. R. Bousso, The holographic principle. Rev. Mod. Phys. **74**, 825–874 (2002)
16. A. Brasil, A.G. Colares, On constant mean curvature spacelike hypersurfaces in Lorentz manifolds. Mat. Contemp. **17**, 99–136 (1999)
17. D. Brill, F. Flaherty, Isolated maximal surfaces in spacetime. Commun. Math. Phys. **50**, 157–165 (1984)
18. M. Caballero, A. Romero, R.M. Rubio, Uniqueness of maximal surfaces in Generalized Robertson-Walker spacetimes and Calabi-Bernstein type problems. J. Geom. Phys. **60**, 394–402 (2010)
19. M. Caballero, A. Romero and R.M. Rubio, New Calabi-Bernstein results for some nonlinear equations. Anal. Appl.**11**(1), 1350002, 13 (2013)

20. M. Caballero, A. Romero, R.M. Rubio, Constant mean curvature spacelike hypersurfaces in Lorentzian manifolds with a timelike gradient conformal vector field. Class. Quantum Grav. **28**, 145009–145022 (2011)
21. E. Calabi, Examples of Bernstein problems for some nonlinear equations. Proc. Symp. Pure Math. **15**, 223–230 (1970)
22. S.Y. Cheng, S.T. Yau, Maximal space-like hypersurfaces in the Lorentz-Minkowski spaces. Ann. Math. **104**, 407–419 (1976)
23. Eigenvalues in Riemannian Geometry, *Pure and Applied Mathematics*, vol. 115 (Academic Press, New York, 1984)
24. S.S. Chern, Simple proofs of two theorems on minimal surfaces. Enseign. Math. **15**, 53–61 (1969)
25. H.Y. Chiu, A cosmological model of universe. Ann. Phys. **43**, 1–41 (1967)
26. Y. Choquet-Bruhat, Quelques proprits des sousvarits maximales d'une varit lorentzienne. Comptes Rend. Acad. Sci. (paris) Serie A **281**, 577–580 (1975)
27. Y. Choquet-Bruhat, Quelques propriétés des sousvariétés maximales d'une variété lorentzienne. C R Acad. Sci. Paris Ser. A B **281**, 577–580 (1975)
28. E.J. Copeland, M. Sami, S. Tsujikawa, Dynamics of dark energy. Int. J. Mod. Phys. D **15**, 1753–1935 (2006)
29. F.J. Estudillo, A. Romero, On the Gauss curvature of maximal surfaces in the 3-dimensional Lorentz-Minkowski space. Comment. Math. Helv. **69**, 1–4 (1994)
30. W.H. Fleming, On the oriented Plateau problem. Rend. Circ. Mat. Palermo **11**, 69–90 (1962)
31. I. Fernndez, P. Mira, Complete maximal surfaces in static Robertson-Walker 3-spaces. Gen. Relat. Gravit. **39**, 2073–2077 (2007)
32. E. de Giorgi, Una estensione del teorema di Bernstein. Ann. Scuola Norm. Sup. Pisa **19**, 79–85 (1965)
33. A. Grigor'yan, Analytic and geometric background of recurrence and non-explosion of the Brownian motion on Riemannian manifolds. Bull. Am. Math. Soc. **36**, 135–249 (1999)
34. D. Hoffman, H. Karcher, Complete embedded minimal surfaces of finite total curvature, in *Geometry V, Encyclopaedia of Mathematical Sciences*, vol. 90 (Springer, 1997), pp. 5–93
35. E. Hopf, On S. Bernstein's theorem on surfaces $z(x, y)$ of nonpositive curvature. Proc. Am. Math. Soc. **1**, 80–85 (1950)
36. J.L. Jaramillo, J.A.V. Kroon, E. Gourgoulhon, From geometry to numerics: interdisciplinary aspects in mathematical and numerical relativity. Class. Quantum Grav. **25**, 093001 (2008)
37. J.L. Kazdan, Parabolicity and the Liouville property on complete Riemannian manifolds, in *Seminar on New Results in Nonlinear Partial Differential Equations (Bonn, 1984)*, ed. By A.J. Tromba, H. Friedr. Aspects of Mathematics E10, (Vieweg and Sohn, Bonn, 1987), pp. 153–166
38. O. Kobayashi, Maximal surfaces in the 3-dimensional Minkowski space $\mathbb{L}^3$. Tokyo J. Math. **6**(2), 297–309 (1983)
39. J.M. Latorre, A. Romero, New examples of Calabi-Bernstein problem for some nonlinear equations. Diff. Geom. Appl. **15**, 153–163 (2001)
40. P. Li, Curvature and function theory in Riemannian manifolds, *Surveys in Differential Geometry*, vol. VII (International Press, 2000) pp. 375–432. Surveys in Differential Geometry: Papers dedicated to Atiyah, Bott, Hirzebruch, and Singer, Vol. VII, (International Press, 2000), pp. 375–432
41. A. Lichnerowicz, L' integration des équations de la gravitation relativiste et le problème des n corps. J. Math. Pures et Appl. **23**, 37–63 (1944)
42. J.E. Marsden, F.J. Tipler, Maximal hypersurfaces and foliations of constant mean curvature in General Relativity. Phys. Rep. **66**, 109–139 (1980)
43. S. Nishikawa, On maximal spacelike hypersurfaces in a Lorentzian manifold. Nagoya Math. J. **95**, 117–124 (1984)
44. H. Omori, Isometric immersions of Riemannian manifolds. J. Math. Soc. Jpn. **19**, 205–214 (1967)
45. B. O'Neill, *Semi-Riemannian Geometry with Applications to Relativity* (Academic Press, 1983)
46. R. Osserman, *A Survey of Minimal Surfaces* (Van Nostrand, New York, IOS9)

47. R. Osserman, On the Gauss curvature of minimal surfaces. Trans. Am. Math. Soc. **12**, 115–128 (1960)
48. J.A. Pelegrín, A. Romero, R.M. Rubio, On maximal hypersurfaces in Lorentz manifolds admitting a parallel lightlike vector field. Class. Quantum Grav. **33**, 055003 (2016)
49. J.A. Pelegrín, A. Romero, R.M. Rubio, Uniqueness of complete maximal hypersurfaces in spatially open $(n+1)$-dimensional Robertson-Walker spacetimes with flat fiber. Gen. Relat. Gravit.**48** (2016)
50. J. Pyo, M. Rodríguez, Simply connected minimal surfaces with finite total curvature in $\mathbb{H}^2 \times \mathbb{R}$. Int. Math. Res. Not. IMRN (2013). https://doi.org/10.1093/imrn/rnt017
51. H. Ringström, *The Cauchy Problem in General Relativity*. ESI Lectures in Mathematics and Physics, 2009
52. A. Romero, Simple proof of Calabi-Bernstein's theorem on maximal surfaces. Proc. Am. Math. Soc. **124**(4), 1315–1317 (1996)
53. A. Romero, R.M. Rubio, New proof of the Calabi-Bernstein theorem. Geom. Dedicata **147**, 173–176 (2010)
54. A. Romero, R.M. Rubio, On maximal surfaces in certain Non-Flat 3-dimensional Robertson-Walker spacetimes. Math. Phys. Anal. Geom. **15**, 193–202 (2012)
55. A. Romero, R.M. Rubio, J.J. Salamanca, Uniqueness of complete maximal hypersurfaces in spatially parabolic Generalized Robertson-Walker spacetimes. Class. Quantum Grav.**30**(2013), 115007 (13pp) (2013)
56. A. Romero, R.M. Rubio, J. Juan, Salamanca, A new approach for uniqueness of complete maximal hypersurfaces in spatially parabolic GRW spacetimes. J. Math. Anal. Appl. **419**, 355–372 (2014)
57. A. Romero, R.M. Rubio, J.J. Salamanca, complete maximal hypersurfaces in certain spatially open generalized Robertson-Walker spacetimes. Rev. R. Acad. Cienc. Exactas Fís. Nat. Ser. A Math. RACSAM 1–10 (2014)
58. R.M. Rubio, Complete constant mean curvature spacelike hypersurfaces in the Einstein-de Sitter spacetime. Rep. Math. Phys. **74**, 127–133 (2014)
59. R.M. Rubio, J.J. Salamanca, Maximal surface equation on a Riemannan 2-manifold with finite total curvature. J. Geom. Phys. **92**, 140–146 (2015)
60. R.K. Sachs, H. Wu, *General Relativity for Mathematicians*. Graduate Texts in Mathematics (Springer, New York, 1977)
61. J. Simons, Minimal varieties in Riemannian manifolds. Ann. of Math. **88**, 62–105 (1968)
62. Y.J. Suh, Generalized maximum principles and their applications to submanifolds and S. S. Chern's conjectures, in *Proceedings of the Eleventh International Workshop on Differential Geometry* (Kyungpook National University, Taegu, 2007), pp. 135–152
63. R.G. Vishwakarma, Einstein-de Sitter model re-examined for newly discovered Type Ia supernovae. Mon. Not. Astron. Soc. **361**, 1382–1386 (2005)
64. S.T. Yau, Harmonic functions on complete Riemannian manifolds. Commun. Pure Appl. Math. **28**, 201–228 (1975)

Null Hypersurfaces on Lorentzian Manifolds and Rigging Techniques

Benjamín Olea

Abstract We introduce the concept of rigging for a null hypersurface in a Lorentzian manifold, which allows us to induce all the necessary geometric objects in a null hypersurface and also to define a Riemannian metric on it, called rigged metric. This metric can be used as an auxiliary tool to study the null hypersurface. Its Levi-Civita connection is called rigged connection and, in general, it will not coincide with the induced connection on the null hypersurface. We show a necessary and sufficient condition for this to happen and we give some examples. Since both rigged connection as induced connection depend on the rigging, we investigate if they can coincide for a suitable choice of the rigging.

Keywords Null hypersurfaces · Rigging · Rigged vector field · Rigged metric Induced connection · Rigged connection

1 Introduction

An hypersurface L in a Lorentzian manifold (M, g) is null if the metric tensor g is degenerate on it. In this case, there exists an unique tangent null direction and all other directions are spacelike and orthogonal to the null direction. In other words, a null hypersurface contains its orthogonal direction. This is why null hypersurfaces can not be studied as spacelike or timelike hypersurfaces, since we can not decompose the ambient tangent space as the direct sum of the tangent space to the null hypersurface and its orthogonal direction.

To solve this problem, it is chosen a (locally defined) null section $\xi \in \mathfrak{X}(L)$ and a screen distribution $\mathcal{S}$, which is just a complementary distribution to the null direction on TL. Once we have made these two arbitrary choices, it is induced a null transverse vector field N (locally defined) on L, which is orthogonal to the screen distribution $\mathcal{S}$ and it holds $g(N, \xi) = 1$. This vector field N allows us to decompose the ambient

B. Olea (✉)
Departamento de Matemática Aplicada, Universidad de Málaga, Málaga, Spain
e-mail: benji@uma.es

M.A. Cañadas-Pinedo et al. (eds.), *Lorentzian Geometry and Related Topics*,
Springer Proceedings in Mathematics & Statistics 211,
DOI 10.1007/978-3-319-66290-9_13

tangent space in an analogous way as for spacelike or timelike hypersurfaces and it plays the role of normal vector field to L. Indeed, since $T_x M = T_x L \oplus span(N_x)$ for all $x \in L$, then we can decompose any vector as a tangent part to L and another part in the direction of N. If we take $U, V \in \mathfrak{X}(L)$, we decompose

$$\nabla_U V = \nabla^L_U V + B(U, V)N,$$

where $\nabla^L_U V \in \mathfrak{X}(L)$ is the induced connection on L and B is a symmetric tensor called null second fundamental form of L.

The induced connection ∇^L depends on the chosen screen distribution and the null section. It is a symmetric connection and the choosen null section ξ is pre-geodesic for it, but, in general, it is not compatible with the metric g and it does not seem an adequate connection to study the null hypersurface. On the other hand, the null second fundamental form only depends on the null section, since $B(U, V) = -g(\nabla_U \xi, V)$. In the same manner, the null mean curvature, which is the trace of the null second fundamental form, only depends on the null section.

Despite the arbitrary election of the screen distribution and the null section, there are concepts that do not depend on them, such as being a totally geodesic or umbilic null hypersurface, having zero mean curvature or being a parallel null hypersurface. Recall that L is called totally umbilic if $B = \rho g$ for certain function $\rho \in C^\infty(L)$ and it is parallel if $(\nabla_U B)(V, W) = -\tau(U)B(V, W)$ for all $U, V, W \in \mathfrak{X}(L)$ and certain one-form τ, [9].

Instead of choosing a null section and a screen distribution independently, we can make only one arbitrary choice: a transverse vector field defined on L, which we call a rigging for L. It will induce a null section (called rigged vector field), a screen distribution and a null transverse vector field, which are all the necessary geometric data to study a null hypersurface. Now, all these objects are coupled and they are related to each other. Moreover, we can choose a suitable rigging to exploit the symmetries of the ambient space. For example, if we choose a closed rigging, then the corresponding induced screen distribution is integrable and if we choose a conformal rigging, then the corresponding rigged vector field is geodesic.

A rigging also induces a Riemannian metric $\widetilde{g}$ (the rigged metric) on the null hypersurface L, which coincides with g on the screen distribution and declares unitary the rigged vector field. The Levi-Civita connection of the rigged metric provides us another connection $\widetilde{\nabla}$ on L (called rigged connection), which does not coincide with the induced connection ∇^L in general. Both the rigged metric as the rigged connection depend on the chosen rigging, but they can be used as an auxiliary tool to study a null hypersurface. For example, using them we can prove two result concerning null hypersurfaces. The first one gives us a curvature condition for a compact totally umbilic null hypersurface to be totally geodesic.

Theorem 1 ([7]) *Let M be an orientable Lorentzian manifold with dimension $n > 2$ which holds the reverse null convergence condition. If there exists a timelike conformal vector field on M, then any compact totally umbilic null hypersurface is totally geodesic.*

The second one relates the multiplicity of the first conjugate point along a null geodesic with a geometric property of the nullcone containing it.

Theorem 2 ([7]) *Let M be a Lorentzian manifold and $\gamma : [0, a] \to M$ a null geodesic such that $\gamma(a)$ is the first conjugate point to $\gamma(0)$ along γ. If the null cone with vertex at $\gamma(0)$ containing γ is totally umbilic, then $\gamma(a)$ has maximum multiplicity.*

Although the rigged connection is not a natural nor canonical object in a null hypersurface, in some situations it seems a more adequate connection than the induced connection. For example, if L is a totally umbilic null hypersurface and the screen distribution is integrable, then the leaves of the screen distribution are not totally umbilical co-dimension two submanifolds of the ambient space in general, but they are totally umbilical hypersurfaces of the Riemannian manifold $(L, \widetilde{g})$.

2 Rigging Vector Fields

In this section, we introduce the notion of rigging vector field for a null hypersurface L and its associated geometric objects, as the rigged vector field, the rigged connection and the rigged metric on L.

Definition 3 A rigging for L is a vector field ζ defined on an open set containing L such that $\zeta_p \notin T_pL$ for each $p \in L$.

Call $i : L \to M$ the canonical inclusion, α the metrically equivalent one-form to ζ and $\omega = i^*(\alpha)$. The tensor $g + \alpha \otimes \alpha$ may not be a Riemannian metric on the domain of definition of ζ, but its pullback by i,

$$\widetilde{g} = i^*(g) + \omega \otimes \omega,$$

defines a Riemannian metric on L, which will be called rigged metric induced from ζ. The Levi-Civita connection $\widetilde{\nabla}$ of $\widetilde{g}$ will be called rigged connection on L.

Definition 4 The rigged vector field induced from the rigging ζ is the $\widetilde{g}$-metrically equivalent vector field to the one-form ω and it is denoted by ξ.

The rigged vector field is a null vector field defined on L which is $\widetilde{g}$-unitary, pre-geodesic for the ambient connection ∇ and it holds $g(\zeta, \xi) = 1$. We consider the screen distribution given by $\mathcal{S} = TL \cap \zeta^{\perp}$ and the null transverse vector field

$$N = \zeta - \frac{1}{2} g(\zeta, \zeta)\xi.$$

Observe that $g(N, \xi) = 1$ and $\mathcal{S}$ is g-orthogonal to N. Moreover, given $U \in \mathfrak{X}(L)$, we have

$$\omega(U) = g(\zeta, U) = g(N, U).$$

Now, given $U, V \in \mathfrak{X}(L)$ and $X \in \mathcal{S}$ we decompose

$$\nabla_U V = \nabla^L_U V + B(U, V)N, \tag{1}$$
$$\nabla_U N = \tau(U)N - A(U), \tag{2}$$
$$\nabla_U \xi = -\tau(U)\xi - A^*(U), \tag{3}$$
$$\nabla^L_U X = \nabla^*_U V + C(U, X)\xi, \tag{4}$$

where $\nabla^L_U V \in \mathfrak{X}(L), \nabla^*_U X, A^*(U), A(U) \in \mathcal{S}$ and τ is certain one-form. The induced connection ∇^L is symmetric and it holds

$$\left(\nabla^L_U g\right)(V, W) = B(U, V)\omega(W) + B(U, W)\omega(V)$$

for all $U, V \in \mathfrak{X}(L)$. The tensors B and C hold

$$B(U, V) = g(A^*(U), V) = -g(\nabla_U \xi, V),$$
$$C(U, X) = g(A(U), X) = -g(\nabla_U N, X),$$
$$B(\xi, V) = 0,$$
$$\omega([X, Y]) = C(X, Y) - C(Y, X)$$

for all $U \in \mathfrak{X}(L)$ and $X, Y \in \mathcal{S}$. The tensor B is the null second fundamental form of L. It is a symmetric tensor and its trace is the null mean curvature, $H_p = \sum_{i=1}^{n-2} B(e_i, e_i)$, being $\{e_1, \ldots, e_{n-2}\}$ an orthonormal basis of $\mathcal{S}_p$. The null hypersurface is totally geodesic if $B = 0$ and totally umbilic if $B = \rho g$ for some function $\rho \in C^\infty(L)$. Both definitions are independent on any election.

It is easy to show that the tensor C is symmetric if and only if $\mathcal{S}$ is integrable. In this case, the second fundamental form of the leaves of $\mathcal{S}$ as co-dimension two submanifolds of (M, g) is given by

$$\mathbb{I}(X, Y) = C(X, Y)\xi + B(X, Y)N \tag{5}$$

and ∇^* is the induced Levi-Civita connection from the ambient space. If the leaves of the screen distribution are totally umbilic (geodesic) submanifolds of (M, g), then L is also totally umbilic (geodesic), but the converse does not hold, in general, as it can be easily checked from formula 5. However, we can ensure the converse if we choose an adequate rigging (see comments below Proposition 5).

The following proposition relates above geometric objects.

Proposition 5 *Given $U, V, W \in \mathfrak{X}(L)$ and $X, Y \in \mathcal{S}$ it holds*

1. $d\alpha(U, X) + \left(L_\zeta g\right)(U, X) + g(\zeta, \zeta)B(U, X) + 2C(U, X) = 0.$
2. $\left(L_\xi \widetilde{g}\right)(X, Y) = -2B(X, Y)$. *In particular, $H = -\widetilde{div}\,\xi$, where $\widetilde{div}$ is the divergence respect to $\widetilde{g}$.*
3. $\tau(U) = -g(\nabla_U \zeta, \xi)$.
4. $\widetilde{\nabla}_X Y = \nabla^*_X Y - \widetilde{g}(\widetilde{\nabla}_X \xi, Y)\xi$.

If we choose a conformal rigging, then $\tau(\xi) = 0$, i.e., the rigged vector field ξ is geodesic for the ambient connection ∇ and the induced connection ∇^L. On the other hand, from point 1, if there exists a closed and conformal rigging for L, then $-2C = g(\zeta, \zeta)B + 2\lambda g$, where $\lambda \in C^\infty(L)$ is a function coming from the conformality condition. So, the screen distribution is integrable and its leaves are totally umbilic (geodesic) co-dimension two submanifold of (M, g) if and only if L is totally umbilic (geodesic) null hypersurface of (M, g).

Call $D = \nabla - \widetilde{\nabla}$ the difference of the Levi-Civita connections of the ambient space (M, g) and the rigged connection. It is a symmetric tensor defined on $T_pL \times T_pL$, which gives vectors in T_pM and it holds the following.

Proposition 6 *Given $U, V, W \in \mathfrak{X}(L)$ and $X, Y \in \mathcal{S}$ it holds*

1. $g(D(U, V), W) = -\frac{1}{2}\left(\omega(W)\left(L_\xi \widetilde{g}\right)(U, V) + \omega(U)d\omega(V, W) + \omega(V)\right.$ $\left.d\omega(U, W)\right)$.
2. $\tau(U) = -\widetilde{g}(D(U, \xi), \xi)$.
3. $D(X, Y) = \left(C(X, Y) + g(\widetilde{\nabla}_X \xi, Y)\right)\xi + B(X, Y)N$.

If we choose a closed rigging, then $d\omega = 0$ and from point 1 of Proposition 5 it follows that C is a symmetric tensor and so the induced screen distribution is integrable. Note that the converse does not hold in general, but if we suppose that the screen distribution is integrable and the rigged vector field is geodesic for the rigged connection, then it holds $d\omega = 0$.

Proposition 7 *Let ζ be a closed rigging for a null hypersurface L and take $U, V, W \in \mathfrak{X}(L)$ and $X, Y \in \mathcal{S}$. Then*

1. *The rigged vector field ξ is geodesic for the rigged connection.*
2. $\left(L_\xi \widetilde{g}\right)(U, V) = -2B(U, V)$.
3. $A^*(U) = -\widetilde{\nabla}_U \xi$.
4. $\widetilde{\nabla}_X Y = \nabla^*_X Y + B(X, Y)\xi$.
5. $\widetilde{g}(\widetilde{\nabla}_U V, W) = g(\nabla_U V, W) + \omega(W)U(\omega(V))$.
6. *The second fundamental form of the leaves of $\mathcal{S}$ in $(L, \widetilde{g})$ is $\widetilde{\mathrm{II}}(X, Y) = B(X, Y)\xi$.*

Some remarks should be made to this proposition. First, observe that locally it always exists a closed rigging for any null hypersurface. The rigged vector field is always pre-geodesic for the ambient connection ∇ and the induced connection ∇^L, but not for the rigged connection, unless the rigging is chosen closed. Finally, point 2 of above proposition holds for all $X, Y \in \mathcal{S}$ even if the rigging is not closed (Proposition 5), but if the rigging is closed it holds for all $U, V \in \mathfrak{X}(L)$.

An immediate consequence of the above proposition is the following.

Proposition 8 *Let L be a null hypersurface and ζ a closed rigging for it.*

1. *L is a totally geodesic null hypersurface if and only if the rigged vector field ξ is parallel for the rigged connection.*
2. *L is a totally geodesic null hypersurface (resp. umbilic) if and only if each leaf of $\mathcal{S}$ is a totally geodesic (resp. umbilic) hypersurface of the Riemannian manifold $(L, \widetilde{g})$.*

Remark that, as it was said before Proposition 5, the leaves of the screen distribution do not need to be totally geodesic or umbilic co-dimension two submanifold of (M, g) even if L is a totally geodesic or umbilic null hypersurface.

In constant curvature, we can describe all the totally geodesic null hypersurfaces. In the Minkowski space $\mathbb{R}^n_1$ they are degenerate hyperplanes. In $\mathbb{S}^n_1 \subset \mathbb{R}^{n+1}_1$ and $\mathbb{H}^n_1 \subset \mathbb{R}^{n+1}_2$, totally geodesic null hypersurfaces are obtained intersecting it with degenerate planes of the ambient space. Important examples of null hypersurfaces in any Lorentzian manifold are local nullcones. In constant curvature, they are the unique totally umbilic null hypersurfaces, [1, 8]. In Robertson-Walker spaces, local nullcones are totally umbilic null hypersurfaces and, under suitable conditions, they are also the unique ones. For example, above holds in $I \times_f \mathbb{S}^n$ with $\int_I \frac{1}{f} > \pi$ and, in particular, in the closed Friedmann Cosmological model, [8].

We can give the local structure of the rigged metric in a totally umbilic null hypersurface if we choose a closed rigging.

Theorem 9 *Let (M, g) be a Lorentzian manifold, L a totally umbilic null hypersurface and ζ a closed rigging for it. For each point $p \in L$, there exists a neighborhood $\Theta \subset L$ of p such that $(\Theta, \widetilde{g})$ is isometric to a twisted product $(I \times S, dr^2 + \lambda^2(r, x) g|_S)$, where the rigged vector field ξ is identified with ∂_r, S is the leaf through p of the screen distribution and*

$$\lambda(r, x) = \exp\left(-\int_0^r \frac{H(\phi_s(x))}{n-2} ds\right)$$

being ϕ the flow of ξ and H the mean curvature of L. Moreover, if L is simply connected and ξ is complete, then above decomposition is global.

In particular, if L is a totally geodesic null hypersurface, then $(L, \widetilde{g})$ is locally isometric to a direct product with one-dimensional base. The existence of a closed rigging is not a strong hypothesis, since it always exists at least locally and the decomposition given in the theorem is also local.

Example 10 Consider the Minkowski space $\mathbb{L}^{n+1}_1 = \left(\mathbb{R}^{n+1}, -dx_0^2 + \cdots + dx_n^2\right)$, the future nullcone with vertex at the origin $C_0^+ = \{(x_0, \ldots, x_n) : -x_0^2 + \cdots + x_n^2 = 0, x_0 > 0\}$ and the rigging $\zeta = -\partial x_0$. The rigged vector field is $\xi = \frac{1}{x_0} P$, where P is the position vector field, and the second fundamental form is $B = -\frac{1}{x_0} g$. If we take $p = (1, 1, \ldots, 0) \in C_0^+$, then the leaf through p of the screen distribution is a $(n-1)$-dimensional euclidean sphere of radius 1 and the integral curve of ξ with initial condition p is $\gamma(t) = (t+1)p$. Applying above theorem, the Riemannian manifold $\left(C_0^+, \widetilde{g}\right)$ is isometric to the warped product given by

$$\left((-1, \infty) \times \mathbb{S}^{n-1}, dr^2 + (1+r)^2 g_{\mathbb{S}^{n+2}}\right),$$

which coincides with the usual metric on the nullcone induced from the euclidean space $\mathbb{R}^{n+1}$.

Example 11 Consider the pseudosphere $\mathbb{S}_1^n = \{x \in \mathbb{L}^{n+1} : -x_0^2 + \cdots + x_n^2 = 1\}$ and decompose $-\partial_{x_0} = \zeta_x + x_0 P_x$ for each point $x \in \mathbb{S}_1^n$, where $\zeta \perp P$. The vector field $\zeta \in \mathfrak{X}(\mathbb{S}_1^n)$ is a timelike, closed and conformal vector field on $\mathbb{S}_1^n$. We suppose it is past-directed.

The future nullcone of $\mathbb{S}_1^n$ with vertex at $p = (0, \ldots, 1) \in \mathbb{S}_1^n$ is given by $\mathcal{C}_p^+ = \mathbb{S}_1^n \cap C_p^+$, where C_p^+ is the future nullcone of $\mathbb{L}^{n+1}$ with vertex at p. Therefore, $\mathcal{C}_p^+$ is a hypersurface of C_p^+ that can be obtained intersecting C_p^+ and the hyperplane $x_n = 0$. If we consider the rigging ζ, the rigged vector field is $\frac{1}{x_0}P$ and so, the rigged metric on $\mathcal{C}_p^+$ coincide with the induced metric from the euclidean cone $(C_p^+, \widetilde{g})$. Thus, $\mathcal{C}_p^+$ is also a $(n-1)$-dimensional euclidean cone.

Example 12 Let m be a positive constant and consider $Q = \{(u, v) \in \mathbb{R}^2 : uv > \frac{-2m}{e}\}$ the Kruskal plane with metric $2F(r(u, v))dudv$, where F and r are certain functions. In the Kruskal spacetime $Q \times_r \mathbb{S}^2$, the hypersurfaces $L_{u_0} = \{(u, v, x) \in Q \times \mathbb{S}^2 : u = u_0\}$ are totally umbilic null hypersurfaces (totally geodesic if $u_0 = 0$). If we consider the closed rigging $\zeta = \frac{1}{F(r)}\partial_u$, then the rigged vector field is ∂_v and $B = -\frac{r_v}{r}g$. From Theorem 9, $\left(L_{u_0}, \widetilde{g}\right)$ is isometric to the warped product $\left(\frac{-2m}{u_0 e}, \infty\right) \times_{\frac{r(u_0, v)}{2m}} \mathbb{S}^2$ if $u_0 \neq 0$ and to the direct product $\mathbb{R} \times \mathbb{S}^2$ is $u_0 = 0$.

The induced curvature tensor is defined as

$$R^L_{UV}W = \nabla^L_U \nabla^L_V W - \nabla^L_V \nabla^L_U W - \nabla^L_{[U,V]}W$$

and it satisfies

$$R^L_{UV}\xi = R_{UV}\xi,$$

where R is the ambient curvature tensor. It also holds the following Gauss-Codazzi equations.

$$\begin{aligned} g(R_{UV}W, X) &= g(R^L_{UV}W, X) + B(U, W)g(A(V), X) \\ &\quad - B(V, W)g(A(U), X), \\ g(R_{UV}W, \xi) &= \left(\nabla^L_U B\right)(V, W) - \left(\nabla^L_V B\right)(U, W) + \tau(U)B(V, W) \\ &\quad - \tau(V)B(U, W), \\ g(R_{UV}W, N) &= g(R^L_{UV}W, N), \end{aligned} \tag{6}$$

where $U, V, W \in \mathfrak{X}(L)$ and $X \in \mathcal{S}$. From these equations it can be deduced the following ones.

$$\begin{aligned} g(R_{UV}X, N) &= \left(\nabla^{*L}_U C\right)(V, X) - \left(\nabla^{*L}_V C\right)(U, X) + \tau(V)C(U, X) \\ &\quad - \tau(U)C(V, X), \\ g(R_{UV}\xi, N) &= C(V, A^*(U)) - C(U, A^*(V)) - d\tau(U, V), \end{aligned}$$

where $\nabla^{*L}_U C$ is defined as

$$\left(\nabla^{*L}_U C\right)(V, X) = U(C(V, X)) - C(\nabla^L_U V, X) - C(V, \nabla^*_U X).$$

Using Eq. 6, we can compute the null sectional curvature respect to ξ of a null plane $\Pi = span(X, \xi)$, where $X \in \mathcal{S}$ is unitary,

$$\mathcal{K}_\xi(\Pi) = \left(\nabla^L_\xi B\right)(X, X) - \left(\nabla^L_X B\right)(\xi, X) + \tau(\xi) B(X, X).$$

In particular, if L is totally geodesic, then we have $\mathcal{K}_\xi(\Pi) = 0$ for any null tangent plane Π to L and if L is totally umbilic, then

$$\mathcal{K}_\xi(\Pi) = \xi(\rho) - \rho^2 + \tau(\xi)\rho,$$

where $B = \rho g$.

Using Gauss-Codazzi equations, we can relate the sectional curvature of the ambient space and the Riemmanian manifold $(L, \widetilde{g})$ for a tangent plane to the screen distribution.

Theorem 13 *Let M be a Lorentzian manifold, L a null hypersurface and ζ a rigging for it. If $\Pi = span(X, Y)$, being $X, Y \in \mathcal{S}$ unitary and orthogonal vectors, then*

$$\begin{aligned} K(\Pi) - \widetilde{K}(\Pi) &= -C(Y, Y)B(X, X) - C(X, X)B(Y, Y) \\ &+ (C(X, Y) + C(Y, X))\, B(X, Y) \\ &+ B(X, X)B(Y, Y) - B(X, Y)^2 + \frac{3}{4} d\omega(X, Y)^2. \end{aligned}$$

In the case of a totally umbilic null hypersurface, the null sectional curvature of a null plane and the sectional curvature in $(L, \widetilde{g})$ are related as follows.

Theorem 14 *Let M be a Lorentzian manifold, L a totally umbilic null hypersurface and ζ a rigging for L. If $\Pi = span(X, \xi)$, where $X \in \mathcal{S}$ is a unitary vector, then*

$$\begin{aligned} \mathcal{K}_\xi(\Pi) - \widetilde{K}(\Pi) &= \tau(\xi)B(X, X) - \widetilde{g}(\widetilde{\nabla}_X \widetilde{\nabla}_\xi \xi, X) + \widetilde{g}(X, \widetilde{\nabla}_\xi \xi)^2 \\ &+ \frac{1}{2}\left(\widetilde{g}(S^2(X), X) - \widetilde{g}(S(X), S(X))\right), \end{aligned}$$

where $S(U) = \widetilde{\nabla}_U \xi$.

If we take a closed rigging, then we can give an explicit relation of the induced curvature tensor and the curvature tensor of the rigged metric $\widetilde{g}$.

Theorem 15 *Let M be a Lorentzian manifold, L a null hypersurface and ζ a closed rigging for it. Take $U, V \in \mathfrak{X}(L)$ and $X \in \mathcal{S}$. Then*

$$\begin{aligned} R^L_{UV}X - \widetilde{R}_{UV}X &= \left(g(R_{UV}X, N) - g(R_{UV}X, \xi)\right)\xi \\ &\quad + C(U,X)A^*(V) - C(V,X)A^*(U) \\ &\quad + B(U,X)\nabla_V\xi - B(V,X)\nabla_U\xi, \\ R^L_{UV}\xi - \widetilde{R}_{UV}\xi &= g(R_{UV}\xi, N)\xi - \tau(U)A^*(V) + \tau(V)A^*(U). \end{aligned}$$

In particular,

1. $\mathcal{K}_\xi(\Pi) = \widetilde{K}(\Pi) + \tau(\xi)\frac{B(X,X)}{g(X,X)}$.
2. $Ric(\xi) = \widetilde{Ric}(\xi) + \tau(\xi)H$.

To end this section, observe that we can also induce a Riemannian metric on a null hypersurface in another way. Given a unitary timelike vector field $E \in \mathfrak{X}(M)$ we define the standard canonical variation as $h = g + 2g(E,\cdot)g(E,\cdot)$ (also called flip metric by some authors). It is a Riemannian metric on the whole M that, in particular, induces a Riemannian metric on any null hypersurface. We consider a null section ξ on L such that $g(E,\xi) = \frac{1}{\sqrt{2}}$ and the screen distribution $\mathcal{S} = TL \cap E^\perp$. The vector field $N = \sqrt{2}E + \xi$ is null and it holds $g(N,\xi) = 1$, so it is the transverse null vector field to L. Moreover, it is h-unitary and h-orthogonal to L, so it is also the unitary normal vector field to L as a hypersurface of the Riemannian manifold (M,h).

Call $\mathbb{I}^h$ the second fundamental form and H^h the mean curvature of L inside (M,h). It holds the following.

Proposition 16 ([10]) *Given $X, Y \in \mathcal{S}$, it holds*

$$\begin{aligned} \mathbb{I}^h(X,Y) &= \left(B(X,Y) - \frac{1}{\sqrt{2}}\left(L_E g\right)(X,Y)\right)N, \\ \mathbb{I}^h(X,\xi) &= -g(\nabla_{E+\sqrt{2}\xi}E, X)N, \\ \mathbb{I}^h(\xi,\xi) &= -\big(2g(\xi,\nabla_E E) + \tau(\xi)\big)N. \end{aligned}$$

In particular, $H^h = H - \sqrt{2}divE + \tau(\xi)$.

As the rigged metric, the standard canonical variation can be also used as an auxiliary tool to prove some results concerning null hypersurfaces. For example, we can give curvature conditions for a compact null hypersurface to be totally geodesic.

Theorem 17 ([10]) *Let (M,g) be a Lorentzian metric furnished with a timelike unitary Killing vector field E. If L is a compact null hypersurface such that:*

1. *The null mean curvature has sign.*
2. *$0 \le Ric(\xi,N) + K(span(\xi,N))$, where ξ is a null section with $g(E,\xi) = \frac{1}{\sqrt{2}}$ and N a null transverse vector field to L (concretely, the symmetric of $-\xi$ respect to E).*

Then L is totally geodesic.

However, the construction of the standard canonical variation h is too rigid to study a null hypersurface since it is a Riemannian metric on the whole ambient space, not only on the null hypersurface. Most important disadvantages of the standard canonical variation are:

- The null pre-geodesics of the null hypersurface L are not pre-geodesics in the Riemannian manifold $(L, i^*(h))$.
- Strong hypotheses are needed to obtain geometric information on L.
- If L is a totally umbilic null hypersurface of (M, g), then it is not a totally umbilic hypersurface in (M, h), even if ζ is parallel.

3 Coincidence of the Induced and the Rigged Connection

Once we have chosen a rigging ζ for a null hypersurface L, we can construct two connections on it: the induced connection ∇^L and the rigged connection $\widetilde{\nabla}$. Both depend on the chosen rigging and they do not have to coincide. We are interested in knowing what conditions on the null hypersurface ensure that these connections are the same. For this, call $D^L = \nabla^L - \widetilde{\nabla}$ which is a symmetric tensor which holds $D - D^L = B \cdot N$.

Following theorem gives necessary and sufficient conditions for the coincidence of the induced and the rigged connection (compare with [3, Theorem 4.1]).

Theorem 18 *Suppose that L is a null hypersurface and take ζ a riggging for it. The rigged and the induced connections coincide if and only if*

$$\begin{aligned} C(U, X) &= B(U, X), \\ \tau(U) &= 0. \end{aligned}$$

for all $U \in TL$ and $X \in \mathcal{S}$. Moreover, in this case $d\omega = 0$ and thus the screen distribution is integrable and the rigged vector field ξ is geodesic for the ambient connection ∇, the induced connection ∇^L and the rigged connection $\widetilde{\nabla}$.

Proof If $D^L = 0$, then from point 2 of Proposition 6 we have $\tau = 0$. In particular, ξ is geodesic for ∇ and $\widetilde{\nabla}$. Now, given $X, Y \in \mathcal{S}$ we have $0 = g(\nabla^L_X \xi - \widetilde{\nabla}_X \xi, Y) = -B(X, Y) - g(\widetilde{\nabla}_X \xi, Y)$, but using point 3 of Proposition 6 we get $B(X, Y) = C(X, Y)$. In particular, $\mathcal{S}$ is symmetric and since ξ is $\widetilde{g}$-geodesic, it holds $d\omega = 0$. Finally,

$$C(\xi, X) = \widetilde{g}(\nabla^L_\xi X, \xi) = \widetilde{g}(\widetilde{\nabla}_\xi X, \xi) = -\widetilde{g}(\widetilde{\nabla}_\xi \xi, X) = 0.$$

Suppose now that $C(U, X) = B(U, X)$ and $\tau(U) = 0$ for all $U \in TL$ and $X \in \mathcal{S}$. From point 2 of Proposition 5 have

$$C(X, X) = B(X, X) = -\widetilde{g}(\widetilde{\nabla}_X \xi, X).$$

Since D^L is a symmetric tensor, from point 3 of Proposition 6 we have $D^L(X, Y) = 0$ for all $X, Y \in \mathcal{S}$. Moreover, using point 1 of Proposition 5 we get $d\alpha(\xi, X) + \left(L_\zeta g\right)(\xi, X) = 0$, which is equivalent to $d\omega(\xi, X) + g(\nabla_\xi \zeta, X) = 0$. But $g(\nabla_\xi \zeta, X) = -g(\zeta, \nabla_\xi X) = 0$ because $\widetilde{\nabla}_\xi X \in \mathcal{S}$ since $C(\xi, X) = 0$. Thus, $d\omega(\xi, X) = 0$ and it follows that ξ is geodesic for $\widetilde{\nabla}$ and $D^L(\xi, \xi) = 0$. Moreover, since $d\omega = 0$, from point 3 of Proposition 7 and formula 4 it also holds $D^L(\xi, X) = 0$ for all $X \in \mathcal{S}$. □

The screen distribution $\mathcal{S}$ is called distinguished if the corresponding one-form τ vanishes. Under some conditions, we can ensure the existence of a distinguished screen distribution. Concretely, if $-\tau(\xi)$ is not an eigenvalue of $A^* : \mathcal{S} \to \mathcal{S}$, then there exists a distinguished screen, [5]. On the other hand, the screen is called conformal if $A(U) = \varphi A^*(U)$ for all $U \in \mathfrak{X}(L)$ and certain function $\varphi \in C^\infty(L)$, or equivalently $C(U, X) = \varphi B(U, X)$ for all $U \in TL$ and $X \in \mathcal{S}$, [4]. Thus, fixed a rigging, the induced and the rigged connection coincide if and only if the screen distribution is distinguished and conformal with constant factor one.

It is possible that the induced and the rigged connection do not coincide for a rigging ζ, but they can coincide for another election of the rigging. Recall that if we choose another rigging, both the induced connection ∇^L and the rigged connection $\widetilde{\nabla}$ change, since the screen distribution and the null section change.

Consider ζ' another rigging for the null hypersurface L and decompose it as

$$\zeta' = \Phi N + U_0,$$

where $U_0 \in \mathfrak{X}(L)$ and $\Phi \in C^\infty(L)$ never vanishes. The corresponding rigged and transverse vector field are given by

$$\begin{aligned} \xi' &= \frac{1}{\Phi}\xi, \\ N' &= \Phi N + V_0, \end{aligned}$$

where $V_0 = U_0 - \frac{1}{2\Phi} g(\zeta', \zeta')\xi \in \mathfrak{X}(L)$. The geometric objects derived from the rigging ζ' are related to those derived from ζ as follows, [2].

$$B'(U, V) = \frac{1}{\Phi} B(U, V),$$

$$\tau'(U) = \tau(U) + \frac{1}{\Phi} d\Phi(U) + \frac{1}{\Phi} B(V_0, U), \tag{7}$$

$$C'(U, X) = \Phi C(U, X) - g(\nabla_U V_0, X) + \tau'(U) g(V_0, X), \tag{8}$$

$$\nabla'^L_U V = \nabla^L_U V - \frac{1}{\Phi} B(U, V) V_0.$$

for all $U, V \in \mathfrak{X}(L)$ and $X \in \mathcal{S}$.

Using above transformation formulas, it can be proven the following corollaries.

Corollary 19 *Let ζ be a rigging for a null hypersurface such that*

$$C(U,X) = \Phi^2 B(U,X),$$
$$\tau(U) = d\ln\Phi,$$

for all $U, V \in \mathfrak{X}(L)$ and certain positive function $\Phi \in C^\infty(L)$. Then the induced and the rigged connection for the rigging $\zeta' = \Phi\zeta + \left(\frac{1}{2\Phi} - \frac{\Phi}{2}g(\zeta,\zeta)\right)\xi$ coincide.

Corollary 20 *Let ζ be a rigging for a totally geodesic null hypersurface such that $C = 0$ and $d\tau = 0$. If L is simply connected, then the induced and the rigged connection coincide for the rigging $\zeta' = \Phi\zeta + \left(\frac{1}{2\Phi} - \frac{\Phi}{2}g(\zeta,\zeta)\right)\xi$, where $\Phi = e^{-f}$ and $\tau = df$ for certain $f \in C^\infty(L)$.*

Corollary 21 *Let ζ be a rigging for a totally geodesic null hypersurface such that $d\tau \neq 0$. Then, the induced and the rigged connection do not coincide for any election of the rigging.*

Example 22 Consider a Lorentzian manifold (M, g) furnished with a timelike unitary parallel vector field E. If L is any null hypersurface, then E is a rigging for it and from Proposition 5 we have

$$C(U,X) = \frac{1}{2}B(U,X)$$
$$\tau(U) = 0,$$

for all $U, V \in \mathfrak{X}(L)$ and $X \in \mathcal{S}$. Thus, using Corollary 19, the induced and the rigged connection on L coincide but for the rigging $\zeta = \sqrt{2}E$.

Moreover, if we consider the standard canonical variation $h = g + 2g(E,\cdot)g(E,\cdot)$, then the induced connection on L from the ambient Riemannian manifold (M, h) also coincides with the rigged and the induced connection induced on L by the rigging ζ. This is because the Levi-Civita connection of g and h concide, since E is parallel, and the null transverse vector field

$$N = \zeta - \frac{1}{2}g(\zeta,\zeta)\xi = \sqrt{2}E + \xi$$

also coincides with the h-unitary normal vector field to L.

Example 23 Consider a plane fronted wave $(M, g) = (Q \times \mathbb{R}^2, g_Q + 2du dv + H(x,u)du^2)$, where (Q, g_Q) is a Riemannian manifold and $H \neq 0$ is a function defined on Q. The vector field ∂_v is null and parallel, so $L = \{(x,u,v) : u = u_0\}$, where u_0 is a constant, is a totally geodesic null hypersurface. We consider the rigging $\zeta = \frac{1}{2}\partial_u$. Using Proposition 5 we have $\tau(U) = C(U,X) = 0$ for all $U \in \mathfrak{X}(L)$ and $X \in \mathcal{S}$. Thus, the induced and the rigged connection on L coincide.

Example 24 Consider $K \subset \mathbb{R}^2$ an open set and two functions $F, r : K \to \mathbb{R}$ such that they never vanish and there exists a constant $u_0 \in \mathbb{R}$ such that

$$r_v(u_0, v) = \frac{r_u(u_0, v)}{F(u_0, v)}$$

for all $(u_0, v) \in K$. Take the Lorentzian surface (K, g_K), where $g_K = 2F(u, v)dudv$ and the warped product $(M, g) = K \times_r Q$, being (Q, g_Q) a Riemannian manifold. Take the null hypersurface $L = \{(u, v, x) : u = u_0\}$ and consider the rigging $\zeta = \frac{1}{F}\partial_u$. The null transverse vector field is $N = \frac{1}{F}\partial_u$, the rigged vector field $\xi = \partial_v$ and the screen distribution $\mathcal{S} \approx TQ$. If $X, Y \in \mathfrak{X}(Q)$ then

$$\nabla_X Y = -\frac{r_v}{r} g(X, Y)N - \frac{r_u}{F \cdot r} g(X, Y)\xi + \nabla_X^Q Y,$$

thus $B(X, Y) = -\frac{r_v}{r} g(X, Y)$ and $C(X, Y) = -\frac{r_u}{F \cdot r} g(X, Y)$. On the other hand, $\nabla_\xi X \perp \zeta$, thus $C(\xi, X) = 0$. Therefore, $B = C$. Moreover, it is easy to check that $\tau(X) = 0$ for all $X \in \mathcal{S}$ and $\tau(\xi) = -\frac{F_v}{F}$. So $d\tau = 0$ and using Corollary 20, the induced and the rigged connection on L coincide.

Example 25 Take $(\mathbb{R}^2, 2dudv)$ and (Q, g_Q) any Riemannian manifold. Consider the twisted product $(M, g) = \left(Q \times \mathbb{R}^2, g_Q + 2f^2(x, u, v)dudv\right)$, being $f \in C^\infty(Q \times \mathbb{R}^2)$ a positive function with $w\left(\frac{f_v}{f}\right) \neq 0$ for some $w \in T_{x_0}Q$. We have that $L = \{(x, u, v) : u = u_0\}$ is a null hypersurface and $\zeta = \frac{1}{f^2}\partial_u$ is a rigging for it. The corresponding rigged vector field is $\xi = \partial_v$ and the screen distribution is $\mathcal{S} \approx TQ$. If $X \in TQ$, we have

$$\nabla_\xi \xi = \frac{2f_v}{f}\xi,$$
$$\nabla_X \xi = \frac{X(f)}{f}\xi,$$

thus, $\tau(\xi) = -\frac{2f_v}{f}$, $\tau(X) = -\frac{X(f)}{f}$ and L is totally geodesic. Moreover, the one-form τ is not closed. Indeed, we have

$$d\tau(\xi, w) = \xi(\tau(X)) - w(\tau(\xi)) = w\left(\frac{f_v}{f}\right) \neq 0.$$

Therefore, using Corollary 21, the induced connection and the rigging connection do not coincide for any election of the rigging.

We can give a family of generalized Robertson-Walker spaces where, as in above example, the rigging and the induced connection do not coincide for any election of the rigging. Totally umbilic null hypersurfaces in generalized Robertson-Walker spaces are determined by twisted decompositions of the fiber.

Theorem 26 ([8]) *Let $I \times_f F$ be a GRW space and consider the rigging $\zeta = f\partial_t$. If L is a totally umbilic null hypersurface, then for each $(t_0, x_0) \in L$ there exists a decomposition of F in a neighborhood of x_0 as a twisted product with one-dimensional base*

$$\left(J \times K, ds^2 + \mu(s,z)^2 g_K\right),$$

where x_0 is identified with $(0, z_0)$ for some $z_0 \in K$ and L is given by

$$\{(t,s,z) \in I \times J \times K : s = \int_{t_0}^{t} \frac{1}{f(r)} dr\}.$$

Moreover, if H is the null mean curvature of L, then

$$\mu(s,z) = \frac{f(t_0)}{f(t)} exp\left(\int_0^s \frac{H(t,r,z) f(t)^2}{n-2} dr\right)$$

for all $(t,s,z) \in L$.

Conversely, if F admits a twisted decomposition in a neighborhood of x_0 as above, then $L = \{(t,s,z) \in I \times J \times K : s = \int_{t_0}^{t} \frac{1}{f(r)} dr\}$ is a totally umbilic null hypersurface with null mean curvature

$$H = \frac{n-2}{f(t)^2}\left(f'(t) + \frac{\mu_s(s,z)}{\mu(s,z)}\right).$$

Take $0 \in I$, $J \subset \mathbb{R}$ two intervals, (Q, g_Q) a Riemannian manifold and consider

$$(M,g) = \left(I \times J \times Q, -dt^2 + f(t)^2\left(ds^2 + \mu(s)^2 g_Q\right)\right),$$

where f is some function with $f(0) = 1$, $\mu(s) = \frac{1}{f(h(s))}$ and $h(s)$ is the function such that $\int_0^{h(s)} \frac{1}{f(r)} dr = s$. Using above theorem, the hypersurface

$$L = \{(t,s,z) \in M : t = h(s)\}$$

is a totally geodesic null hypersurface. Since the rigging $\zeta = f\partial_t$ is closed and conformal, it holds $\nabla_U \zeta = f'U$ for all $U \in \mathfrak{X}(M)$ and using point 1 of Proposition 5 we get $\tau = 0$ and $C = -f'g$. Observe that the screen distribution $\mathcal{S}$ induced from ζ is integrable and the leaf S through a point $(0, 0, z_0) \in L$ is isometric to $\left(Q, g_Q\right)$.

Suppose that there exists another rigging ζ' such that the corresponding induced and rigging connection coincide and call $N' = \Phi N + V_0$ the associated null transverse vector field induced from ζ', where $\Phi \in C^\infty(L)$ and $V_0 \in \mathfrak{X}(L)$. From Eqs. 7 and 8 we get $d\Phi = 0$ and $g(\nabla_U V_0, X) = \Phi f' g(U, X)$ for all $U \in \mathfrak{X}(L)$ and $X \in \mathcal{S}$. Since L is totally geodesic, if we decompose $V_0 = a\xi + X_0$, where $X_0 \in \mathcal{S}$, then $g(\nabla_U V_0, X) = g(\nabla_U X_0, X)$ for all $U \in \mathfrak{X}(L)$ and $X \in \mathcal{S}$. If we restrict the vector field X_0 to the leaf S through $(0, 0, z_0)$, then we get a vector field $E \in$

$\mathfrak{X}(K)$ which holds $g_K(\nabla^K_X E, Y) = \Phi f'(0) g_K(X, Y)$ for all $X, Y \in \mathcal{X}(K)$. Therefore, the Riemannian manifold (Q, g_Q) locally decomposes as a warped product $\big((-\varepsilon, \varepsilon) \times P, du^2 + (\Phi f'(0)u + 1)^2 g_P\big)$, where E is identified with $(\Phi f'(0)u + 1)\partial_u$, [6].

Summarizing, in the Lorentzian manifold $I \times_f \big(J \times_\mu Q\big)$, the existence of a rigging for L such that the induced and the rigged connection coincide is codified by the local decompositions as warped product with one-dimensional base of (Q, g_Q). If it does not admit such local decomposition, then the induced and rigged connection do not coincide for any rigging. For example, the Riemannian direct product $\mathbb{S}^2 \times \mathbb{S}^2$ does not admit any local warped product decomposition as before, since for any vector we can find two planes containing it with different sectional curvature. So the induced and the rigged connection do not coincide for any election of the rigging if $Q = \mathbb{S}^2 \times \mathbb{S}^2$.

References

1. M.A. Akivis, V.V. Goldberg, On some methods of construction of invariant normalizations of lightlike hypersurfaces. Differ. Geom. Appl. **12**, 121–143 (2000)
2. C. Atindogbe, Normalization and prescribed extrinsic scalar curvature on lightlike hypersurfaces. J. Geom. Phys. **60**, 1762–1770 (2010)
3. C. Atindogbe, J.P. Ezin, T. Tossa, Pseudo-inversion of degenerate metrics. Int. J. Math. Math. Sci. **55**, 3479–3501 (2003)
4. C. Atindogbe, K.L. Duggal, Conformal screen on lightlike hypersurfaces. Int. J. Pure Appl. Math. **11**, 421–442 (2004)
5. K.L. Duggal, A. Giménez, Lightlike hypersurfaces of Lorentzian manifolds with distinguished screen. J. Geom. Phys. **55**, 107–222 (2005)
6. M. Gutiérrez, B. Olea, Global decomposition of a Lorentzian manifold as a generalized Robertson-Walker space. Differ. Geom. Appl. **27**, 146–156 (2009)
7. M. Gutiérrez, B. Olea, Induced Riemannian structures on a null hypersurface. Math. Nachr. **289**, 1219–1236 (2016)
8. M. Gutiérrez, B. Olea, Totally umbilic null hypersurfaces in generalized Robertson-Walker spaces. Differ. Geom. Appl. **42**, 15–30 (2015)
9. M. Navarro, O. Palmas, D.A. Solis, Null hypersurfaces in generalized Robertson-Walker spacetimes. J. Geom. Phys. **106**, 256–267 (2016)
10. B. Olea, Canonical variation of a Lorentzian metric. J. Math. Anal. Appl. **419**, 156–171 (2014)

Surfaces With Light-Like Points In Lorentz-Minkowski 3-Space With Applications

Masaaki Umehara and Kotaro Yamada

Abstract With several concrete examples of zero mean curvature surfaces in the Lorentz-Minkowski 3-space $\boldsymbol{R}^3_1$ containing a light-like line recently having been found, here we construct all real analytic germs of zero mean curvature surfaces by applying the Cauchy-Kovalevski theorem for partial differential equations. A point where the first fundamental form of a surface degenerates is said to be *light-like*. We also show a theorem on a property of light-like points of a surface in $\boldsymbol{R}^3_1$ whose mean curvature vector is smoothly extendable. This explains why such surfaces will contain a light-like line when they do not change causal types. Moreover, several applications of these two results are given.

Keywords Maximal surface · Mean curvature · Type change · Zero mean curvature · Lorentz-Minkowski space

2010 Mathematics Subject Classification Primary 53A10 · Secondary 53B30, 35M10

1 Introduction

In this paper, we denote by $\boldsymbol{R}^3_1$ the Lorentz-Minkowski 3-space of inner product $\langle\ ,\ \rangle$ of signature $(-++)$, and write the canonical coordinate system of $\boldsymbol{R}^3_1$ as (t, x, y).

Umehara was partially supported by the Grant-in-Aid for Scientific Research (A) No. 26247005, and Yamada by (C) No. 26400066 from Japan Society for the Promotion of Science.

M. Umehara
Department of Mathematical and Computing Sciences, Tokyo Institute of Technology, Tokyo 152-8552, Japan
e-mail: umehara@is.titech.ac.jp

K. Yamada (✉)
Department of Mathematics, Tokyo Institute of Technology, Tokyo 152-8551, Japan
e-mail: kotaro@math.titech.ac.jp

M.A. Cañadas-Pinedo et al. (eds.), *Lorentzian Geometry and Related Topics*, Springer Proceedings in Mathematics & Statistics 211,
DOI 10.1007/978-3-319-66290-9_14

Klyachin [13] showed that a zero mean curvature C^3-immersion $F\colon U \to \boldsymbol{R}^3_1$ of a domain $U \subset \boldsymbol{R}^2$ into the Lorentz-Minkowski 3-space $\boldsymbol{R}^3_1$ containing a light-like point o satisfies one of the following two conditions:

(a) *There exists a null curve σ (i.e., a regular curve in $\boldsymbol{R}^3_1$ whose velocity vector field is light-like) on the image of F passing through $F(o)$ which is nondegenerate (i.e., its projection into the xy-plane is a locally convex plane curve, cf. Definition* 4.1). *Moreover, the causal type of the surface changes from time-like to space-like across the curve.*

(b) *There exists a light-like line segment passing through $F(o)$ consisting of the light-like points of F. Zero mean curvature surfaces which change type across a light-like line belong to this class.*

The case (a) is now well understood (cf. [9, 12, 13]). In fact, under the assumption that F is real analytic, the surface in the class (a) can be reconstructed from the null curve σ as follows:

$$F(u,v) := \begin{cases} \dfrac{\sigma(u+i\sqrt{v})+\sigma(u-i\sqrt{v})}{2} & (v \geq 0), \\[2ex] \dfrac{\sigma(u+\sqrt{|v|})+\sigma(u-\sqrt{|v|})}{2} & (v < 0), \end{cases} \tag{1.1}$$

where $i = \sqrt{-1}$, and we extend the real analytic curve σ as a complex analytic map into $\boldsymbol{C}^3$.

We call the point o as in the case (a) a *nondegenerate light-like point* of F. A typical example of such a surface is obtained by a null curve $\gamma(u) = (u, \cos u, \sin u)$ and the resulting surface is a *helicoid*, which is a zero mean curvature surface (i.e., ZMC-surface) in $\boldsymbol{R}^3_1$ as well as in the Euclidean 3-space. The Ref. [6] is an expository article of this subject. Moreover, an interesting connection between type change of ZMC-surfaces and 2-dimensional fluid mechanics was also given in [6]. The existence and properties of entire ZMC-graphs in $\boldsymbol{R}^3_1$ with nondegenerate light-like points are discussed in [3]. Embedded ZMC-surfaces with nondegenerate light-like points with many symmetries are given in [4, 8].

On the other hand, several important ZMC-surfaces satisfying (b) are given in [1, 5]. In contrast to the case (a), these examples of surfaces do not change causal types across the light-like line.[1] A family of surfaces constructed in [7] satisfying (b) and also changes its causal type. In spite of this progress, there was still no machinery available to find surfaces of type (b) and no simple explanation for why only two cases occur at light-like points.

In this paper, we clarify such phenomena as follows: We denote by $\mathcal{Y}^r$ $(r \geq 3)$ the set of germs of C^r-differentiable immersions in $\boldsymbol{R}^3_1$ whose mean curvature vector field can be smoothly extended at a light-like point. We prove a property of regular

[1] In this paper, we say that *a surface changes its causal types across the light-like line* if the causal type of one-side of the line is space-like and the other-side is time-like. If the causal type of the both sides of the line coincides, we say that the surface *does not change its causal type across the light-like line*.

surfaces in the class $\mathcal{Y}^3$, which contains the above Klyachin's result as a special case. Our approach is different from that of [13]: we use the uniqueness of ordinary differential equations to prove the assertion. We also show a general existence of real analytic ZMC-surfaces and surfaces in $\mathcal{Y}^\omega$ using the Cauchy-Kovalevski theorem for partial differential equations. As a consequence, new examples of ZMC-surfaces which change type along a given light-like line are obtained.

2 Preliminaries

We denote by $\mathbf{0} := (0, 0, 0)$ (resp. $o := (0, 0)$) the origin of the Lorentz-Minkowski 3-space $\boldsymbol{R}^3_1$ of signature $(- + +)$ (resp. the plane $\boldsymbol{R}^2$) and denote by (t, x, y) the canonical coordinate system of $\boldsymbol{R}^3_1$. An immersion $F : U \to \boldsymbol{R}^3_1$ of a domain $U \subset \boldsymbol{R}^2$ into $\boldsymbol{R}^3_1$ is said to be *space-like* (resp. *time-like*, *light-like*) at p if the tangent plane of the image $F(U)$ at $F(p)$ is space-like (resp. time-like, light-like), that is, the restriction of the metric $\langle\ ,\ \rangle$ to the tangent plane is positive definite (resp. indefinite, degenerate). We denote by $\widetilde{\mathcal{I}}^r$ $(r \geq 2)$ the set of germs of C^r-immersions into $\boldsymbol{R}^3_1$ which map the origin o in the uv-plane to the origin $\mathbf{0}$ in $\boldsymbol{R}^3_1$. ($F \in \widetilde{\mathcal{I}}^\omega$ means that F is real analytic.) Let $F : (U, o) \to \boldsymbol{R}^3_1$ be an immersion in the class $\widetilde{\mathcal{I}}^r$. We denote by U_+ (resp. U_-) the set of space-like (resp. time-like) points, and set

$$U_* := U_+ \cup U_-.$$

A point $p \in U$ is light-like if $p \notin U_*$. We denote by $\widetilde{\mathcal{I}}^r_L (\subset \widetilde{\mathcal{I}}^r)$ the set of germs of C^r-immersion such that o is a light-like point.

If $F \in \widetilde{\mathcal{I}}^r_L$, the tangent plane of the image of F at o contains a light-like vector and does not contain time-like vectors. Thus, we can express the surface as a graph

$$F = (f(x, y), x, y), \tag{2.1}$$

where $f(x, y)$ is a C^r-function defined on a certain neighborhood of the origin of the xy-plane. Let

$$B_F := 1 - f_x^2 - f_y^2 \qquad \left(f_x = \frac{\partial f}{\partial x},\quad f_y = \frac{\partial f}{\partial y}\right). \tag{2.2}$$

Then the point of the graph (2.1) is space-like (resp. time-like) if and only if $B_F > 0$ (resp. $B_F < 0$) at the point. Since $F \in \widetilde{\mathcal{I}}^r_L$, the origin $o = (0, 0)$ is light-like, that is, $B_F(0, 0) = 0$. Hence there exists $\theta \in [0, 2\pi)$ such that

$$f_x(0, 0) = \sin\theta, \qquad f_y(0, 0) = \cos\theta.$$

So by a rotation about the t-axis, we may assume

$$f_x(0,0) = 0, \qquad f_y(0,0) = 1 \tag{2.3}$$

without loss of generality.

We denote by $\mathcal{I}_L^r (\subset \widetilde{\mathcal{I}}_L^r)$ the set of germs of C^r-immersion F with properties (2.1) and (2.3). Then

$$\iota : \mathcal{I}_L^r \ni F \mapsto (\iota_F :=) f \in \{f \in C_o^r(\boldsymbol{R}^2)\,;\ f(0,0) = f_x(0,0) = 0,\ f_y(0,0) = 1\}, \tag{2.4}$$

which maps F to the function f as in (2.1), is a bijection, where $C_o^r(\boldsymbol{R}^2)$ is the set of C^r-function germs on a neighborhood of $o \in \boldsymbol{R}^2$.

For $F \in \mathcal{I}_L^r$,

$$U_+ = \{p \in U\,;\ B_F(p) > 0\}, \qquad U_- = \{p \in U\,;\ B_F(p) < 0\} \tag{2.5}$$

hold, where B_F is the function as in (2.2). We let

$$A_F := (1 - f_x^2) f_{yy} + 2 f_x f_y f_{xy} + (1 - f_y^2) f_{xx}. \tag{2.6}$$

Then the mean curvature function of F

$$H_F := \frac{A_F}{2|B_F|^{3/2}} \tag{2.7}$$

is defined on U_* (cf. [10, Lemma 2.1]). We first remark the following:

Proposition 2.1 (cf. Klyachin [13, Example 4]) *If B_F vanishes identically, then so does A_F.*

Proof Since $B_F = 0$, we have $1 - f_x^2 = f_y^2$. By differentiating this, we get $f_x f_{xy} = -f_y f_{yy}$, and

$$(1 - f_x^2) f_{yy} = f_y (f_y f_{yy}) = -f_x f_y f_{xy}. \tag{2.8}$$

Similarly, we have

$$(1 - f_y^2) f_{xx} = f_x^2 f_{xx} = -f_x f_y f_{xy}. \tag{2.9}$$

By (2.8) and (2.9), we get the identity $A_F = 0$. □

We denote by Λ^r the set of germs of immersions $F \in \mathcal{I}_L^r$ with identically vanishing B_F, that is, Λ^r is the set of germs of *light-like immersions*. We denote by $C_o^\omega(\boldsymbol{R}, 0_2)$ the set of real analytic functions φ satisfying $\varphi(0) = d\varphi(0)/dx = 0$. Then the following assertion holds:

Proposition 2.2 *The map*

$$\lambda : \Lambda^\omega \ni F \mapsto (\lambda_F :=) f(x, 0) \in C_o^\omega(\boldsymbol{R}, 0_2)$$

is bijective, where $f = \iota_F$ (cf. (2.4)).

Proof Since B_F vanishes identically, taking in account of (2.3), we can write

$$f_y = \sqrt{1 - f_x^2}.$$

This can be considered as a normal form of a partial differential equation under the initial condition

$$f(x, 0) := \psi(x) \qquad (\psi \in C_o^\omega(\boldsymbol{R}, 0_2)), \tag{2.10}$$

because of the condition $f_x(0, 0) = \psi'(0) = 0$. So we can apply the Cauchy-Kovalevski theorem (cf. [15]) and show the uniqueness and existence of the solution f satisfying (2.10). □

Remark 2.3 The above proof of the existence of a light-like surface $F(x, y)$ satisfying (2.10) is local, that is, it is defined only for small $|y|$. Later, we will show that F is a ruled surface and has an explicit expression, see Corollary 4.7 and (4.8).

Example 2.4 The light-like plane $F(x, y) = (y, x, y)$ belongs to the class Λ^ω such that $\lambda_F = 0$.

Example 2.5 The light-cone $F(x, y) = (\sqrt{x^2 + (1 + y)^2} - 1, x, y)$ is a light-like surface satisfying $\lambda_F = \sqrt{1 + x^2} - 1$.

3 Surfaces with Smooth Mean Curvature Vector Field

Let $F : (U, o) \to (\boldsymbol{R}^3_1, \boldsymbol{0})$ be an immersion of class $\mathcal{I}^r_L$ $(r \geq 3)$ such that U_* is open and dense in U, and fix a C^{r-2}-function φ. We say that F is *φ-admissible* if

$$A_F - \varphi B_F^2 = 0 \tag{3.1}$$

holds, where A_F and B_F are as in (2.6) and (2.2), respectively. We denote

$$\mathcal{Y}^r_\varphi := \{F \in \mathcal{I}^r_L\,;\ F \text{ is } \varphi\text{-admissible}\}. \tag{3.2}$$

An immersion germ $F \in \mathcal{I}^r_L$ is called *admissible* if it is φ-admissible for a certain $\varphi \in C^{r-2}_o(\boldsymbol{R}^2)$. The set

$$\mathcal{Y}^r := \bigcup_{\varphi \in C^{r-2}_o(\boldsymbol{R}^2)} \mathcal{Y}^r_\varphi$$

consists of all germs of φ-admissible immersions. The following assertion explains why the class $\mathcal{Y}^r$ is important.

Proposition 3.1 ([10]) *Let $F : (U, o) \to (\boldsymbol{R}^3_1, \boldsymbol{0})$ be an immersion in the class $\mathcal{I}^r_L$ for $r \geq 3$. Then the mean curvature vector field $\boldsymbol{H}_F$ can be C^{r-2}-differentially extended on a neighborhood of o if and only if F belongs to the class $\mathcal{Y}^r$.*

Proof This assertion follows from the fact that

$$H_F = \frac{A_F}{2(B_F)^2}(F_x \times F_y)$$

on U^*, where $\times$ denotes the vector product in $\boldsymbol{R}^3_1$. □

Since $\varphi = 0$ implies $H_F = 0$,

$$\mathcal{Z}^r := \{F \in \mathcal{I}^r_L\,;\ A_F = 0\}(= \mathcal{Y}^r_0) \tag{3.3}$$

is the set of germs of zero mean curvature immersions in $\boldsymbol{R}^3_1$ at the light-like point o. By definition, we have (cf. Proposition 2.1)

$$\Lambda^r \subset \mathcal{Z}^r \subset \mathcal{Y}^r \qquad (r \geq 3). \tag{3.4}$$

Surfaces in the class $\mathcal{Y}^r$ are investigated in [10], and an entire graph in $\mathcal{Y}^r$ which is not a ZMC-surface was given. In this section, we shall show a general existence result of surfaces in the class $\mathcal{Y}^\omega$. We fix a germ of a real analytic function $\varphi \in C^\omega_o(\boldsymbol{R}^2)$, and take an immersion $F \in \mathcal{Y}^\omega_\varphi$.

Definition 3.2 Let $f := \iota_F$ be the function associated with $F \in \mathcal{Y}^\omega_\varphi$ (cf. (2.4)). Then

$$\gamma_F(x) := (f(x,0),\ f_y(x,0)) \tag{3.5}$$

is a real analytic plane curve, which we call the *initial curve* associated with F.

We denote by $C^\omega_{(0,1)}(\boldsymbol{R}, \boldsymbol{R}^2)$ the set of germs of C^ω-maps $\gamma : (\boldsymbol{R}, 0) \to (\boldsymbol{R}^2, (0,1))$. By definition (cf. (2.3)), $\gamma_F(x) \in C^\omega_{(0,1)}(\boldsymbol{R}, \boldsymbol{R}^2)$ holds. We prove the following assertion:

Theorem 3.3 *For $\varphi \in C^\omega_0(\boldsymbol{R}^2)$, the set $\mathcal{Y}^\omega_\varphi$ is non-empty. More precisely, the map*

$$\mathcal{Y}^\omega_\varphi \ni F \mapsto \gamma_F \in C^\omega_{(0,1)}(\boldsymbol{R}, \boldsymbol{R}^2)$$

is bijective. Moreover, the base point o is a light-like point of F satisfying $\nabla B_F \neq 0$ (resp. $\nabla B_F = 0$) if $\dot{\gamma}_F(0) \neq (0,0)$ (resp. $\dot{\gamma}_F(0) = (0,0)$), where "dot" denotes d/dx and $\nabla B_F := ((B_F)_x, (B_F)_y)$ is the gradient vector of the function B_F.

Proof Suppose that $F \in \mathcal{Y}^\omega_\varphi$. Since $A_F - \varphi B_F^2$ vanishes identically (cf. (3.1)), $f = \iota_F$ satisfies

$$f_y = g, \quad g_y = -\frac{2f_x g g_x + (1-g^2) f_{xx} - (1 - f_x^2 - g^2)^2 \varphi}{1 - f_x^2}, \tag{3.6}$$

which is the normal form for partial differential equations. So we can apply the Cauchy-Kovalevski theorem (cf. [15]) for a given initial data

$$(f(x,0), g(x,0)) := \gamma(x) \qquad (\gamma \in C^{\omega}_{(0,1)}(\boldsymbol{R}, \boldsymbol{R}^2)).$$

Then the solution (f, g) of (3.6) is uniquely determined. Obviously, the resulting immersion $F_\gamma := (f(x, y), x, y)$ gives a surface in $\mathcal{Y}^{\omega}_{\varphi}$ whose initial curve is γ. The second assertion follows from the fact that $\dot{\gamma}(0) = (0, 0)$ if and only if $\nabla B_F(0, 0) = \mathbf{0}$, where $\nabla B_F := ((B_F)_x, (B_F)_y)$. □

When $\varphi = 0$, we get the following:

Corollary 3.4 *The map* $\mathcal{Z}^{\omega} \ni F \mapsto \gamma_F \in C^{\omega}_{(0,1)}(\boldsymbol{R}, \boldsymbol{R}^2)$ *is bijective.*

The following is a direct consequence of this corollary and Theorem 3.3.

Corollary 3.5 *In the above correspondence, it holds that*

$$\Lambda^{\omega} = \left\{ F \in \mathcal{Z}^{\omega}\,;\ \gamma_F = \left(\psi, \sqrt{1 - \dot{\psi}^2}\right),\ \ \psi \in C^{\omega}_{o}(\boldsymbol{R}, 0_2) \right\}.$$

4 A Property of Light-Like Points

Definition 4.1 Let I be an open interval, and $\sigma : I \to \boldsymbol{R}^3_1$ a regular curve of class C^r $(r \geq 3)$. The space curve σ is called *null* if $\sigma'(t) = d\sigma/dt$ is light-like. Moreover, σ is called *nondegenerate* if $\sigma''(t)$ is not proportional to $\sigma'(t)$ for each $t \in I$.

The orthogonal projection of a nondegenerate null curve into the xy-plane is a locally convex plane curve. The following assertion is a generalization of Klyachin's result in the introduction, since ZMC-surfaces are elements of $\mathcal{Y}^3$.

Theorem 4.2 *Let* $F : (U, o) \to \boldsymbol{R}^3_1$ *be an immersion of class* $\mathcal{Y}^3$*. Then, one of the following two cases occurs:*

(a) ∇B_F *does not vanish at* o*, and the image of the level set* $F(\{B_F = 0\})$ *consists of a nondegenerate null regular curve in* $\boldsymbol{R}^3_1$*, where* $f = \iota_F$*.*
(b) ∇B_F *vanishes at* o*, and the image of the level set* $F(\{B_F = 0\})$ *contains a light-like line segment in* $\boldsymbol{R}^3_1$ *passing through* $F(o)$*.*

Proof The first assertion (a) was proved in [10, Proposition 3.5]. So it is sufficient to prove (b). We may assume that $F \in \mathcal{Y}^3_{\varphi}$ and $\varphi \in C^1_o(\boldsymbol{R}^2)$. Let $f := \iota_F$. Since f is of class C^3, applying the division lemma (Lemma A.1 in Appendix A) for $g(x, y) := 2(f(x, y) - f(0, y) - x f_x(0, y))$, there exists a C^1-function h such that

$$f(x, y) = a_0(y) + a_1(y)x + \frac{h(x, y)}{2}x^2 \quad \big(a_0(y) := f(0, y),\ a_1(y) := f_x(0, y)\big). \tag{4.1}$$

By (2.3), it holds that

$$a_0(0) = 0, \quad a_0'(0) = 1, \qquad a_1(0) = 0. \tag{4.2}$$

Moreover, since ∇B_F vanishes at o, we have

$$a_1'(0) = 0. \tag{4.3}$$

We set (cf. (2.6) and (3.1))

$$\tilde{A} := A_F - \varphi B_F^2.$$

Since $F \in \mathcal{Y}_\varphi^3$, $\tilde{A}$ vanishes identically. We set

$$h_0(y) := h(0, y),\ h_1(y) := h_x(0, y),\ h_2(y) := h_y(0, y),$$
$$\varphi_0(y) := \varphi(0, y),\ \varphi_1(y) := \varphi_x(0, y),$$

Then we have

$$\begin{aligned} 0 = \tilde{A}|_{x=0} =& \varphi_0((a_0')^2 + a_1^2 - 1)^2 + h_0\left(1 - (a_0')^2\right) \\ &+ \left(1 - a_1^2\right) a_0'' + 2a_1 a_0' a_1', \end{aligned} \tag{4.4}$$

$$\begin{aligned} 0 = \tilde{A}_x|_{x=0} =& -2a_1 h_0 a_0'' + 4\varphi_0\left(1 - (a_0')^2 - a_1^2\right)\left(a_1 h_0 + a_0' a_1'\right) \\ &+ 2a_1 h_2 a_0' - \varphi_1\left(-(a_0')^2 - a_1^2 + 1\right)^2 \\ &- h_1\left((a_0')^2 - 1\right) - \left(a_1^2 - 1\right) a_1'' + 2a_1(a_1')^2. \end{aligned} \tag{4.5}$$

These two identities (4.4) and (4.5) can be rewritten in the form

$$\left(1 - a_1^2\right) a_0'' = \Psi_1(x, y, a_0, a_0', a_1, a_1'),$$
$$-2a_1 h_0 a_0'' + \left(a_1^2 - 1\right) a_1'' = \Psi_2(x, y, a_0, a_0', a_1, a_1'),$$

where Ψ_1 and Ψ_2 are continuous functions of five variables. Since $1 - a_1^2(0) = 1$, this gives a normal form of a system of ordinary differential equations with unknown functions a_0 and a_1. Moreover, this system of differential equations satisfies the local Lipschitz condition, since Ψ_1 and Ψ_2 are polynomials in a_0, a_0', a_1 and a_1'. Here,

$$(a_0, a_1) = (y, 0) \tag{4.6}$$

gives a solution of this system of equations. Then the uniqueness of the solution with the initial conditions (4.2) and (4.3) implies that (4.6) holds for F. As a consequence, we have $F(0, y) = (y, 0, y)$, proving the assertion. $\square$

Remark 4.3 If F is of nonzero constant mean curvature H, then $A_F - 2H|B_F|^{3/2}$ vanishes identically. In this case, we also get the relations $A_F|_{x=0} = (A_F)_x|_{x=0} = 0$, which can be considered as a system of ordinary equations like as in the above proof. However, this does not seem to satisfy the local Lipschitz condition, and the above proof does not work directly in this case. Fortunately, an analogue of Theorem 4.2

can be proved for real analytic surfaces with nonzero constant mean curvature by modifying the above argument (see [16] for details).

As a corollary, we immediately get the following:

Corollary 4.4 *Any light-like points on a surface in the class $\mathcal{Y}^3$ are not isolated.*

If a light-like point o is nondegenerate, then B_F changes sign, that is, F changes causal type. So the following corollary is also obtained.

Corollary 4.5 *If an immersion $F \in \mathcal{Y}^3$ does not change its causal type at o, then there exists a light-like line L passing through $f(o)$ such that the set of the light-like points of F is a regular curve γ and the image of $F \circ \gamma$ lies on the line L.*

We next give an application of Theorem 4.2 for light-like surfaces:

Definition 4.6 Let $\sigma(t)$ $(t \in I)$ be a space-like C^∞-regular curve defined on an interval I. Since the orthogonal complement of $\sigma'(t)$ is Lorentzian, there exists a nonvanishing vector field $\xi(t)$ along σ such that $\xi(t)$ points the light-like direction which is orthogonal to $\sigma'(t) = d\sigma(t)/dt$, that is, it holds that

$$\langle \xi(t), \xi(t) \rangle = \left\langle \xi(t), \sigma'(t) \right\rangle = 0,$$

where $\langle \ , \ \rangle$ means the canonical Lorentzian inner product in $\boldsymbol{R}^3_1$. The possibility of such vector fields $\xi(t)$ are essentially two up to a multiplication of nonvanishing smooth functions. Then the map

$$F(t,s) := \sigma(t) + s\,\xi(t) \qquad (t \in I, |s| < \epsilon) \tag{4.7}$$

gives a light-like immersion if $\epsilon > 0$ is sufficiently small (This representation formula was given in Izumiya-Sato [11]). We call such a F a *light-like ruled surface* associated to the space-like curve σ.

For example, consider an ellipse $\sigma(t) = (0, a\cos t, \sin t)$ on the xy-plane in $\boldsymbol{R}^3_1$, where $a > 0$ is a constant. Then an associated light-like surface is given by (4.7) by setting

$$\xi(t) := \left(\sqrt{a^2 \sin^2 t + \cos^2 t}, \cos t, a \sin t\right).$$

Figure 1 presents the resulting light-like surface for $a = 2$.

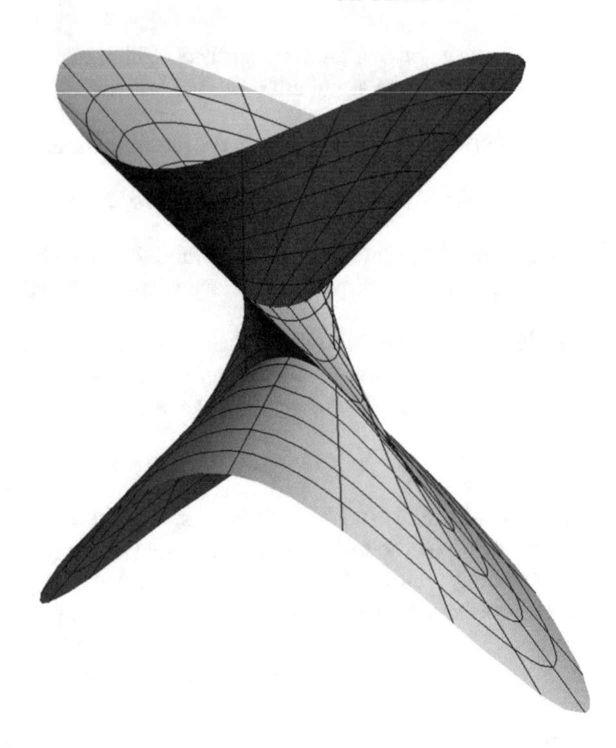

Fig. 1 The light-like surface associated to an ellipse

The following corollary asserts that light-like regular surfaces are locally regarded as ruled surfaces.

Corollary 4.7 *A light-like surface germ $F \in \Lambda^\infty$ can be parametrized by a light-like ruled surface along a certain space-like regular curve.*

Proof Let F be a light-like surface such that $\iota_F(x, 0) = \psi(x)$ as in (2.10), where $\psi(0) = \dot{\psi}(0) = 0$ ($\dot{} = d/dx$). Then it holds that $\sigma(x) := F(x, 0) = (\psi(x), x, 0)$ is a space-like curve for sufficiently small x. There are $\pm$-ambiguity of light-like vector fields

$$\xi^\pm(x) := \left(1, \dot{\psi}(x), \pm\sqrt{1 - \dot{\psi}(x)^2}\right)$$

along the curve $\sigma(x)$ perpendicular to $\dot{\sigma}(x)$. By Corollary 4.5, F must be a ruled surface foliated by light-like lines. Since the light-like line in the image of F passing through the origin is $y \mapsto (y, 0, y)$, the light-like ruled surface

$$(t, s) \mapsto \sigma(t) + s\,\xi^+(t) \tag{4.8}$$

gives a new parametrization of the surface F, that proves the assertion. □

It should be remarked that the property of light-like points (cf. Theorem 4.2) can be generalized for surfaces in arbitrarily given Lorentzian 3-manifolds, see [16]. It is well-known that space-like ZMC-surfaces have nonnegative Gaussian curvature. Regarding this fact, we prove the following assertion on the sign of Gaussian curvature at nondegenerate light-like points:

Proposition 4.8 *Let F be an immersion in the class $\mathcal{Y}^r$ $(r \geq 3)$. Suppose that o is a nondegenerate light-like point. Then the Gaussian curvature function K diverges to ∞ at o.*

When F is of ZMC, the assertion was proved in Akamine [2].

Proof Let $f = \iota_F$ is the function associated with $F \in \mathcal{Y}^r$ (cf. (2.4)) with (4.1) and set

$$C_F := f_{xx} f_{yy} - (f_{xy})^2.$$

Then the Gaussian curvature of F is given by

$$K := -\frac{C_F}{(B_F)^2}, \tag{4.9}$$

where B_F is the function as in (2.2). Since the function A_F as in (2.6) satisfies $A_F(o) = a_0''(0) = 0$, we have

$$C_F(o) = h(o)a_0''(0) - a_1'(0)^2 = -a_1'(0)^2.$$

Here $a_1'(0) \neq 0$ since o is a nondegenerate light-like point. Thus $C_F(o) \neq 0$ holds, and we get the conclusion because of the fact $B_F(o) = 0$. □

5 Properties of Surfaces in $\mathcal{Y}^\omega$

We denote by $\mathcal{Y}^r_a$ (resp. $\mathcal{Y}^r_b$) the subset of $\mathcal{Y}^r$ consisting of surfaces such that the origin o is a nondegenerate (resp. degenerate) light-like point. Then

$$\mathcal{Y}^r := \mathcal{Y}^r_a \cup \mathcal{Y}^r_b$$

holds. We next define two subsets of $\mathcal{Z}^r$ (cf. (3.3)) as

$$\mathcal{Z}^r_a := \{F \in \mathcal{Z}^r \,;\, \gamma_F(0) \neq (0,0)\} = \{F \in \mathcal{Z}^r \,;\, \nabla B_F(0,0) \neq \mathbf{0}\} = \mathcal{Y}^r_a \cap \mathcal{Z}^r,$$
$$\mathcal{Z}^r_b := \{F \in \mathcal{Z}^r \,;\, \gamma_F(0) = (0,0)\} = \{F \in \mathcal{Z}^r \,;\, \nabla B_F(0,0) = \mathbf{0}\} = \mathcal{Y}^r_b \cap \mathcal{Z}^r,$$

where B_F is the function as in (2.2).

As explained in the introduction, surfaces in $\mathcal{Z}_a^r$ can be constructed using the formula as in (1.1). On the other hand, to get ZMC-surfaces in $\mathcal{Z}_b^\omega$, we can apply Corollary 3.4. However, it is only a general existence result, and is not useful if one would like to know the precise behavior of the surfaces along the degenerate light-like lines. Here, we focus on the class $\mathcal{Z}_b^\omega$. We first show that $\mathcal{Y}_b^\omega$ and will sow that surfaces in the class $\mathcal{Y}_b^\omega$ have quite similar properties as zero mean curvature surfaces in the class $\mathcal{Z}_b^\omega$: Each surface $F \in \mathcal{Y}_b^\omega$ is expressed as the graph of a function

$$f(x, y) := y + \frac{\alpha_F(y)}{2}x^2 + \frac{\beta_F(y)}{3}x^3 + h(x, y)x^4, \tag{5.1}$$

where α_F, β_F and h are certain real analytic functions. In fact, as seen in the proof of Theorem 4.2, $a_1(y) = 0$ holds when o is a degenerate light-like point, where $a_1(y) := f_x(0, y)$. We call $\alpha_F(y)$ and $\beta_F(y)$ in (5.1) the *second approximation function* and the *third approximation function* of F, respectively. These functions give the following approximation of F:

$$f(x, y) \approx y + \frac{\alpha_F(y)}{2}x^2 + \frac{\beta_F(y)}{3}x^3. \tag{5.2}$$

Proposition 5.1 *For $F \in \mathcal{Y}_b^\omega$, there exists a real number μ_F (called the* characteristic *of F) such that α_F and β_F satisfy*

$$\alpha_F' + \alpha_F^2 + \mu_F = 0, \tag{5.3}$$
$$\beta_F'' + 4\alpha_F\beta_F' = 0. \tag{5.4}$$

Moreover, if $\mu_F > 0$ (resp. $\mu_F < 0$) then F has no time-like points (resp. no space-like points). In particular, if F changes causal type, the $\mu_F = 0$.

When $F \in \mathcal{Z}_b^\omega$, this assertion for $\alpha := \alpha_F$ was proved in [5]. So the above assertion is its generalization.

Proof Since F in $\mathcal{Y}^\omega$, there exists a function $\varphi \in C_o^\omega(\boldsymbol{R}^2)$ such that $F \in \mathcal{Y}_\varphi^\omega$. Let $f := \iota_F$ and set $A := A_F$ and $B := B_F$ as in (2.2) and (2.6), respectively. Then (5.1) implies that

$$B(0, y) = B_x(0, y) = 0. \tag{5.5}$$

Since $\varphi = A/B^2$ is a smooth function, the L'Hospital rule yields that

$$\begin{aligned}\varphi(0, y) &= \lim_{x\to 0}\frac{A(x, y)}{B(x, y)^2} = \lim_{x\to 0}\frac{A_x(x, y)}{2B_x(x, y)B(x, y)}\\ &= \lim_{x\to 0}\frac{A_{xx}(x, y)}{2B_{xx}(x, y)B(x, y) + 2B_x(x, y)B_x(x, y)}\\ &= \lim_{x\to 0}\frac{A_{xxx}(x, y)}{2B_{xxx}(x, y)B(x, y) + 6B_{xx}(x, y)B_x(x, y)}.\end{aligned}$$

Thus, (5.5) yields that we have

$$0 = A(0, y) = A_x(0, y) = A_{xx}(0, y) = A_{xxx}(0, y).$$

Here $A|_{x=0} = A_x|_{x=0} = 0$ does not produce any restrictions for $\alpha := \alpha_F$ and $\beta := \beta_F$. On the other hand, we have

$$0 = A_{xx}|_{x=0} = 2\alpha\alpha' + \alpha'', \tag{5.6}$$

where the prime means the derivative with respect to y. Hence $\alpha' + \alpha^2$ is a constant function, and get the first relation. We then get

$$0 = A_{xxx}|_{x=0} = 4\alpha\beta' + \beta'', \tag{5.7}$$

which yields the second assertion. The last assertion follows from the fact that $B_{xx}(0, y) = -2(\alpha' + \alpha^2) = 2\mu_F$. □

We can find the solution $\alpha = \alpha_F$ of the ordinary differential equation (5.3) under the conditions

$$\alpha(0) = \ddot{u}(0), \qquad \mu_F = -\big(\ddot{u}(0)^2 + \dddot{v}(0)\big)$$

for a given initial curve $\gamma(x) = (u(x), v(x))$. By a homothetic change

$$\widetilde{F}(x, y) := (\tilde{f}(x, y), x, y) \qquad \left(\tilde{f}(x, y) := \frac{1}{m} f(mx, my),\ m > 0\right),$$

one can normalize the characteristic μ_F to be -1, 0 or 1. In fact, as shown in [5],

$$\alpha^+ := -\tan(y + c) \qquad (|c| < \pi/2)$$

is a general solution of (5.3) for $\mu_F = 1$,

$$\alpha_I^0 := 0 \;\text{ and }\; \alpha_{II}^0 := \frac{1}{y + c} \qquad (c \in \boldsymbol{R} \setminus \{0\})$$

are the solutions for $\mu_F = 0$, and

$$\begin{aligned} \alpha_I^- &:= \tanh(y + c)\ (c \in \boldsymbol{R}), \\ \alpha_{II}^- &:= \coth(y + c)\ (c \in \boldsymbol{R} \setminus \{0\}), \\ \alpha_{III}^- &:= \pm 1 \end{aligned}$$

are the solutions for $\mu_F = -1$. Thus, as pointed out in [5], $\mathcal{Y}_b^\omega$ consists of the following six subclasses:

$$\mathcal{Y}^+,\ \mathcal{Y}_I^0,\ \mathcal{Y}_{II}^0,\ \mathcal{Y}_I^-,\ \mathcal{Y}_{II}^-,\ \mathcal{Y}_{III}^-.$$

Remark 5.2 To find surfaces in the class $\mathcal{Y}_{III}^{-}$, we may set $\alpha_{III}^{-} := 1$ without loss of generality. In fact, if $F \in \mathcal{Y}_b^{\omega}$ satisfies $\alpha_F = -1$, then we can write

$$F = \big(f(x, y), x, y\big), \qquad f(x, y) = y - \frac{x^2}{2} + k(x, y)x^3,$$

where $k(x, y)$ is a C^{ω}-function defined on a neighborhood of the origin. If we set $X := -x$, $Y := y$ and set $\tilde{F} := -F$, then we have

$$\tilde{F}(X, Y) = \left(Y + \frac{X^2}{2} + k(-X, Y)X^3, X, Y\right).$$

Thus $\alpha_{\tilde{F}} = 1$.

Like as in the case of $\alpha = \alpha_F$, we can find the solution $\beta := \beta_F$ of the ordinary differential equation (5.4) with the given initial condition $2(\beta(0), \beta'(0)) = \dddot{\gamma}_F(0)$. In fact, β can be written explicitly for each $\alpha = \alpha^+, \alpha_I^0, \alpha_{II}^0, \alpha_I^-, \alpha_{II}^-, \alpha_{III}^-$ as follows:

$$\begin{aligned}
\beta^+ &= c_1\left(2 + \sec^2(y + c)\right)\tan(y + c) + c_2, \\
\beta_I^0 &= c_1 y + c_2, \\
\beta_{II}^0 &= \frac{c_1}{(y + c)^3} + c_2, \\
\beta_I^- &= c_1\left(2 + \operatorname{sech}^2(y + c)\right)\tanh(y + c) + c_2, \\
\beta_{II}^- &= c_1\left(2 - \operatorname{csch}^2(y + c)\right)\coth(y + c) + c_2, \\
\beta_{III}^- &= c_1 e^{\pm 4y} + c_2.
\end{aligned}$$

In particular, we get the following assertion:

Proposition 5.3 *The second and the third approximation functions of each $F \in \mathcal{Y}_b^{\omega}$ can be written in terms of elementary functions.*

The following assertion implies that our approximation for $F \in \mathcal{Y}^{\omega}$ itself is an element of $\mathcal{Y}^{\omega}$:

Proposition 5.4 *Let α and β be analytic functions satisfying*

$$\alpha'' + 2\alpha\alpha' = 0, \qquad \beta'' + 4\alpha\beta' = 0. \tag{5.8}$$

Then the immersion

$$F_{\alpha,\beta}(x, y) := \left(y + \frac{\alpha(y)}{2}x^2 + \frac{\beta(y)}{3}x^3, x, y\right)$$

belongs to the class $\mathcal{Y}^{\omega}$.

This assertion follows from the proof of Proposition 5.1 immediately. Moreover, the following assertion holds:

Proposition 5.5 *We fix an analytic function $\varphi \in C^{\omega}_{o}(\boldsymbol{R}^2)$. Let α, β be analytic functions satisfying* (5.8)*. Then there exists an immersion $F \in \mathcal{Y}^{\omega}_{\varphi}$ such that $\alpha = \alpha_F$ and $\beta = \alpha_F$. In the case of $\varphi = 0$, this implies the existence of $F_{\alpha,\beta} \in \mathcal{Z}^{\omega}_{b}$.*

Proof We set

$$\gamma(x) = (1, 0) + \frac{x^2}{2}(\alpha(0), \beta(0)) + \frac{x^3}{3}(\alpha'(0), \beta'(0)).$$

By Theorem 3.3, there exists a unique immersion $F \in \mathcal{Y}^{\omega}_{\varphi} \cap \mathcal{Y}^{\omega}_{b}$ such that $\gamma_F = \gamma$. Then we have $\alpha_F = \alpha$ and $\beta_F = \beta$, proving the assertion. □

Finally, we prove the following assertion for the sign of the Gaussian curvature near a degenerate light-like point. (cf. Proposition 4.8 for the nondegenerate case.)

Proposition 5.6 *Let F be an immersion in the class $\mathcal{Y}^{\omega}_{b}$. Then the Gaussian curvature function K diverges to ∞ at a degenerate light-like point if $\mu_F > 0$. On the other hand, if $\mu_F = 0$ and $\delta_F \neq 0$, then $K(x, y)$ diverges to $+\infty$ (resp. $-\infty$) on the domain of $B_F(x, y) > 0$ (resp. $B_F(x, y) < 0$) as $(x, y) \to (0, 0)$, where*

$$\delta_F := \beta'(0) + 3\alpha(0)\beta(0) \tag{5.9}$$

and $\alpha = \alpha_F$, $\beta = \beta_F$.

Proof Recall that the Gaussian curvature K is expressed as (cf. (4.9))

$$K = -\frac{C_F}{(B_F)^2} \qquad (C_F = f_{xx} f_{yy} - f_{xy}^2),$$

where $f = \iota_F$ and B_F is the function defined as in (2.2). Since $F \in \mathcal{Y}^{\omega}_{b}$, can be expanded as (5.1). Then we have

$$\begin{aligned} C_F &= f_{xx} f_{yy} - (f_{xy})^2 \\ &= \left(\frac{1}{2}\alpha\alpha'' - (\alpha')^2\right) x^2 + \left(\beta\alpha'' - 2\alpha'\beta' + \frac{1}{3}\alpha\beta''\right) x^3 + \text{(higher order terms)}. \end{aligned}$$

Since $F \in \mathcal{Y}^{r}_{b}$, the relations (5.3) and (5.4) hold, and then we have

$$C_F = \alpha' \mu_F x^2 - \frac{2}{3}\left(3\alpha'\beta' + 3\alpha\beta\alpha' + 2\alpha^2\beta'\right) x^3 + \text{(higher order terms)}.$$

If $\mu_F > 0$, then $\alpha'(0) < 0$ by (5.3), so we get the conclusion. We next assume $\mu_F = 0$, then $\alpha' = -\alpha^2$ and we have

$$C_F(x,0) = \frac{2\alpha(0)^2\delta_F}{3}x^3 + (\text{higher order terms}).$$

In this situation, it holds that

$$B_F(x,0) = -\frac{2\delta_F}{3}x^3 + (\text{higher order terms}).$$

Thus the sign of the Gaussian curvature $K(x,0)$ coincides with that of $B_F(x,0)$, proving the assertion. □

6 Examples

In this section, we give several examples of zero mean curvature surfaces: We now give here a recipe to give more refined approximate solutions as follows: For $F \in \mathcal{Z}_b^\omega$, we can expand the function $f = \iota_F$ as

$$f(x,y) = y + \sum_{k=2}^{\infty} \frac{a_k(y)}{k} x^k. \tag{6.1}$$

We call each function $a_k(y)$ as the *k-th approximation function* of F. Remark that a_2 and a_3 coincide with α_F and β_F in (5.1), respectively:

$$\alpha_F = a_2, \qquad \beta_F = a_3. \tag{6.2}$$

We give here several examples:

Example 6.1 The light-like plane (cf. Example 2.4) $F(x,y) = (y,x,y)$ belongs to $\Lambda^\omega \cap \mathcal{Z}_I^0$ such that $\gamma_F = (0,1)$ and $\alpha_F = \beta_F = 0$.

Example 6.2 The light-cone (cf. Example 2.5)

$$F(x,y) = (\sqrt{x^2 + (1+y)^2} - 1, x, y)$$

belongs to $\Lambda^\omega \cap \mathcal{Z}_{II}^0$ such that

$$\gamma_F = (\sqrt{1+x^2} - 1, 1/\sqrt{1+x^2})$$

and $\alpha_F = 1/(1+y)$, $\beta_F = 0$.

Example 6.3 The surface $F(x,y) = (y + x^2/2, x, y)$ is a zero mean curvature surface in $\mathcal{Z}_{III}^-$, which satisfies $\gamma_F = (x^2/2, 0)$ and

$$\alpha_F = 1, \qquad \beta_F = 0.$$

Example 6.4 Recall the space-like Scherk surface $\{(t, x, y) \in \boldsymbol{R}^3_1;\ \cos t = \cos x \cos y\}$ (cf. [5, Example 3]). We replace (t, x, y) by $(t + (\pi/2), x, (\pi/2) - y)$, we have the expression

$$F(x, y) = \left(-\arccos(\cos x \sin y) - \frac{\pi}{2}, x, y\right)$$

satisfying (2.3). This is a zero mean curvature surface in $\mathcal{Z}^+$, which satisfies $\gamma_F = (-\pi, \cos x)$ and

$$\alpha_F = -\tan y, \qquad \beta_F = 0.$$

Example 6.5 The time-like Scherk surface of the first kind (cf. [5, Example 4]) can be normalized as

$$F(x, y) = \left(\operatorname{arccosh}(\cosh x \cosh(y + 1)) - 1, x, y\right),$$

which is a zero mean curvature surface in $\mathcal{Z}^-_I$, This satisfies

$$\gamma_F = \left(-1 + \operatorname{arccosh}(\cosh x \cosh 1), \frac{\sinh 1 \cosh x}{\sqrt{(\cosh 1 \cosh x)^2 - 1}}\right)$$

and

$$\alpha_F = \coth y, \qquad \beta_F = 0.$$

Example 6.6 The time-like Scherk surface of the second kind (cf. [5, Example 5])

$$F(x, y) = \left(\operatorname{arcsinh}(\cosh x \sinh y), x, y\right)$$

is a zero mean curvature surface in $\mathcal{Z}^-_{II}$, which satisfies $\gamma_F = (0, \cosh x)$ and

$$\alpha_F = \tanh y, \qquad \beta_F = 0$$

hold.

For $F \in \mathcal{Z}^\omega_b$, it holds for $k \geq 4$ that

$$\left.\frac{d^k A}{dx^k}\right|_{x=0} = 0 \qquad (A := A_F), \tag{6.3}$$

which can be considered as an ordinary differential equation of the k-th approximation function a_k as in (6.1). As shown in [7], (6.3) is equivalent to

$$a_k'' + 2(k-1)a_2 a_k' + k(3-k)a_2' a_k + k(P_k + Q_k - R_k) = 0, \tag{6.4}$$

where P_k, Q_k, R_k are terms written using $\{a_s\}_{s<k}$ as follows:

$$P_k := \sum_{m=3}^{k-1} \frac{2(k-2m+3)}{k-m+2} a_m a'_{k-m+2},$$

$$Q_k := \sum_{m=2}^{k-2} \sum_{n=2}^{k-m} \frac{3n-k+m-1}{mn} a'_m a'_n a_{k-m-n+2},$$

$$R_k := \sum_{m=2}^{k-2} \sum_{n=2}^{k-m} \frac{a_m a_n a''_{k-m-n+2}}{k-m-n+2}.$$

When $k = 4$, (6.4) reduces to

$$a''_4 + 6a_2 a'_4 - 4a'_2 a_4 + 3a_2(a'_2)^2 - 2(a_2)^2 a''_2 + \frac{8}{3} a_3 a'_3 = 0.$$

Using (6.4) and (6.1), one can get an appropriate approximation for F.

Finally, we remark on existence results of zero mean curvature surface using Corollary 3.4: For $F \in \mathcal{Z}^\omega$, we set

$$\gamma_F = (0, 1) + (0, v_1)x + \sum_{n=2}^{4} (u_n, v_n) \frac{x^n}{n} + \text{(higher order terms)}.$$

Then $F \in \mathcal{Z}^\omega_b$ if and only if $v_1 = 0$. Under the assumption $v_1 = 0$, the characteristic (cf. (5.3)) μ_F and the constant δ_F in (5.9) satisfy

$$\mu_F = -(u_2^2 + v_2) \qquad \text{and} \qquad \delta_F = 3u_2 u_3 + v_3.$$

We set

$$\Delta_F := 4u_3^2 + 8u_2 u_4 + v_2^2 + 2v_4.$$

Proposition 6.7 *A surface $F \in \mathcal{Z}^\omega$ belongs to $\mathcal{Z}^0_I$ (resp. $\mathcal{Z}^0_{II}$)* if $\mu_F = 0$ and $u_2 = 0$ (resp. $\mu_F = 0$ and $u_2 \neq 0$). Suppose that $\mu_F = 0$. Then

(1) F changes causal type if $\delta_F \neq 0$, and
(2) F has no time-like (resp. space-like) part if $\delta_F = 0$ and $\Delta_F < 0$ (resp. $\Delta_F > 0$).

Proof The causal type of F depends on the sign of $B := B_F$. As shown in [5], $B|_{x=0} = B_x|_{x=0} = 0$. Moreover, one can easily see that

$$B(x, 0) = \mu_F x^2 - \frac{2\delta_F}{3} x^3 - \frac{\Delta_F}{4} x^4 + \text{(higher order terms)}. \tag{6.5}$$

So we get the conclusion. □

Example 6.8 In [7], $F \in \mathcal{Z}^{\omega}$ satisfying

$$\gamma_F(x) := (0, 1 + 3cx^3)$$

is constructed, which belongs to the class $\mathcal{Z}^{\omega}_{b}$ and changes its causal type. Although the existence of this F is obtained by applying Corollary 3.4, the advantage of the method in [7] is that we can get the explicit approximation for F at the same time.

Until now, the existence of zero mean curvature surfaces (i.e., ZMC-surfaces) in the following three cases was unknown (cf. the footnote of [6, P. 194]);

(i) ZMC-surfaces in $\mathcal{Z}^0_I$ without space-like part,
(ii) ZMC-surfaces in $\mathcal{Z}^0_I$ without time-like part,
(iii) ZMC-surfaces in $\mathcal{Z}^0_{II}$ which changes causal type.

We can show the existence of the above remaining cases:

Corollary 6.9 *There exist ZMC-immersions satisfying (i), (ii), and (iii), respectively.*

Proof We set $\mu_F = 0$. If $u_2 \neq 0$ and $\delta_F \neq 0$, then $F \in \mathcal{Z}^0_{II}$ which changes causal type (i.e., it gives the case (iii)). On the other hand, if $u_2 = \delta_F = 0$ and $\Delta_F < 0$ (resp. $\Delta_F > 0$), then it gives the case (ii) (resp. (i)). □

Acknowledgements The authors thank Shintaro Akamine, Udo Hertrich-Jeromin, Wayne Rossman and Seong-Deog Yang for valuable comments.

Appendix A. Division Lemma

Lemma A.1 *Let g be a C^r-function ($r \geq 1$) defined on a convex domain U of the xy-plane including the origin o, satisfying*

$$g(0, y) = \frac{\partial g}{\partial x}(0, y) = \frac{\partial^2 g}{\partial x^2}(0, y) = \cdots = \frac{\partial^k g}{\partial x^k}(0, y) = 0 \qquad \big((0, y) \in U\big) \quad \text{(A.1)}$$

for a nonnegative integer $k < r$. Then there exists a C^{r-k-1}-function h defined on U such that

$$g(x, y) = x^{k+1} h(x, y) \qquad \big((x, y) \in U\big). \quad \text{(A.2)}$$

Proof We shall prove by an induction in k. Since

$$g(x, y) = \int_0^1 \frac{dg(tx, y)}{dt}\, dt = \int_0^1 x g_x(tx, y) dt = x \int_0^1 g_x(tx, y)\, dt,$$

the conclusion follows for $k = 0$, by setting

$$h(x, y) := \int_0^1 g_x(tx, y)\,dt.$$

Assume that the statement holds for $k - 1$. If g satisfies (A.1), there exists a C^{r-k}-function $\varphi(x, y)$ defined on U such that

$$g(x, y) = x^k \varphi(x, y) \qquad \big((x, y) \in U\big). \tag{A.3}$$

Differentiating this k-times in x, we have

$$0 = \frac{\partial^k g}{\partial x^k}(0, y) = k!\varphi(0, y)$$

because of (A.1). Hence, by the case $k = 0$ of this lemma, there exists C^{k-r-1}-function $h(x, y)$ defined on U such that $\varphi(x, y) = xh(x, y)$. The function h is the desired one. □

References

1. S. Akamine, *Causal characters of zero mean curvature surfaces of Riemann-type in the Lorentz-Minkowski 3-space*. Kyushu J. Math. **71**, 211–249 (2017)
2. S. Akamine, *Behavior of the Gaussian curvature of timelike minimal surfaces with singularities*, preprint, arXiv:1701.00238
3. S. Fujimori, Y. Kawakami, M. Kokubu, W. Rossman, M. Umehara, K. Yamada, Zero mean curvature entire graphs of mixed type in Lorentz-Minkowski 3-space. Quart. J. Math. **67**, 801–837 (2016)
4. S. Fujimori, Y. Kawakami, M. Kokubu, W. Rossman, M. Umehara, K. Yamada, Analytic extension of Jorge-Meeks type maximal surfaces in Lorentz-Minkowski 3-space. Osaka J. Math. **54**, 249–272 (2017)
5. S. Fujimori, Y.W. Kim, S.-E. Koh, W. Rossman, H. Shin, H. Takahashi, M. Umehara, K. Yamada, S.-D. Yang, Zero mean curvature surfaces in $\mathbf{L}^3$ containing a light-like line. C. R. Acad. Sci. Paris. Ser. I. **350**, 975–978 (2012)
6. S. Fujimori, Y.W. Kim, S.-E. Koh, W. Rossman, H. Shin, M. Umehara, K. Yamada, S.-D. Yang, Zero mean curvature surfaces in Lorentz-Minkowski 3-space and 2-dimensional fluid mechanics. Math. J. Okayama Univ. **57**, 173–200 (2015)
7. S. Fujimori, Y.W. Kim, S.-E. Koh, W. Rossman, H. Shin, M. Umehara, K. Yamada, S.-D. Yang, Zero mean curvature surfaces in Lorentz-Minkowski 3-space which change type across a light-like line, Osaka J. Math., 52 (2015), 285–297. Erratum to the article "Zero mean curvature surfaces in Lorentz-Minkowski 3-space which change type across a light-like line". Osaka J. Math. **53**, 289–293 (2016)
8. S. Fujimori, W. Rossman, M. Umehara, K. Yamada, S.-D. Yang, Embedded triply periodic zero mean curvature surfaces of mixed type in Lorentz-Minkowski 3-space. Michigan Math. J. **63**, 189–207 (2014)
9. C. Gu, The extremal surfaces in the 3-dimensional Minkowski space. Acta. Math. Sinica. **1**, 173–180 (1985)

10. A. Honda, M. Koiso, M. Kokubu, M. Umehara, K. Yamada, Mixed type surfaces with bounded mean curvature in 3-dimensional space-times. Differ. Geom. Appl. **52**, 64–77 (2017). https://doi.org/10.1016/j.difgeo.2017.03.009
11. S. Izumiya, T. Sato, Lightlike hypersurfaces along spacelike submanifolds in Minkowski space-time. J. Geom. Phys. **71**, 30–52 (2013)
12. Y.W. Kim, S.-E. Koh, H. Shin, S.-D. Yang, Spacelike maximal surfaces, timelike minimal surfaces, and Björling representation formulae. J. Korean Math. Soc. **48**, 1083–1100 (2011)
13. V.A. Klyachin, Zero mean curvature surfaces of mixed type in Minkowski space. Izv. Math. **67**, 209–224 (2003)
14. O. Kobayashi, Maximal surfaces in the 3-dimensional Minkowski space L^3. Tokyo J. Math. **6**, 297–309 (1983)
15. S. G. Krantz, H. R. Parks, A Primer of Real Analytic Functions, 2nd edn. (Birkhäuser, USA, 2002)
16. M. Umehara, K. Yamada, *Hypersurfaces with light-like points in Lorentzian manifolds*, in preparation

Printed in the United States
By Bookmasters